1982

Coded Character Sets, History and Development

Coded Character Sets, History and Development

CHARLES E. MACKENZIE
IBM Corporation

ADDISON-WESLEY PUBLISHING COMPANY
Reading, Massachusetts • Menlo Park, California
London • Amsterdam • Don Mills, Ontario • Sydney

Library of Congress Cataloging in Publication Data

Mackenzie, Charles E
Coded-character sets.

Includes index.
1. Coding theory. I. Title.
QA268.M27 519.4 77-90165
ISBN 0-201-14460-3

 Printed in the United States of America. Published simultaneously in Canada. Library of Congress Catalog Card No. 77-90165.

ISBN 0-201-14460-3
BCDEFGHIJK-HA-89876543210

Dedicated to Dr. Louis Robinson
without whose insistence and persistence
this book would not have been written.

THE SYSTEMS PROGRAMMING SERIES

*The Program Development Process Part I—The Individual Programmer	Joel D. Aron
The Program Development Process Part II—The Programming Team	Joel D. Aron
*The Structure and Design of Programming Languages	John E. Nicholls
*Mathematical Foundations of Programming	Frank Beckman
Structured Programming: Theory and Practice	Richard C. Linger Harlan D. Mills Bernard I. Witt
*The Environment for Systems Programs	Frederic G. Withington
*Coded Character Sets; History and Development	Charles E. Mackenzie
*An Introduction To Database Systems, Second Edition	C. J. Date
Interactive Computer Graphics	James Foley Andries Van Dam
*Sorting and Sort Systems	Harold Lorin
*Compiler Design Theory	Philip M. Lewis II Daniel J. Rosenkrantz Richard E. Stearns
*Communications Architecture for Distributed Systems	R. J. Cypser

*Recursive Programming Techniques	William Burge
Conceptual Structures: Information Processing in Mind and Machines	John F. Sowa
*Modeling and Analysis: An Introduction to System Performance Evaluation Methodology	Hisashi Kobayashi

*Published

Foreword

The field of systems programming primarily grew out of the efforts of many programmers and managers whose creative energy went into producing practical, utilitarian systems programs needed by the rapidly growing computer industry. Programming was practiced as an art where each programmer invented his own solutions to problems with little guidance beyond that provided by his immediate associates. In 1968, the late Ascher Opler, then at IBM, recognized that it was necessary to bring programming knowledge together in a form that would be accessible to all systems programmers. Surveying the state of the art, he decided that enough useful material existed to justify a significant publication effort. On his recommendation, IBM decided to sponsor The Systems Programming Series as a long term project to collect, organize, and publish principles and techniques that would have lasting value throughout the industry.

The Series consists of an open-ended collection of text-reference books. The contents of each book represent the individual author's view of the subject area and do not necessarily reflect the views of the IBM Corporation. Each is organized for course use but is detailed enough for reference. Further, the Series is organized in three levels: broad introductory material in the foundation volumes, more specialized material in the software volumes, and very specialized theory in the computer science volumes. As such, the Series meets the needs of the novice, the experienced programmer, and the computer scientist.

The Editorial Board

Preface

The word "code" is a word of broad meaning and application. Legal codes, fire safety codes, building construction codes, a code of ethics, and so on, exemplify the use of the word in some of its dictionary meanings, "a system of rules or regulations on any subject." A dictionary meaning that comes closer to the context of this book is "a system of signals."

From early beginnings, humans have used many methods to convey information over a distance. Indians (of North America) used a set of smoke signals for sending messages. A semaphore, a vertical post with one or more arms moving in a vertical plane, was and is used to send messages over line-of-sight distances.

The method that comes close to the meaning used in this book is the Morse Code, an alphabet in which the letters are expressed as dots and dashes. This method can be used visibly with short and long flashes of light, audibly with short and long bursts of sound, electrically with short and long pulses of current, and so on. The interesting aspect of the Morse Code is that it is based on two possible states—dot or dash, short or long, and so on—that is to say, it is binary in nature. Standing aside from the spaces between dots and dashes, and between letters, the Morse Code may be regarded as a binary code.

Analogously to the Morse Code, the set of alphabetic, numeric and special (such as period, comma, plus sign, minus sign) symbols processed by a computer are associated with a set of particular binary representations. Such a set of graphic symbols and binary representations is called a coded character set, or, more familiarly, a code.

The binary aspect of a coded character set stems naturally from the binary, or two-state, nature of many mechanisms, components, or processes of a computer. A switch is on or off, a relay is normal or transferred, a vacuum tube is or is not passing current, a condenser is or is not charged, a magnetic pole is north or south, a voltage is positive or negative or is equal to or less than a reference voltage, and so on. Relays, vacuum tubes, transistors, magnetic cores, diodes, as used in computer circuits, are binary in nature.

In the decimal number system, there are ten digits—0, 1, 2, 3, 4, 5, 6, 7, 8, 9. In the binary number system, there are two digits—0 and 1. Very early in the history of computing, the words "binary digit" were contracted to the word "bit"; "a bit may be 0, or 1," means "a binary digit may be 0 or 1." A discrete grouping of contiguous bits, 1001011 for example, is called a bit pattern.

A coded character set, or code, is a set of meanings associated with a set of bit patterns. For a particular code, the number of bits is generally a fixed number; all bit patterns in a particular code have five bits, or all bit patterns in a particular code have six bits, and so on. This aspect of a fixed number of bits in the bit patterns of a particular code is frequently used to characterize a code as a 5-bit code, or as a 6-bit code, and so on. In this respect, the Morse Code, which has different numbers of bits for different letters, although it continues to be used for sending messages, was deemed not to be satisfactory for computing purposes.

The number of different possible bit patterns in a particular code depends on the fixed number of bits of that code. In consequence, the number of different possible meanings that may be associated on a one-to-one basis with the different bit patterns of a code depends on the number of bits of a code. Reasoning in the opposite direction suggests that the number of different meanings required in the code of a computer may be a determining factor in the number of bits in a code.

Perhaps the most famous code in the history of computing was that invented by Dr. Herman Hollerith of the United States Census Bureau in the late nineteenth century. His code was a decimal code based on the position of a punched hole across a paper card—ten digits, ten punching positions. His code was actually a twelve-position code—ten positions for digits, two positions for other purposes (positive or negative, for example). Today, more than seven decades later, Dr. Hollerith's twelve-position code is fundamental in the punched card code used by many/most computers.

A number of different codes have evolved in the computing and data communication fields: different codes evolved because different requirements emerged as computing and data communication evolved. Many factors shaped the different codes. This book describes those factors and how they either led to or mandated decisions in the development of some codes. This book is not a definitive book on all computer or data communication codes. Discussion is limited to those codes which have evolved, have been developed, or have been used in the author's personal experience.

Mainly, the factors discussed are of a technical nature, but some of the factors are of an economic or cost nature. For example, in computers, bit patterns are stored in registers. In early computers, registers were implemented in vacuum-tube technology. The number of bits to be stored in a register bore a relation to the number of tubes needed in the register—8-bit registers required more tubes than 6-bit registers. The manufacturing cost of a register was related to the number of tubes in the register. In this sense, a 6-bit code was considered to be more "economical" than an 8-bit code.

Two processes have shaped the evolution and development of codes. One process is the process of developing computing and communication products and systems, a process of individual manufacturers. The other process is the developing of standards for the data processing industry, a process of both manufacturers and users, in concert.

With respect to the first process, during the 1960s, two great technological evolutions were occurring in the data processing field. On one hand, computing systems were evolving from an architecture of six bits to an architecture of eight bits. (Many people consider this to have been more of a revolution than an evolution.) On the other hand, communications systems were evolving from five-bit codes to six-, seven-, and eight-bit codes.

With respect to the second process, during the 1960s, there was a quite remarkable development of standards in the field of data processing. One particular area of standardization was the area of coded character sets and their representation on physical media—magnetic tape, paper tape, punched cards, data transmission, tape cassettes, and so on. This standardization effort was exerted on both the national and international level. In the United States alone during the 1960s, some twenty standards in this area were started, and most were completed.

As might be supposed, the interaction between these two processes was considerable. One characteristic of codes is very interesting. In the data processing industry over the last twenty years, older computing and communications products and systems have not infrequently been replaced with newer, more economically efficient products and systems. But old codes do not die, nor do they fade away. A 5-bit telegraph communications code standardized in 1931 is still in wide use although a 7-bit communications code was standardized in 1963, and many products implementing the 7-bit code are available. A 6-bit computer code developed in 1962 continues in wide use, although 8-bit computers with an 8-bit computing code have largely replaced the 6-bit computers. Codes have the characteristic of continuity and long-life expectancy due to user's application demands.

A problem that has to be faced in a technical book such as this is the existence of the specialized jargon used by professionals in the subject. Words or terms that make up the jargon came from two sources. The first source is words with a general meaning or meanings in the English language that are given a very specialized meaning in the jargon. Such specialized meanings are not in common use and will not be found in common dictionaries. An example is the word "track." In railroading, "track" means one thing; in fur trapping, it means something else; and in horse racing, it means yet something else. These meanings will likely be found in common dictionaries. But in the field of magnetic tape engineering, "track" has a meaning most unlikely to be found in common dictionaries, although it is likely to be found in technical dictionaries for the field of data processing. The second source of jargon is new words or terms invented by the professionals. An example here is "bit." The meaning "binary digit," from which "bit" was contracted, is not likely to be found in common dictionaries, although its meaning is well known in the data processing field.

Technical jargon must be used in a book on a technical subject. Early in this book some terms and concepts very necessary to an understanding of the field of coded character sets are defined and explained; the glossary of this book is devoted to a comprehensive set of definitions of terms.

Just as letters, digits, and special symbols make up a language in which humans intercommunicate, the letters, digits, and special symbols with associated bit patterns of a coded character set make up the language in which information is passed, interchanged, and processed by computers. A complete knowledge of the art of computing, which includes both the manufacture and use of computers, requires a knowledge of the art of coded character sets. This book describes some of that art.

The author would like to express his appreciation to Mrs. Helena Russo, Mrs. Janet Palome, and Mrs. Betty Birdsall, who did the lengthy and frequently very difficult typing of the manuscript of this book.

Poughkeepsie, New York C.E.M.
January 1980

Contents

CHAPTER 6
THE SIZE AND STRUCTURE OF PTTC

CHAPTER 7
THE STRUCTURE OF EBCDIC

CHAPTER 8
THE SEQUENCE OF EBCDIC

CHAPTER 9
THE DUALS OF EBCDIC

CHAPTER 10
THE GRAPHIC SUBSETS OF EBCDIC

CHAPTER 14
THE SEQUENCE OF ASCII

CHAPTER 15
WHICH BIT FIRST?

CHAPTER 16
DECIMAL ASCII

CHAPTER 20
ASCII IN AN 8-BIT INTERCHANGE ENVIRONMENT

CHAPTER 21
THE ALPHABETIC EXTENDER PROBLEM

CHAPTER 22
GRAPHIC SUBSETS FOR THE GOVERNMENT

CHAPTER 23
WHICH ASCII?

1
The Standards Process

Most of the codes discussed in this book have been developed in the context of developing data processing standards of one kind or another. These standards may be categorized as being either public or company standards. Public standards are those developed by governmental, national, or international organizations. Company standards are those developed by a company. Many company standards are well known outside the developing company, and in many instances are used by companies or organizations other than the developing company. Although the discussion of company standards is intended to be of a general nature, it does draw primarily on the author's experience in the IBM Corporation.* Also, most of the national standards discussed in this book are those developed in the United States of America, again by reason of the author's familiarity. Equivalent national standards have been developed in many other countries.

1.1 THE PUBLIC COMMITTEE PROCESS

The suggestion to standardize in a particular subject area may originate anywhere; an individual, a company, a government agency/department, a society/association, a standards committee, and so on.

Public standards are developed by committees—committees established specifically for the process of developing the standard, or standards, and staffed with professionals from the field of the subject.

* The views expressed in this book are those of the author and not necessarily those of the IBM Corporation.

Generally speaking, the organization is as follows. At the top will be an administrative body, whose functions are to establish the procedural rules for developing standards, to monitor adherence to these rules, to determine that any particular standard is not in technical conflict with other standards, and to publish and distribute the standards. In the case of national standards, the administrative body will generally be the national standards institute or association of the country.

Reporting to the administrative body will be one or more managerial committees, each dedicated to a particular subject area of standardization. The area of standardization assigned to the managerial committee is generally divided into subareas. Technical subcommittees are established to develop standards for the subareas. One main function of such managerial committees is to direct and coordinate the activities of technical subcommittees who do the actual work of developing and drafting the standards. The other main function is to assess the economic (and sometimes social) implications of draft standards.

Usually some organization will serve as secretariat for the committee and subcommittees. The secretariat distributes to the members, and keeps on file, the minutes, papers, and other correspondence of the committee and subcommittees.

The committees and subcommittees function very similarly. There will be a chairman, usually a vice-chairman (sometimes called chairperson and vice-chairperson today), and a secretary. Minutes of the meetings are kept. Members submit papers of a technical, economic, or social nature. The papers, and the subject matter of the standard(s), are discussed at meetings. Decisions on points of issue and points of agreement are taken by votes or ballots, under various rules of majority, consensus, or unanimity. The meetings are conducted under parliamentary rules of procedure. Draft standards are (generally) subjected to some form of public review before final approval.

In the case of national managerial committees, members are companies, governmental units, and professional societies or associations. In the case of national technical subcommittees, members are professionals knowledgeable in the subject area of the standard(s). In the case of international committees and subcommittees, members are countries, with actual attendees at meetings being delegations selected by the countries. Not unexpectedly, the individuals on country delegations are usually selected from the members of national committees and subcommittees.

1.2 THE COMPANY PROCESS

Company standards are generally developed by the same procedures and methods the company uses to manage itself and to develop its products.

1.3 DECISION PROCESSES

Usually, national and international standards are derived from and based on well-established industrial practices or techniques. The task of a standards committee developing a standard in such instances is to describe completely, consistently, and unambiguously what already exists, removing or smoothing any incompletenesses, inconsistencies, and ambiguities.

In some cases, standards committees foresee the need to develop a standard where practices or techniques are not well established, or do not exist at all. Such standards are called anticipatory standards. The main problem for standards committees in such instances is to try to guess or anticipate what the needs of users will be. These guesses are always speculative and judgmental, and frequently controversial. Sometimes, the most controversial aspect of such guesses is whether a standard is actually needed before users build up experience, practices, and techniques over a period of time and a range of applications.

The development processes for public and company standards are in some respects the same. A group of professionals knowledgeable in the subject area is called together, a chairman or coordinator is appointed, and the group is charged with the responsibility to develop a standard for the subject area. The group reviews the subject area, reviews relevant technical facts, and drafts the standard.

Inevitably, on one or more aspects of the standard, technical alternatives will emerge, and decisions for one of the alternatives must be made. If, after review of the alternatives, the group is unanimous in selection of a particular one, the matter is resolved. But if the group is not unanimous in opinion, a decision must be made. It is in respect of such technical decisions that the process in a company is quite different from the process of a standards committee.

In the company, if the group is not unanimous, a management decision must be made. It may be made by the group coordinator. Or it may be referred to a higher level of management or to a series of management levels. But in all cases, the decision will be made by a single person. It is made after that person reviews the alternatives, and the pros and cons, and makes a decision based on personal judgment.

In a public standards committee, the decision is not made by a single person. It is made by taking a vote or ballot, the outcome of the voting process being determined by pre-established rules of majority or consensus for the particular committee. That is to say, the decision is a reflection of the combined personal judgments of all committee members, each committee member's judgment being given an equal weight. In theory, it should be possible to follow the company approach of letting the most knowledgeable person on the committee make the decision. In practice, it is not possible to determine who of the committee members is the most knowledgeable. The equal-weight voting approach is the only practical and workable one for a committee.

In a particular situation when the pros and cons of alternatives are based purely on technical aspects, the committee is not likely to have difficulty in arriving at a decision. The decision can be made purely on technical merit, and it is simply a question of determining the relative technical merits of the alternatives. The professionals on the committee are very well qualified to make such determinations.

An interesting situation that sometimes arises, (more likely in the development of an anticipatory standard than in the standardization of an established industry practice) is that two technical alternatives face the committee, and each alternative would be equally satisfactory. In such situations, the *act* of making the decision is more important than the technical *matter* of the decision. For example, standardization in the area of data communications eventually faced the question of order of transmission of the bits of a byte—should transmission be low-order bit first or high-order bit first? A priori, there were arguments in favor of each of the alternatives, and the arguments were clearly of equal technical weight. It did not matter, a priori, which choice was made, but it was necessary to make the choice.

A posteriori, once the choice was made, and implementations emerged, it did matter, because then the fact of implementation for the particular choice was a weighty argument.

Intuitively, it would seem that, for a particular subject area, one standard, which is to say one technique or one practice, best serves the interests of the data processing industry. Thus, if a card code is to be standardized, only one card code (whatever it may be) should be standardized. Two card codes would result in conflicts and confusions. Many standards associations, as a cardinal principle, forbid the approval of conflicting standards in any area.

But there are situations where more than one standard, a family of standards, is a viable solution, each member of the family serving a particular purpose in the general subject area. For example, in the area of data transmission, standards specifying different speeds or rates of transmission have been developed. In the area of magnetic tape, standards specifying different densities of recording have been developed. Such families of standards reflect the practical economics that exist. Thus, in general, the lower the density of recording, the lower the cost of the magnetic tape drive. A low density of recording may be quite satisfactory in some data processing applications, and then the user will appreciate the lower cost of tape drives. Other data processing applications may require a higher density of recording, and for such situations, the user accepts the higher cost of tape drives.

1.4 ECONOMIC CONSIDERATIONS

Frequently, factors other than technical, such as economic and sometimes social, are involved, and then the committee's decision process becomes much more difficult. A standard committee, when developing standards in a particular subject area, may face a number of possible situations.

Situation 1. There is a single, uniform practice in the subject area.

Situation 2. There is essentially a single practice in the subject area, but with slight individual variations.

Situation 3. There are a number of different practices in the area, with much in common but with appreciable differences.

Situation 4. There are a number of different practices in the subject area, with little if anything in common.

Situation 1 is the simplest for the committee. All that is needed is to draft a standard which accurately describes the established practice. Of course, there may be some question on the accuracy of the description, but the committee members are well qualified to resolve just such questions.

Situations 2, 3, and 4 become increasingly more difficult for the committee members to resolve. The difficulty is the same kind for these three situations, but different in degree. The difficulty is that the practices under review are in use in the industry, and the final decision of the standard will make some current practices standard, while making other current practices nonstandard. Then, if those who are using the just-defined nonstandard practice want to use the just-defined standard practice, they will have to change what they are doing, or the way they are doing it. Such changes will generally involve cost to the user.

In such situations, then, economic as well as technical factors affect the decision process. Indeed, there are situations where the economic factors are more, sometimes much more, significant than the technical factors. And, while the technical factors can be determined with some degree of precision, it will generally be difficult or impossible to determine the economic factors with any degree of precision.

1.5 NAMES OF STANDARDS

National and international standards take their titles (which lead to their names) from the organizations under which they were developed, and to some extent, from the purpose for which the standard was developed. Company standards often take their titles from the purpose for which the code was developed (the Paper Tape and Transmission Code, for example).

The international organization responsible for standards in the data processing field (as well as in many other fields) is the International Organization for Standardization (ISO). Until recently, "standards" developed under ISO were not called "standards," but were called "Recommendations." The intent of such documents was vested in the name, "Recommendation." It was recommended when national standards bodies developed their own national standards that such standards be based on the ISO Recommendations. Recently, ISO decided to call their documents ISO Standards in name as well as in fact. Another international organization, responsible for all matters pertaining to worldwide telegraph and telephone communications, is the International Telegraph and Telephone Consultative Committee. Its acronym, CCITT, comes from the equivalent French name for the organization (Commité Consultatif International Telegraphique et Telephonique). A European organization that develops data processing standards is the European Computer Manufacturers Association (ECMA).

In the United States, the national standards organization has gone through a number of changes of name. Organized in 1918 as the American Engineering Standards Committee, it became the American Standards Association (ASA) in 1928. In 1966, it was re-named the United States of America Standards Institute (USASI) and in 1969 it took its present name, the American National Standards Institute (ANSI).

A 5-bit code was standardized in 1931 by CCITT for telegraph communications purposes. It is designated CCITT #2, and is still in worldwide use.

The U.S. Army developed a 7-bit code for data communications that became a U.S. Military Standard in 1960. Its developers coined for it the name FIELDATA.

A 7-bit code described in this book has been standardized by a number of national and international standards organizations:

a) In 1963, under ASA, it became the American Standard Code for Information Interchange, acronym ASCII (pronounced 'ass-key). When ASA became USASI in 1966, the code was called the United States of America Standard Code for Information Interchange, with acronym USASCII (pronounced you-'sass-key). However, the previous acronym ASCII, prominent in the literature, was officially designated as an acceptable alternative acronym. When USASI became ANSI in 1969, the code was called the American National Standard for Information Interchange. Needless to say ANSCII was proposed as a new acronym, but the standards committee rejected further name changes, and ANSCII as an acronym was rejected. ASCII was then designated as the preferable acronym. (USASCII is an acceptable alternative acronym, but has fallen into disuse.)

b) In 1967, it was incorporated into the ECMA Standard for a 7-Bit Input/Output Character Code, ECMA-6.

c) In 1967, it was incorporated into an ISO Recommendation, the 6 and 7-Bit Coded Character Set for Information Processing Interchange. In that context, it is referred to as the ISO 7-bit code.

d) In 1969, it was incorporated into the Japanese Industrial Standard Code for Information Interchange (JISCII).

e) In 1968, it was incorporated into a CCITT standard designated CCITT #5.

These 7-bit codes are essentially the same. They differ in graphic symbols which reflect different national requirements. This similarity is not coincidental; it is intentional—the result of professionals in different countries working together to achieve that result.

The original twelve character (ten numerics and two special symbols) code invented by Dr. Herman Hollerith in the late nineteenth century grew to include alphabetics and special symbols. It also was incorporated into national and international standards, specifying either 128 or 256 characters:

a) In 1969, 128 characters were incorporated into the American National Standard Hollerith Punched Card Code. This standard took its name from the original inventor of the card code. It is now referred to as the Hollerith Card Code.

b) In 1970, 128 characters were incorporated into an ISO Recommendation, Representation of ISO 7-Bit Coded Character Set on 12-Row Punched Cards. It is referred to as the ISO 12-Row Card Code.

c) In 1970, the American Standard was extended to incorporate 256 characters, retaining the same name.

d) In 1971, another ISO Recommendation incorporated 256 characters, Representation of 8-Bit Patterns on 12-Row Punched Card. It also is referred to as the ISO 12-Row Card code.

Items (a) and (b) are identical; items (c) and (d) are identical. The 128 characters of (a) and (b) are a subset of the 256 characters of (c) and (d). As with the 7-bit code standards, this is intentional, not coincidental.

Four codes developed in IBM are discussed in this book. Two of these codes (described in more detail in Chapter 2) were named in consequence of a particular aspect of codes; namely, that the decimal numbers 0 through 9, when represented in a binary code, have particular binary bit-patterns which are called binary coded decimal in the literature. The acronym, BCD, is well understood in the data processing industry to characterize a code whose decimal numbers are in the binary coded decimal representation.

The first code developed in IBM, formalized in 1962, is a 6-bit code called the BCD Interchange Code, with acronym BCDIC (pronounced bee-see-dick). An 8-bit code adopted within IBM in 1964 is called the Extended BCD Interchange Code with acronym EBCDIC (pronounced ebb-see-dick).

Two other IBM standard codes were developed for use in perforated tape and transmission products. These 6-bit codes were originally named Perforated Tape and Transmission Code for use in 6-Bit BCD Environments, with acronym PTTC/6, and Perforated Tape and Transmission Code for use in 8-Bit BCD Environments, with acronym PTTC/8. These names turned out to be confusing. People thought that PTTC/6 meant that it was a 6-bit code, and PTTC/8 meant it was an 8-bit code. The former was correct, the latter was incorrect. Therefore, PTTC/6 was renamed the Perforated Tape and Transmission Code for use in BCDIC Environments, with acronym PTTC/BCD, and PTTC/8 was renamed the Perforated Tape and Transmission Code for use in EBCD Environments, with acronym PTTC/EBCD. Whether the confusion was reduced is moot, but the second set of names has remained.

Reference is made in this book to various American National Standards and ISO Recommendations:

1. The American National Standard Code for Information Interchange, X3.4-1968, referred to in this book as ASCII.
2. ISO Recommendation, 6 and 7-Bit Coded Character Sets for Information Processing Interchange, ISO/R646-1967, referred to in this book as the ISO 7-Bit code.
3. The American National Standard Bit Sequencing of the American National Standard Code for Information Interchange in Serial-by-Bit Data Transmission, X3.15-1966.
4. The American National Standard Hollerith Punched Card Code X3.26-1970, referred to in this book as the Hollerith Card Code.

5. ISO Recommendation, ISO 7-Bit Coded Character Set on 12-Row Punched Cards, ISO/R1679-1970, referred to in this book as the ISO 12-Row Card code.
6. ISO Recommendation, Representation of 8-Bit Patterns on 12-Row Punched cards, ISO/R2021-1971, referred to in this book as the ISO 12-Row Card Code.

Copies of these American National Standards and ISO Recommendations are available from the American National Standards Institute, 1430 Broadway, New York, New York 10018.

The 7-bit bit codes of items (1) and (2) above are similar. When there is no need to distinguish between them, they are referred to generically as the 7-Bit Code in this book. When distinction is necessary, one is referred to as ASCII, the other as the ISO 7-Bit Code.

The 256-character card codes of items (2) and (5) above are equivalent. When it is necessary to distinguish between them, one is referred to as the Hollerith Card Code, the other as the ISO 12-Row Card Code.

2 Terms and Concepts

There are some basic terms which should be understood at the onset of reading this book. These are grouped in this chapter for convenience. (A lengthy set of terms and definitions is found in the Glossary.)

A fundamental concept involved in data processing products is the binary, or two-state, nature of many mechanisms, devices, and processes:

- A relay is transferred or normal.
- A switch is on or off.
- A condenser is charged or discharged.
- A light is on or off.
- A diode is, or is not, conducting current.
- A vacuum tube is, or is not, conducting current.
- A magnetic pole is North or South.
- A punching position on a paper card or on paper tape is punched or unpunched; which is to say, in a punching position, a hole is present or absent.
- At a point in an electrical circuit, the voltage is positive or negative, or is zero or negative, or is zero or positive, or is high or low, and so on.

The decimal number system has the familiar ten digits 0, 1, 2, 3, 4, 5, 6, 7, 8, 9. The binary number system has two digits, 0 and 1. The representation of physical, electrical, or magnetic two-state situations such as those above by binary digits is the analytic process of representing a physical situation by a mathematical model. In the literature, the term "binary digit" soon came to be contracted to "bit."

2.1 BIT

A bit is a binary digit, either 0 or 1.

2.2 BIT PATTERN

A bit pattern is an ordered set of bits, usually of a fixed length.

Example 1 101011, a bit pattern of 6 bits

Example 2 1100011, a bit pattern of 7 bits

Example 3 10011100, a bit pattern of 8 bits

A bit pattern of n bits is called an n-bit bit pattern. Thus we speak of 6-bit bit patterns, 7-bit bit patterns, 8-bit bit patterns, and so on.

2.3 BYTE

A byte is a bit pattern of fixed length. Thus we speak of 8-bit bytes, 6-bit bytes, and so on.

2.4 BINARY VARIABLE

A binary variable is a variable which can take two possible values or represent two possible states.

Three major conventions for representing bit patterns of binary variables have developed.

- The first convention is the obvious one, a string of 0s and 1s; thus 10100, 1001111, 10010101, and so on.
- The second convention is based on the realization that, for a binary variable, call it A, we have either A or the inverse of A; we have either A or "not A." The convention is to represent "not A" (or the inverse of A) as $\bar{A}$ (A overlined). Thus for a set of three binary variables, A, B, C, we may have eight possible states:

Example 4

$$\bar{A}\bar{B}\bar{C}$$
$$\bar{A}\bar{B}C$$

$$\bar{A}B\bar{C}$$
$$\bar{A}BC$$
$$A\bar{B}\bar{C}$$
$$A\bar{B}C$$
$$AB\bar{C}$$
$$ABC$$

- The third convention is based on a presence/absence concept and the naming of the specific bit positions within a bit pattern.

Example 5

The four bit positions of a 4-bit bit pattern are named 8, 4, 2, 1; these are the decimal equivalents of 2^3, 2^2, 2^1, 2^0, respectively. Then the sixteen 4-bit bit patterns are represented as in Fig. 2.1, sometimes in a columnar form as at the left and sometimes in a compact form as at the right.

	8	4	2	1	
0					No bits
1				1	1
2			2		2
3			2	1	21
4		4			4
5		4		1	41
6		4	2		42
7		4	2	1	421
8	8				8
9	8			1	81
10	8		2		82
11	8		2	1	821
12	8	4			84
13	8	4		1	841
14	8	4	2		842
15	8	4	2	1	8421

Fig. 2.1 8421 notation

Under the second convention, A and $\bar{A}$ are equated to 1 and 0, respectively. Under the third convention, presence and absence are equated to 1 and 0, respectively.

Example 6

Figure 2.2 shows the sixteen possible states of a 4-bit bit pattern represented under the three conventions, using A, B, C, D as variables for the second convention.

Convention 1	Convention 2	Convention 3
0000	$\bar{A}\bar{B}\bar{C}\bar{D}$	No bits
0001	$\bar{A}\bar{B}\bar{C}D$	1
0010	$\bar{A}\bar{B}C\bar{D}$	2
0011	$\bar{A}\bar{B}CD$	21
0100	$\bar{A}B\bar{C}\bar{D}$	4
0101	$\bar{A}B\bar{C}D$	41
0110	$\bar{A}BC\bar{D}$	42
0111	$\bar{A}BCD$	421
1000	$A\bar{B}\bar{C}\bar{D}$	8
1001	$A\bar{B}\bar{C}D$	81
1010	$A\bar{B}C\bar{D}$	82
1011	$A\bar{B}CD$	821
1100	$AB\bar{C}\bar{D}$	84
1101	$AB\bar{C}D$	841
1110	$ABC\bar{D}$	842
1111	$ABCD$	8421

Fig. 2.2 Conventions for binary notation

The first and second conventions lead to a uniform, fixed-length representation. The third convention leads to a compact, variable-length representation.

2.5 BIT NUMBERING AND BIT NAMING

For purposes of reference, the bit positions of the bit patterns of a code are numbered, or named:

- For a 7-bit code (Fig. 2.26) the seven bits are numbered b7, b6, b5, b4, b3, b2, b1, from high- to low-order significance.
- For an 8-bit representation based on that 7-bit code (Fig. 2.27) the eight bits are numbered a8, a7, a6, a5, a4, a3, a2, a1, from high- to low-order significance.
- For the code table of Fig. 2.28, which is an 8-bit code (structured differently from the 8-bit representation in Fig. 2.27), the eight bits

are numbered 0, 1, 2, 3, 4, 5, 6, 7, from high- to low-order significance.

- For 6-bit codes (Fig. 2.29), the six bits are named B, A, 8, 4, 2, 1, from high- to low-order significance. This bit-naming convention for the four low-order bits is based on the 8421 convention previously described.

2.6 BIT STRING

A bit string is a contiguous sequence of bits, usually not a fixed length. In data processing applications, bit patterns of variable length are generally called bit strings.

2.7 CARD HOLE PATTERNS

The twelve vertical punching rows of a punched card are called the 12-row, the 11-row, the 0-row, the 1-row, . . . , the 9-row (see Fig. 2.3). The vertical punching rows of a card give their names to hole punches in those rows. Thus a hole punch in the 12-row is called a 12-punch, a hole punch in the 11-row is called an 11-punch, a hole punch in the 0-row is called a 0-punch, and so on. (The numeric designators may also be spelled out, twelve-row, eleven-row, twelve-punch, eleven-punch, etc.)

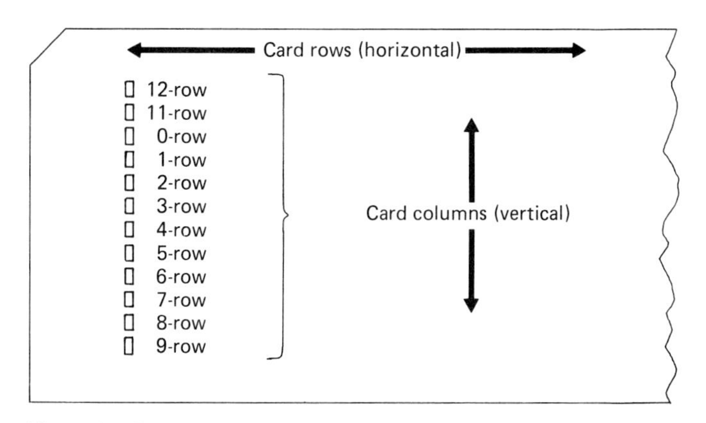

Fig. 2.3 Punched card

2.7.1 Hole Pattern

A hole pattern is a set of punched holes within a single vertical punching column of a card.

In documents, a hole pattern is given as the punches separated by hyphens. Thus 12-8-2, 12-11-3, 12-11-0-8-7 and so on.

2.8 ZONE ROW, ZONE PUNCH

The 12-row and 11-row are called zone rows. The 12-punch and 11-punch are called zone punches. The 9-row and 0-row are sometimes called zone rows, sometimes digit rows (Section 2.9 below). The 9-punch and 0-punch are sometimes called zone punches, sometimes digit punches (Section 2.9 below).

2.9 DIGIT ROW, DIGIT PUNCH

The 1-row, 2-row, 3-row, 4-row, 5-row, 6-row, 7-row, 8-row are called digit rows. The 1-punch, 2-punch, 3-punch, 4-punch, 5-punch, 6-punch, 7-punch, 8-punch are called digit punches. The 9-row and 0-row are sometimes called digit rows, sometimes zone rows (Section 2.8 above). The 9-punch and 0-punch are sometimes called digit punches, sometimes row punches (Section 2.8 above).

2.10 GRAPHIC

A graphic is a particular shape, printed, typed, or displayed, that represents an alphabetic, numeric, or special symbol.

In documents, books, magazines, newspapers, for example, we find three kinds of symbols; letters, numbers, and special symbols used for punctuation, mathematical operations, editorial inserts, and the like. These symbols are called graphic symbols; more commonly, simply graphics.

2.10.1 Alphabetic

An *alphabetic* is a letter in the alphabet of a country. Generally taken to mean a letter of the Latin alphabet but sometimes particularized as, for example, Latin alphabetic, Cyrillic alphabetic, Greek alphabetic, Hebraic alphabetic.

2.10.2 Numeric

A *numeric* is one of the ten decimal digits 0, 1, 2, 3, 4, 5, 6, 7, 8, 9.

2.10.3 Special

A *special* is a graphic symbol indicating a specific purpose.

Special symbols are frequently multi-purpose. Thus "." may be a period or a decimal point; "-" may be a hyphen or a minus sign, or a dash.

Example 7

Some specials commonly found on data processing products are

. , : ; ? ! " ( ) < > + − / *

= | ¬ — @ # % & $ ¢ { } [ ]

2.11 CONTROL MEANING

Control meaning refers to a particular function or operation that controls hardware or software products of systems. Control functions come in many categories. Some of the categories are as follows:

Format effectors. Functions to control the formatting of data on a printed page, or on a display.

Information separators. Functions to separate and block data.

Device controls. Functions to control a device (as "On" or "Off") or to control actions within a device.

Transmission controls. Functions to control intercommunications on data transmission lines.

Mode change. Functions to set or change some particular mode of operation.

Miscellaneous. Functions which do not fall into the above categories.

2.12 CHARACTER

A character is a specific bit pattern and an assigned meaning.

2.12.1 Graphic Character

A graphic character is a specific bit pattern and an assigned graphic meaning.

In order that data processing equipment may process graphic information, specific bit patterns must be assigned to specific graphic meanings. Thus if 100 0001 is assigned to graphic meaning of the alphabetic A, for example, the electrical circuits of a printer will analyze bit patterns, and when it detects 100 0001, the letter A will be printed.

2.12.2 Control Character

A control character is a specific bit pattern and an assigned control meaning.

Data processing products perform certain control functions. For example, a typewriter performs the operations of spacing, backspacing, up shifting, down shifting, tabulation, carriage return. If the typewriter is to operate as a printer, certain bit patterns must be assigned the meaning of control functions.

2.13 DATA STREAM

A data stream is a variable-length string of bit patterns, representing the data of a data processing application.

2.14 CODED CHARACTER SET—CODE

A coded character set is a specific set of bit patterns or hole patterns to which both specific graphic and control meanings have been assigned.

2.14.1 Bit Code

A bit code is a set of bit patterns to which either graphic or control meanings have been assigned.

A code byte in general can be of variable length. The Morse code, for example, has variable-length code bytes. However, codes used in data processing systems invariably have fixed-length bytes.

The code byte prescribes the number of different possible bit patterns in a code—the code byte is generally used to characterize a code. Thus we speak of a 5-bit code, or a 6-bit code, or a 7-bit code, and so on. A n-bit code has 2^n possible different bit patterns. A 4-bit code has $2^4 = 16$ possible different bit patterns. A 5-bit code has $2^5 = 32$ possible different bit patterns. A 6-bit code has $2^6 = 64$ possible different bit patterns. And so on.

Generally, the number of different possible bit patterns of a code prescribes also the number of possible characters in a code. Thus, a 6-bit code has 64 characters, and an 8-bit code has 256 characters. A 6-bit code used in the early days of data processing is shown in Fig. 2.4. It is to be noted that graphic meanings only are assigned and that not all bit patterns have an assigned meaning. This early 6-bit code consisted of 48 characters (64 would be possible)—the Space character, 10 numerics, 26 alphabetics, and 11 specials.

Three concepts (to be explained)—duals, character sequences, and shifted codes—allow the assignment of more meanings to a code than the total possible number of different bit patterns.

2.14.2 Card Code

A card code is a set of hole patterns to which graphic or control meanings have been assigned.

Bit pattern	Graphic	Bit pattern	Graphic
No bits	Space	B	- Hyphen, minus
1	1	B1	J
2	2	B2	K
21	3	B21	L
4	4	B4	M
41	5	B41	N
42	6	B42	O
421	7	B421	P
8	8	B8	Q
81	9	B81	R
82	0	B82	
821	# Number sign	B821	$ Dollar sign
84	@ At sign	B84	* Asterisk
841		B841	
842		B842	
8421		B8421	
A		BA	& Ampersand
A1	/ Slash	BA1	A
A2	S	BA2	B
A21	T	BA21	C
A4	U	BA4	D
A41	V	BA41	E
A42	W	BA42	F
A421	X	BA421	F
A8	Y	BA8	H
A81	Z	BA81	I
A82		BA82	
A821	, Comma	BA821	. Period
A84	% Percent sign	BA84	¤ Lozenge
A841		BA841	
A842		BA842	
A8421		BA8421	

Fig. 2.4 Early code

2.15 REPRESENTATION

Representation refers to the form or manner in which the characters of a coded character set are recorded or transmitted on some medium, such as magnetic tape, magnetic disk, magnetic card, magnetic tape cassette/cartridge, magnetic core, paper tape, punched cards, data transmission lines, etc.

For such media representations, it is necessary to specify a precise relationship between the format characteristics of the medium (rows, columns, tracks, etc.) and the bits of the bit pattern of a character.

Characters may also be represented by graphic shapes either printed on paper or displayed on cathode ray tubes. Such graphic shapes may have a conventional font for human reading or a stylized font for machine reading (optical character recognition, OCR, or magnetic ink character recognition, MICR).

Bit Pattern			A	B	BA
		SP		–	&
1		1	/	J	A
2		2	S	K	B
2 1		3	T	L	C
4		4	U	M	D
4 1		5	V	N	E
4 2		6	W	O	F
4 2 1		7	X	P	G
8		8	Y	Q	H
8 1		9	Z	R	I
8 2		0			
8 2 1		#	,	$	.
8 4		@	%	*	⌑
8 4 1					
8 4 2					
8 4 2 1					

Fig. 2.5 6-bit code table, 8421 convention

A more subtle form of representation is where a sequence of characters is used, as an entity, to represent some single graphic or control meaning (see, for example, Chapter 26, Code Extension).

2.16 CODE TABLE

A code table is a compact matrix form of rows and columns for exhibiting the bit patterns and assigned meanings of a code. The 6-bit code, previously listed in Fig. 2.4, is exhibited in a code table using the 8421 convention (Fig. 2.5). It is also exhibited using the binary convention for representing bit patterns (Fig. 2.6).

The rule for reading these code tables is that the two high-order bits of the 6-bit bit pattern are shown as column headings, and the four low-order bits are shown as row sidings.

Bit Pattern → 6 5 ↓ 4 3 2 1		0 0	0 1	1 0	1 1
0 0 0 0		SP		–	&
0 0 0 1		1	/	J	A
0 0 1 0		2	S	K	B
0 0 1 1		3	T	L	C
0 1 0 0		4	U	M	D
0 1 0 1		5	V	N	E
0 1 1 0		6	W	O	F
0 1 1 1		7	X	P	G
1 0 0 0		8	Y	Q	H
1 0 0 1		9	Z	R	I
1 0 1 0		0			
1 0 1 1		#	,	$	.
1 1 0 0		@	%	*	⌑
1 1 0 1					
1 1 1 0					
1 1 1 1					

Fig. 2.6 6-bit code table, binary convention

Example 8

From the code tables of Figs. 2.5 and 2.6 we derive the following:

Graphic meaning	Bit pattern Fig. 2.5	Bit pattern Fig. 2.6
7	421	00 0111
R	B81	10 1001
E	BA41	11 0101
Space	No bits	00 0000

It is common practice to represent codes in code tables of 16 rows. Thus, a 6-bit code has a code table of 4 columns and 16 rows, a 7-bit code has a code table of 8 columns and 16 rows, an 8-bit code has a code table of 16 columns and 16 rows, and so on.

It is common practice to exhibit control meanings in code tables by either abbreviations or acronyms of the name of the control meaning.

Example 9

Control meaning	Abbreviation or acronym
Space	SP
Segment mark	SM
Record mark	RM
End of Transmission	EOT
Acknowledge	ACK
Negative Acknowledge	NAK
Null	NUL
Bell	BEL

A card code may be exhibited in a code table in the same way that a bit code is exhibited in a code table. The conventions for bit-code code tables are also used for card-code code tables. Zone punch hole patterns are shown as column headings. Digit punch hole patterns are shown as row sidings. The hole pattern for a particular character is made up of the column heading and row siding. A 64-character card code is shown in Fig. 2.7. The Hollerith Card Code is shown in Fig. 2.8.

Example 10

From Fig. 2.7 we derive the following:

Graphic	Hole pattern
Space	No holes
Z	0-9
<	12-8-6

Note. In card-code code tables, there may be exceptions to the general rule of column headings and row siding. These will be designated with small footnote numbers, with the actual hole patterns for such code positions (shown below) appearing in the table.

	Hole Pattern → ↓		0	11	12
		SP	SB [1]	-	&
	1	1	/	J	A
	2	2	S	K	B
	3	3	T	L	C
	4	4	U	M	D
	5	5	V	N	E
	6	6	W	O	F
	7	7	X	P	G
	8	8	Y	Q	H
	9	9	Z	R	I
	0	0	RM [2]	!	?
	8-3	#	,	$	.
	8-4	@	%	*	⌑
	8-5	:	WS	]	[
	8-6	>	\	;	<
	8-7	TM	SM	MC	GM

Hole Patterns:
[1] 8-2
[2] 0-8-2

Control Characters
SP - Space
TM - Tape Mark
SB - Substitute Blank
RM - Record Mark
WS - Word Separator
SM - Segment Mark
MC - Mode Change
GM - Group Mark

Fig. 2.7 Card-code code table

It is possible to exhibit, in one code table, both bit patterns and hole patterns, with zone bits and zone punches as column headings and digit bits and digit punches as row sidings. See, for example, Fig. 2.9. In more complex code tables, such as Figs. 2.8 and 2.10, zone punches for

Hole Pat.	12				12	12		12	12				12	12		12	Hole Pat.
		11				11	11	11		11				11	11	11	
			0		0		0	0			0		0		0	0	
	&	-	0	SP	{	¦	}										8-1
1	A	J	/	1	a	j	~		SOH	DC1							9-1
2	B	K	S	2	b	k	s		STX	DC2		SYN					9-2
3	C	L	T	3	c	l	t		ETX	DC3							9-3
4	D	M	U	4	d	m	u										9-4
5	E	N	V	5	e	n	v		HT		LF						9-5
6	F	O	W	6	f	o	w			BS	ETB						9-6
7	G	P	X	7	g	p	x		DEL		ESC	EOT					9-7
8	H	Q	Y	8	h	q	y			CAN							9
9	I	R	Z	9	i	r	z			EM			NUL	DLE			9-1
8-2	[	]	\	:													9-2
8-3	.	$	,	#					VT								9-3
8-4	<	*	%	@					FF	FS		DC4					9-4
8-5	(	)	_	`					CR	GS	ENQ	NAK					9-5
8-6	+	;	>	=					SO	RS	ACK						9-6
8-7	!	^	?	"					SI	US	BEL	SUB					9-7
Hole Pat.	12				12	12		12	12				12	12		12	
		11				11	11	11		11				11	11	11	
			0		0		0	0			0		0		0	0	
	8	8	8	8	8	8	8	8	8	8	8	8	8	8	8	8	

1	3
2	4

Block	Hole Patterns at:
1	Top and Left
2	Bottom and Left
3	Top and Right
4	Bottom and Right

Fig. 2.8 Hollerith Card Code

Bit Pattern →			A	B	BA
↓	Hole Pattern →		0	11	12
		SP	SB [1]	-	&
1	1	1	/	J	A
2	2	2	S	K	B
2 1	3	3	T	L	C
4	4	4	U	M	D
4 1	5	5	V	N	E
4 2	6	6	W	O	F
4 2 1	7	7	X	P	G
8	8	8	Y	Q	H
8 1	9	9	Z	R	I
8 2	0	0	RM [2]	!	?
8 2 1	8-3	#	,	$	.
8 4	8-4	@	%	*	⌑
8 4 1	8-5	:	WS	]	[
8 4 2	8-6	>	\	;	<
8 4 2 1	8-7	TM	SM	MC	GM

Hole Patterns:
[1] 8-2
[2] 0-8-2

Fig. 2.9 Code table, bit patterns and hole patterns

characters in the top rows of the table are different than they are for characters in the bottom rows of the table, and digit punches for characters in the left columns of the table are different than they are for characters in the right columns of the table. In such a case, zone punches are shown as column headings and column footings and digit punches are shown as left and right row sidings.

A rule for reading hole patterns for such a table must be stated. The table of Fig. 2.10, is blocked into four blocks, as shown below, with the rule for reading as follows:

1	3
2	4

Row	Bit Pat.	Hole Pat.	0	1	2	3	4	5	6	7	8	9	A	B	C	D	E	F	Hole Pat.
	Bit Pat.		00				01				10				11				
			00	01	10	11	00	01	10	11	00	01	10	11	00	01	10	11	
		Hole Pat.	9	9	9	9	9	9	9	9									Hole Pat.
			12				12	12		12	12	12		12	12				
				11				11	11	11		11	11	11		11			
					0		0		0	0	0		0	0			0		
0	0000		NUL [1]	DLE [2]	DS [3]	[4]	SP [5]	& [6]	- [7]	[8]					{ [9]	} [10]	\ [11]	0 [12]	8-1
1	0001	1	SOH	DC1	SOS				/ [13]		a	j	~		A	J	[14]	1	1
2	0010	2	STX	DC2	FS	SYN					b	k	s		B	K	S	2	2
3	0011	3	ETX	TM							c	l	t		C	L	T	3	3
4	0100	4	PF	RES	BYP	PN					d	m	u		D	M	U	4	4
5	0101	5	HT	NL	LF	RS					e	n	v		E	N	V	5	5
6	0110	6	LC	BS	ETB	UC					f	o	w		F	O	W	6	6
7	0111	7	DEL	IL	ESC	EOT					g	p	x		G	P	X	7	7
8	1000	8		CAN							h	q	y		H	Q	Y	8	8
9	1001	8-1		EM						`	i	r	z		I	R	Z	9	9
A	1010	8-2	SMM	CC	SM		¢	!	¦ [15]	:									8-2
B	1011	8-3	VT	CU1	CU2	CU3	.	$	,	#									8-3
C	1100	8-4	FF	IFS		DC4	<	*	%	@									8-4
D	1101	8-5	CR	IGS	ENQ	NAK	(	)	_	'									8-5
E	1110	8-6	SO	IRS	ACK		+	;	>	=									8-6
F	1111	8-7	SI	IUS	BEL	SUB	|	¬	?	"								EO	8-7
		Hole Pat.	9	9	9	9									9	9	9	9	
			12				12				12	12		12	12	12	12	12	
				11				11				11	11	11		11	0	11	
					0				0		0		0	0	0			0	

Hole Patterns:

[1]	9-12-0-8-1	[7]	11	[13]	0-1
[2]	9-12-11-8-1	[8]	12-11-0	[14]	9-11-0-1
[3]	9-11-0-8-1	[9]	12-0	[15]	12-11
[4]	9-12-11-0-8-1	[10]	11-0		
[5]	No Pch	[11]	0-8-2		
[6]	12	[12]	0		

1	3
2	4

Block	Hole Patterns at:
1	Top and Left
2	Bottom and Left
3	Top and Right
4	Bottom and Right

Fig. 2.10 256-character code table

Block 1: Zone punches at top of table, digit punches at left.
Block 2: Zone punches at bottom of table, digit punches at left.
Block 3: Zone punches at top of table, digit punches at right.
Block 4: Zone punches at bottom of table, digit punches at right.

2.16.1 Column Number, Row Number

For purposes of easy reference, the columns and rows of a code table are numbered and named. For the code table of Fig. 2.26, the 8 columns are numbered 0, 1, 2, 3, 4, 5, 6, 7, and the 16 rows are numbered 0, 1, 2, 3, . . . , 14, 15.

For the code table of Fig. 2.10, both the 16 columns and 16 rows are numbered (or named) 0, 1, 2, 3, 4, 5, 6, 7, 8, 9, A, B, C, D, E, F. This notation is called the *hexadecimal* notation.

2.16.2 Code Table Character Position, Code Table Characters Location

The position or location of a character in a code table is stated according to its column and row number. For the tables of Figs. 2.26 and 2.27, the convention is to give the position as x/y, where x is the code table column number and y is the code table row number. For the code table of Fig. 2.10, the hexadecimal convention mn is used, where m is the hexadecimal column number and n is the hexadecimal row number.

Example 11

In the code table of Fig. 2.26, the letter R is in position 5/2.

Example 12

In the code table of Fig. 2.10, the letter R is in position D9.

2.17 CODE NAMES

The following codes, to be discussed in detail later in this book, are used in this chapter to illustrate certain basic characteristics of codes. Their names, the derivation of which will be described later in this book, are used in this chapter. (The term *shifted*, used below, is explained later in this chapter.)

a) CCITT #2 A 58-character, shifted 5-bit code.
b) FIELDATA A 128-character, 7-bit code.
c) ASCII A 128-character, 7-bit code.
d) PTTC A 111-character, shifted 6-bit code.
e) BCDIC A 64-character, 6-bit code and 12-row card code.
f) EBCDIC A 256-character, 8-bit code and 12-row card code.
g) Hollerith A 256-character, 12-row card code.

BASIC CHARACTERISTICS

There are some basic characteristics of coded character sets. Not all of these characteristics will be exhibited by any particular code.

2.18 SHIFTED CODE

Recall that the total number of possible different bit patterns of a code is prescribed by the number of bits in the code byte: a code byte of 5 bits gives rise to $2^5 = 32$ different bit patterns; a code byte of 6 bits gives rise to $2^6 = 64$ different bit patterns; a code byte of 7 bits gives rise to $2^7 = 128$ different bit patterns; etc.

Ordinarily, the number of possible different characters (a character is a bit pattern with an assigned meaning) in a code equals the number of possible different bit patterns. But, by the use of a technique called shifting, the number of characters in a code may be increased beyond the number of bit patterns. Under this technique, the meaning of a bit pattern depends not only on the bit pattern itself, but also on the fact that it has been preceded in the data stream by some other particular bit pattern, which is called a precedence character or a shift character.

In CCITT #2 (Fig. 2.11), for example, there are two characters, Figure Shift (11011) and Letter Shift (11111). The meaning of a bit pattern in a data stream is determined not only by the bit pattern itself but also by which of the two precedence bit patterns has preceded it. By preceded, we do not necessarily mean "immediately" preceded. For example, if the bit pattern 01010 has been preceded by the bit pattern 11011 (Figure Shift), it would mean "4", but if it had been preceded by the bit pattern 11111 (Letter Shift), it would mean "R". A precedence character, when detected in the data stream, establishes a mode which remains in effect until another precedence character is detected, which then disestablishes the previous mode and establishes its own mode, which in its turn remains in effect until the subsequent detection of another precedence character.

The precedence characters are generally called shift characters because they are associated with the mechanism in a serial printer such as a typewriter which shifts from one case to the other.

In the serial printers that implement CCITT #2, the shift keys "lock in" the shift mode of the printing mechanism. Thus when the key or keys are depressed to generate the Figure Shift character, the Figure Shift Case is set for the printing mechanism and it remains set until the key or keys are depressed to generate the Letter Shift character. At that time, the Letter Shift case of the printing mechanism is set and it remains set until the key or keys are depressed to generate the Figure Shift character.

Bit pattern	Letter case	Figure case	Bit pattern	Letter case	Figure case
00000	Not used	Not used	10000	E	3
00001	T	5	10001	Z	+ or "
00010	CR	CR	10010	D	(2)
00011	0	9	10011	B	?
00100	SP	SP	10100	S	'
00101	H	(1)	10101	Y	6
00110	N	,	10110	F	(1)
00111	M	.	10111	X	/
01000	LF	LF	11000	A	-
01001	L	)	11001	W	2
01010	R	4	11010	J	Bell
01011	G	(1)	11011	FS	FS
01100	I	8	11100	U	7
01101	P	0	11101	Q	1
01111	C	:	11110	K	(
01111	V	= or ;	11111	(3)LS	LS

(1) For National Use
(2) Used for Answer Back
(3) Also used for Delete

CR Carriage Return
SP Space
LF Line Feed
FS Figure Shift
LS Letter Shift

Fig. 2.11 CCITT #2

In precedence codes, certain bit patterns, usually those associated with control meanings, are independent of shift. That is to say, the bit pattern of a shift-independent character has the same meaning, regardless of which precedence bit pattern has preceded it in the data stream. In CCITT #2, the control characters Carriage Return, Space, Line Feed, Figure Shift, and Letter Shift are shift-independent. There is a human-factors reason for this. Assume the following:

a) The Space bit pattern operates only in Letter Shift, not in Figure Shift.

b) An operator is transmitting data using a keyboard.

c) The data consists of blocks of numerics, the blocks separated by a Space.

Each time the operator comes to the end of a numeric block and wishes to key the Space, he would first have to depress the Letter Shift key, then the Space key, then the Figure Shift key (to reestablish the Figure Shift mode for the next block of numerics). In short, to generate the Space character he would have to have depressed three keys. Similarly, if we had assumed that the Space bit pattern operated only in Figure Shift (not in Letter Shift) and if the operator was transmitting text (alphabet blocks, separated by a Space), he would have to depress three keys in order to generate the Space character.

In both instances, if the Space key operated in both Letter Shift and Figure Shift, he would have had to depress only one key, the Space key. In short, making the Space character shift-independent increases operator productivity by decreasing the number of key strokes needed. Analysis shows that the other control characters—Carriage Return, Form Feed, Letter Shift, and Figure Shift—should be shift-independent for similar reasons.

If the number of bits in a code byte is x and if the number of shift-independent characters in a code is Y, then

- number of shift-dependent characters $= 2^{x+1} - 2Y$;
- total number of different characters shift-dependent and shift-independent $= 2^{x+1} - Y$.

CCITT #2 is a 5-bit shifted code, with 6 shift-independent characters. The number of shift-dependent characters is 52, and the total number of different characters is 58. PTTC (Fig. 2.30) is a 6-bit shifted code and has 17 shift-independent characters. The number of shift-dependent characters is 84; the total number of different characters is 111.

2.19 BINARY CODED DECIMAL (BCD)

The binary bit patterns for the ten decimal digits, shown in Fig. 2.12 under both the 8421 convention and the binary convention, are called Binary Coded Decimal bit patterns, with acronym BCD.

2.19.1 BCD for Numerics

For a code to have the characteristic of BCD bit patterns for numerics, the low-order four bits of the bit patterns for the numerics must be as shown in Fig. 2.12, and the high-order bits must be the same for all numerics. Figure 2.13 shows excerpts from two codes, ASCII and EBCDIC, with BCD for the numerics.

Decimal digits	Binary Coded Decimal bit patterns	
	8421 convention	Binary convention
0		0000
1	1	0001
2	2	0010
3	2 1	0011
4	4	0100
5	4 1	0101
6	4 2	0110
7	4 2 1	0111
8	8	1000
9	8 1	1001

Fig. 2.12 BCD bit patterns

	Column	0	1	2	3	4	5	6	7					C	D	E	F
	Bit Pat. →													11			
Row	↓	000	001	010	011	100	101	110	111					00	01	10	11
0	0000				0		P										0
1	0001				1	A	Q							A	J		1
2	0010				2	B	R							B	K	S	2
3	0011				3	C	S							C	L	T	3
4	0100				4	D	T							D	M	U	4
5	0101				5	E	U							E	N	V	5
6	0110				6	F	V							F	O	W	6
7	0111				7	G	W							G	P	X	7
8	1000				8	H	X							H	Q	Y	8
9	1001				9	I	Y							I	R	Z	9
10	1010					J	Z										
11	1011					K											
12	1100					L											
13	1101					M											
14	1110					N											
15	1111					O											
		ASCII												EBCDIC			

Fig. 2.13 BCD for numerics and alphabetics

2.19.2 BCD for Alphabetics

For some codes, the alphabetics have bit patterns where the low-order four bits for A to I, for J to R, and for S to Z have BCD bit patterns. In Fig. 2.13 EBCDIC exhibits this characteristic while ASCII does not.

2.20 SEQUENCES OF BIT PATTERNS

2.20.1 Numerics in Numeric Sequence

The natural sequence of numerics is 0, 1, 2, 3, 4, 5, 6, 7, 8, 9. The binary bit patterns of the numerics may be in numeric sequence for a code. In Fig. 2.14, ASCII and EBCDIC exhibit this characteristic; CCITT #2 and BCDIC do not. (BCDIC almost does, since its numerics are in the sequence 1, 2, 3, 4, 5, 6, 7, 8, 9, 0.)

	Column	0	1	2	3									C	D	E	F
	Bit Pat. →													11			
Row	↓	000	001	010	011	00	01	10	11	00	01	10	11	00	01	10	11
0	0000				0		3										0
1	0001				1	5				1							1
2	0010				2					2							2
3	0011				3	9				3							3
4	0100				4					4							4
5	0101				5		6			5							5
6	0110				6					6							6
7	0111				7					7							7
8	1000				8					8							8
9	1001				9		2			9							9
10	1010									0							
11	1011																
12	1100					8	7										
13	1101					0	1										
14	1110																
15	1111																
		ASCII				CCITT#2				BCDIC				EBCDIC			

Fig. 2.14 Numerics, numeric sequence, contiguous sequence

2.20.2 Numerics in Contiguous Sequence

For some codes, the binary bit patterns of the numerics are in contiguous sequence, that is, the sequence of bit patterns is continuous and uninterrupted. In Fig. 2.14, ASCII, BCDIC, and EBCDIC exhibit this characteristic; CCITT #2 does not.

2.20.3 Alphabetics in Alphabetic Sequence

The natural sequence of alphabetics is A, B, C, . . . , X, Y, Z. For some codes, the binary bit patterns of the alphabetics are in the same relative sequence as the alphabetics. Figure 2.15 shows the alphabetics of ASCII, FIELDATA, BCDIC, and EBCDIC. Figure 2.16 shows that the alphabetics of EBCDIC, although not contiguous in the sequence of bit patterns, are nevertheless in relative sequence. By contrast, Fig. 2.17 shows that the alphabetics of BCDIC are not in relative sequence. Figure 2.18 shows that the alphabetics of ASCII are in relative sequence and in contiguous sequence. The alphabetics of FIELDATA can be seen from Fig. 2.15 to be in relative sequence and in contiguous sequence.

	Column	4	5	6	7	4	5	6	7	0	1	2	3	C	D	E	F
														11			
Row	Bit Pat.	100	101	110	111	100	101	110	111	00	01	10	11	00	01	10	11
0	0000		P						K								
1	0001	A	Q						L			J	A	A	J		
2	0010	B	R						M		S	K	B	B	K	S	
3	0011	C	S						N		T	L	C	C	L	T	
4	0100	D	T						O		U	M	D	D	M	U	
5	0101	E	U						P		V	N	E	E	N	V	
6	0110	F	V					A	Q		W	O	F	F	O	W	
7	0111	G	W					B	R		X	P	G	G	P	X	
8	1000	H	X					C	S		Y	Q	H	H	Q	Y	
9	1001	I	Y					D	T		Z	R	I	I	R	Z	
10	1010	J	Z					E	U								
11	1011	K						F	V								
12	1100	L						G	W								
13	1101	M						H	X								
14	1110	N						I	Y								
15	1111	O						J	Z								
		ASCII				FIELDATA				BCDIC				EBCDIC			

Fig. 2.15 Contiguous and noncontiguous alphabetics

1100 0000	
0001	A
0010	B
0011	C
0100	D
0101	E
0110	F
0111	G
1100 1000	H
1001	I
1010	
1011	
1100	
1101	
1110	
1111	
1101 0000	
0001	J
0010	K
0011	L
0100	M
0101	N
0110	O
0111	P
1101 1000	Q
1001	R
1010	
1011	
1100	
1101	
1110	
1111	
1110 0000	
0001	
0010	S
0011	T
0100	U
0101	V
0110	W
0111	X
1110 1000	Y
1001	Z
1010	
1011	
1100	
1101	
1110	
1111	

Fig. 2.16 EBCDIC alphabetics in relative sequence and in non-contiguous sequence

01 0000	
0001	
0010	S
0011	T
0100	U
0101	V
0110	W
0111	X
01 1000	Y
1001	Z
1010	
1011	
1100	
1101	
1110	
1111	
10 0000	
0001	J
0010	K
0011	L
0100	M
0101	N
0110	O
0111	P
10 1000	Q
1001	R
1010	
1011	
1100	
1101	
1110	
1111	
11 0000	
0001	A
0010	B
0011	C
0100	D
0101	E
0110	F
0111	G
11 1000	H
1001	I
1010	
1011	
1100	
1101	
1110	
1111	

Fig. 2.17 BCDIC alphabetics not in relative sequence and in noncontiguous sequence

100 0000	
0001	A
0010	B
0011	C
0100	D
0101	E
0110	F
0111	G
100 1000	H
1001	I
1010	J
1011	K
1100	L
1101	M
1110	N
1111	O
101 0000	P
0001	Q
0010	R
0011	S
0100	T
0101	U
0110	V
0111	W
101 1000	X
1001	Y
1010	Z
1011	
1100	
1101	
1110	
1111	
110 0000	
0001	
0010	
0011	
0100	
0101	
0110	
0111	
110 1000	
1001	
1010	
1011	
1100	
1101	
1110	
1111	

Fig. 2.18 ASCII alphabetics in relative sequence and in contiguous sequence

2.20.4 Alphabetics in Contiguous Sequence

For some codes, the bit patterns of the alphabetics are in contiguous sequence. In Fig. 2.15, ASCII and FIELDATA exhibit this characteristic (ASCII is also shown in Fig. 2.18). BCDIC and EBCDIC do not, as can be seen in Figs. 2.17 and 2.16.

2.20.5 Alphabetics in Noncontiguous Sequence

For some codes, the bit patterns of the alphabetics are not in contiguous sequence. In Figs. 2.17 and 2.16, BCDIC and EBCDIC exhibit this characteristic. ASCII (Fig. 2.18) and FIELDATA (Fig. 2.15) do not.

Note 1. Characteristics described in Sections 2.20.4 and 2.20.5 are, of course, opposite. A full discussion of the significance of contiguity and noncontiguity of the alphabetics is given later in this book.

Note 2. Some codes (for example, that of the IBM 7030 (Stretch) computer) exhibit the characteristic of "interleaved alphabets;" that is, the upper- and lower-case alphabetics are interleaved. This is discussed more fully in Chapter 3.

2.21 SIGNED NUMERICS

It is a common practice in punched card applications to punch the 11-punch in the same card column as a numeric to indicate a negative numeric. Thus 11-0, 11-1, . . . , 11-9 represent −0, −1, . . . , −9, respectively. It is a recognized though little-used practice to punch the 12-punch in the same card column as a numeric to indicate a positive numeric. Thus 12-0, 12-1, ... , 12-9 represent +0, +1, . . . , +9, respectively. And, of course, 0, 1, . . . , 9 punches are used to represent absolute numerics 0, 1, . . . , 9, respectively. This is shown in Sections 1 and 2 of Fig. 2.19.

In the Hollerith card code, the hole patterns 12-0, 12-1, 12-2, . . . , 12-9 are assigned to {, A, B, . . . , I; the hole patterns 11-0, 11-1, 11-2, . . . , 11-9 are assigned to }, J, K, . . . , R; the hole patterns 0-2, 0-3, . . . , 0-9 are assigned to S, T, . . . , Z; and the hole patterns 0, 1, 2, . . . , 9 are assigned to 0, 1, 2, . . . , 9 as shown in Section 1 of Fig. 2.20. For ASCII and EBCDIC, the graphics { and }, the alphabetics A through Z, and numerics 0 through 9 have bit patterns as shown in Sections 2 and 3 of Fig. 2.20.

It is to be noted, therefore, that such over-punched numerics in the card code have a duality of meaning. For example, the hole pattern 12-1 might mean A, or it might mean +1. There is nothing intrinsic to the hole pattern itself that determines which meaning is to be applied. The actual meaning would be determined within the context of a data processing application.

Row	Column								3	4	5	6	7	C	D	E	F
	Bit Pat. →													11			
↓									011	100	101	110	111	00	01	10	11
0	0000	12-0	11-0	0		+0	-0	0	0		-7			+0	-0		0
1	0001	12-1	11-1	1		+1	-1	1	1	+1	-8			+1	-1		1
2	0010	12-2	11-2	2		+2	-2	2	2	+2	-9			+2	-2		2
3	0011	12-3	11-3	3		+3	-3	3	3	+3				+3	-3		3
4	0100	12-4	11-4	4		+4	-4	4	4	+4				+4	-4		4
5	0101	12-5	11-5	5		+5	-5	5	5	+5				+5	-5		5
6	0110	12-6	11-6	6		+6	-6	6	6	+6				+6	-6		6
7	0111	12-7	11-7	7		+7	-7	7	7	+7				+7	-7		7
8	1000	12-8	11-8	8		+8	-8	8	8	+8				+8	-8		8
9	1001	12-9	11-9	9		+9	-9	9	9	+9				+9	-9		9
10	1010									-1							
11	1011									-2			+0				
12	1100									-3							
13	1101									-4			-0				
14	1110									-5							
15	1111									-6							
		Hole Patterns				Equivalent Signed Numerics			ASCII Signed Numerics					EBCDIC Signed Numerics			
		Section 1				Section 2			Section 3					Section 4			

Fig. 2.19 Signed numerics

In consequence of the relationship between positive, negative, and absolute numerics and hole patterns (Sections 1 and 2, Fig. 2.19) and in consequence of the relationship between hole patterns and ASCII and EBCDIC bit patterns (Sections 1, 2, and 3, Fig. 2.20), the positive, negative, and absolute numerics take bit patterns for ASCII and EBCDIC as shown in Sections 3 and 4 of Fig. 2.19.

The signed and absolute numerics for EBCDIC (Section 4, Fig. 2.19) exhibit the following characteristics:

a) For all numerics, signed or absolute, the numerics 0 to 9 have the low-order four bits as BCD bit patterns.

b) For all positive numerics 0 through 9, the four high-order bits are the same.

Column							0	1	2	3	4	5	6	7	C	D	E	F
Bit Pat.							01				10				11			
							00	01	10	11	00	01	10	11	00	01	10	11
Row	Bit Pat.	Hole Pat.	12	11	0													
0	0000	0	{	}		0				0	A	Q			{	}		0
1	0001	1	A	J		1				1	B	R			A	J		1
2	0010	2	B	K	S	2				2	C	S			B	K	S	2
3	0011	3	C	L	T	3				3	D	T			C	L	T	3
4	0100	4	D	M	U	4				4	E	U			D	M	U	4
5	0101	5	E	N	V	5				5	F	V			E	N	V	5
6	0110	6	F	O	W	6				6	G	W			F	O	W	6
7	0111	7	G	P	X	7				7	H	X			G	P	X	7
8	1000	8	H	Q	Y	8				8	I	Y			H	Q	Y	8
9	1001	9	I	R	Z	9				9	J	Z			I	R	Z	9
10	1010										K			{				
11	1011										L							
12	1100										M			}				
13	1101										N							
14	1110										O							
15	1111										P							
			HOLLERITH HOLE PATTERNS SECTION 1				ASCII BIT PATTERNS SECTION 2								EBCDIC BIT PATTERNS SECTION 3			

Fig. 2.20 Alphabetics and numerics

c) For all negative numerics 0 through 9, the four high-order bits are the same.

d) For all absolute numerics 0 through 9, the four high-order bits are the same.

Note. In characteristics (b), (c), and (d) above, the actual four high-order bits are not important. What is important is that for each category—(b), (c), (d)—the four high-order bits are the same.

It is clear that when the arithmetic circuits of a CPU are built around the EBCDIC signed and absolute numerics advantage can be taken of characteristics (a), (b), (c), and (d). It is equally clear, that for ASCII, arithmetic circuits would have to be more complex, since characteristics (a), (b), and (c) are not present. A full discussion of this is given later.

2.22 SPACE CHARACTER HAS "NO PUNCHES" CARD CODE

It is an established card practice for the Space character to generate a "no punches," or "blank column," card code. This characteristic is essential in data processing card applications where fields are left blank on punched cards in the initial keypunching operation—blank fields to be filled with punched data in subsequent card operations.

The Hollerith Card Code, also called the Twelve-Row Card Code, and the EBCDIC Card Code (see Chapters 11, 16, and 17) have this characteristic. The 96-Column Card (see Chapter 27) has this characteristic. During the technical debates in standards committees on binary card codes and on the Decimal ASCII Card Code (Chapter 16, Decimal ASCII), there was a technical controversy as to whether the "no punches" card hole pattern should be assigned to the Space character or to the Null character. This controversy was finally resolved with respect to Decimal ASCII by assigning the "no punches" to the Space character, in accord with de facto practice. It was not resolved for binary card codes, because the standards committee ceased to study binary card codes.

2.23 DUALS

The practice of mapping more than one graphic meaning to a single bit pattern or hole pattern is quite common. The different graphics with the same bit pattern or hole pattern are called duals. Sometimes, more than two graphics are mapped to a single bit pattern or hole pattern.

The duals of BCDIC are shown in Fig. 2.21.

Graphics	Hole pattern	Bit pattern
@ or '	8-4	84
# or =	8-3	8 21
& or +	12	BA
% or (	0-8-4	A84
¤ or)	12-8-4	BA84

Fig. 2.21 BCDIC duals

Some European languages require 29 letters, three more than the 26 letters of the English language. The additional three letters, which occur in both lower- and upper-case alphabetics, are called diacritics. Some codes, EBCDIC and the ISO 7-Bit Code, for example, accommodate this aspect by assigning six code positions for alphabetic extenders (or National Use graphics, as they are sometimes called). The EBCDIC scheme is shown in Fig. 2.22, followed by the ISO scheme in Fig. 2.23.

		GRAPHICS			
Hex position	Bit pattern	U.S.A.	Germany	Norway/ Denmark	Sweden/ Finland
7B	0111 1011	#	Ä	Æ	Ä
7C	0111 1100	@	Ö	Ø	Ö
5B	0101 1011	$	Ü	Å	Å
7F	0111 1111	"	ä	æ	ä
4A	0100 1010	¢	ö	ø	ö
5A	0101 1010	!	ü	å	å

Fig. 2.22 EBCDIC alphabetic extender graphics

		GRAPHICS			
Column row	Bit pattern	U.S.A.	Germany	Norway/ Denmark	Sweden/ Finland
5/11	101 1011	[	Ä	Æ	Ä
5/12	101 1100	/	Ö	Ø	Ö
5/13	101 1101	]	Ü	Å	Å
7/11	111 1011	{	ä	æ	ä
7/12	111 1100	¦	ö	ø	ö
7/13	111 1101	}	ü	å	å

Fig. 2.23 ISO National Use graphics

It is to be noted that the five BCDIC duals (Fig. 2.21) create duals within a country (U.S.A.), while the alphabetic extender duals create duals between countries. The former situation can be very troublesome (if all ten graphics are needed in the same data processing application, for example), while the latter situation does not cause trouble (for example, systems problems) as far as is known today. Duals are not good or bad, per se. Each situation must be examined individually.

There are, theoretically, two kinds of duals.

2.23.1 Many-to-one

Many-to-one refers to different meanings mapped into the same code position. This is the type described above.

2.23.2 One-to-many

One-to-many refers to a single meaning mapped into different code positions. Generally, this is a situation that will arise not within a code but rather between two different codes. For example, the 7-Bit Code has two different control characters, Line Feed and Carriage Return. These two functions are conbined into one EBCDIC Control Character, New Line. There is an obvious problem in trying to determine the translation relationship between these codes with respect to these three characters.

2.24 COLLATING SEQUENCE MATCHES BIT SEQUENCE

The bit sequence of a code is from low (all zero-bits) to high (all one-bits). Thus for EBCDIC, the bit sequence is 00000000, 00000001, 00000010, . . . , 11111101, 11111110, 11111111. In a code, graphic meanings are assigned to some of the bit patterns. For reasons outside the code, there may be an established sequence, from low to high, for these graphics. Such a sequence is called a collating sequence. The collating sequence of the graphics may, or may not, match the bit sequence of the graphics.

In the 64-character, 6-bit BCDIC, for example, the collating sequence does not match the bit sequence. Figure 2.9 shows the 64 characters in bit sequence. Each of the 64 BCDIC characters was assigned a collating number, from 0, low, to 63, high. The 64-characters of Fig. 2.9 are shown reordered into correct collating sequence in Fig. 2.24, with the collating numbers shown in each code table position. Figure 2.25 shows *some* of the BCDIC characters in column (1). Column (2) shows the collating number, and column (3) shows the bit patterns from Fig. 2.9.

The sorting or collating operation in a computer involves putting items in an ordered sequence, the collating sequence. Visualize a sort on a one-character field. Then, for two items, X1 or X2, the following question is asked:

Is X1 greater than, equal to, or less than X2?

When this question is answered, the two items X1 and X2 can then be arranged in correct sequence. Actually, the comparison instruction, which asks the question above, performs a binary subtraction, X1 − X2, and examines the sign and magnitude of the result.

First a binary subtraction is performed:

$$X1 - X2 = Y.$$

		SP 0	ϒ 16	G 32	U 48
		. 1	\ 17	H 33	V 49
		⌑ 2	⧻ 18	I 34	W 50
		[3	ƀ 19	! 35	X 51
		< 4	# 20	J 36	Y 52
		‡ 5	@ 21	K 37	Z 53
		& 6	: 22	L 38	0 54
		$ 7	> 23	M 39	1 55
		* 8	√ 24	N 40	2 56
		] 9	? 25	O 41	3 57
		; 10	A 26	P 42	4 58
		Δ 11	B 27	Q 43	5 59
		– 12	C 28	R 44	6 60
		/ 13	D 29	‡ 45	7 61
		, 14	E 30	S 46	8 62
		% 15	F 31	T 47	9 63

Fig. 2.24 BCDIC collating numbers

Then,

If Y is minus, $X1 < X2$;

or

If Y is zero, $X1 = X2$;

or

If Y is positive, $X1 > X2$.

Performing this binary comparison on the bit patterns of column (3) will not yield the desired result. But if the binary comparison were performed on the pseudo bit patterns of column (4), the desired result would be yielded. In short, if the bit patterns of column (3) are converted into the pseudo bit patterns of column (4) *before* comparison, the graphics of BCDIC can be sorted according to the prescribed collating sequence.

In some BCDIC computers, this conversion before comparison was achieved with a software routine; in other BCDIC computers it was achieved with a hardware comparator. In one instance there was a performance penalty, and in the other instance there was additional hardware cost.

1 Graphic	2 Collating number	3 Bit pattern	4 Pseudo bit pattern
Space	0	00 0000	00 0000
.			
.			
$	7	10 1011	00 0111
*	8	10 1100	00 1000
.			
.			
?	25	11 1010	01 1001
A	26	11 0001	01 1010
B	27	11 0010	01 1011
.			
.			
H	33	11 1000	10 0001
I	34	11 1001	10 0010
.			
.			
J	36	10 0001	10 0100
K	37	10 0010	10 0101
.			
.			
Q	43	10 1000	10 1011
R	44	10 1001	10 1100
.			
.			
S	46	01 0010	10 1110
T	47	01 0011	10 1111
.			
.			
Y	52	01 1000	11 0100
Z	53	01 1001	11 0101
0	54	00 1010	11 0110
1	55	00 0001	11 0111
2	56	00 0010	11 1000
.			
.			
8	62	00 1000	11 1110
9	63	00 1001	11 1111

Fig. 2.25 BCDIC collating sequence

In developing EBCDIC, a primary design factor was collating sequence (see Chapter 8, the Sequence of EBCDIC). The 88 graphics of EBCDIC were assigned 8-bit bit patterns such that the collating sequence matched the bit sequence, thus saving software or hardware costs for customers.

2.25 SUMMARY OF CODE CHARACTERISTICS

Seven codes or representations are given as follows:

Code	Figure
ASCII	2.26
An 8-bit representation	2.27
EBCDIC	2.28
BCDIC	2.29
PTTC	2.30
CCITT #2	2.31
FIELDATA	2.32

These are analyzed below as they do, or do not, exhibit the previous characteristics.

Figure	2.26	2.27	2.28	2.29	2.30	2.31	2.32
Characteristics ↓ / Code	ASCII	8-Bit Rep.	EBCDIC	BCDIC	PTTC	CCITT #2	FIEL DATA
Shifted code	No	No	No	No	Yes	Yes	No
BCD for numerics	Yes	Yes	Yes	No	No	Yes	Yes
BCD for alphabetics	No	No	Yes	Yes	Yes	No	No
Numerics in numeric sequence	Yes	Yes	Yes	No	No	No	Yes
Numerics in contiguous sequence	Yes	Yes	Yes	Yes	Yes	No	Yes
Alphabetics in alphabetic sequence	Yes	Yes	Yes	No	No	No	Yes
Alphabetics in contiguous sequence	Yes	Yes	No	No	No	No	Yes
Alphabetics in noncontiguous sequence	No	No	Yes	Yes	Yes	Yes	No
Signed numerics	No	No	Yes	Yes	Yes	No	No
Collating sequence matches bit sequence	Yes	Yes	Yes	No	No	No	Yes

	b7	0	0	0	0	1	1	1	1
	b6	0	0	1	1	0	0	1	1
	b5	0	1	0	1	0	1	0	1
b4 b3 b2 b1	Row \ Col	0	1	2	3	4	5	6	7
0 0 0 0	0	NUL	DLE	SP	0	@	P		p
0 0 0 1	1	SOH	DC1	!	1	A	Q	a	q
0 0 1 0	2	STX	DC2	"	2	B	R	b	r
0 0 1 1	3	ETX	DC3	#	3	C	S	c	s
0 1 0 0	4	EOT	DC4	$	4	D	T	d	t
0 1 0 1	5	ENQ	NAK	%	5	E	U	e	u
0 1 1 0	6	ACK	SYN	&	6	F	V	f	v
0 1 1 1	7	BEL	ETB	'	7	G	W	g	w
1 0 0 0	8	BS	CAN	(	8	H	X	h	x
1 0 0 1	9	HT	EM	)	9	I	Y	i	y
1 0 1 0	10	LF	SUB	*	:	J	Z	j	z
1 0 1 1	11	VT	ESC	+	;	K	[	k	{
1 1 0 0	12	FF	FS	,	<	L	\	l	¦
1 1 0 1	13	CR	GS	– 11	=	M	]	m	}
1 1 1 0	14	SO	RS	.	>	N	^	n	~
1 1 1 1	15	SI	US	/ 0-1	?	O	_	o	DEL

Fig. 2.26 ASCII

Row	Column	0	1	2	3	4	5	6	7	8	9	10	11	12	13	14	15
	Bit Pat. →	00				01				10				11			
	↓	00	01	10	11	00	01	10	11	00	01	10	11	00	01	10	11
0	0000	NUL	DLE	SP	0	@	P	`	p								
1	0001	SOH	DC1	! ①	1	A	Q	a	q								
2	0010	STX	DC2	"	2	B	R	b	r								
3	0011	ETX	DC3	#	3	C	S	c	s								
4	0100	EOT	DC4	$	4	D	T	d	t								
5	0101	ENQ	NAK	%	5	E	U	e	u								
6	0110	ACK	SYN	&	6	F	V	f	v								
7	0111	BEL	ETB	'	7	G	W	g	w								
8	1000	BS	CAN	(	8	H	X	h	x								
9	1001	HT	EM	)	9	I	Y	i	y								
10	1010	LF	SUB	*	:	J	Z	j	z								
11	1011	VT	ESC	+	;	K	[	k	{								
12	1100	FF	FS	,	<	L	\	l	¦								
13	1101	CR	GS	-	=	M	]	m	}								
14	1110	SO	RS	.	>	N	^ ②	n	~	EC							
15	1111	SI	US	/	?	O	_	o	DEL	BC							EO

① May be "|"
② May be "¬"

Fig. 2.27 An 8-bit representation

Row	Bit Pat.	Hole Pat.	0	1	2	3	4	5	6	7	8	9	A	B	C	D	E	F	Hole Pat.
	Bit Pat.		00				01				10				11				
			00	01	10	11	00	01	10	11	00	01	10	11	00	01	10	11	
		Hole Pat.	9	9	9	9	9	9	9	9									Hole Pat.
			12				12	12		12	12	12		12	12				
				11				11	11	11		11	11	11		11			
					0		0		0	0	0		0	0			0		
0	0000		NUL [1]	DLE [2]	DS [3]	[4]	SP [5]	& [6]	- [7]	[8]					{ [9]	} [10]	\ [11]	0 [12]	8-1
1	0001	1	SOH	DC1	SOS				/ [13]		a	j	~		A	J	[14]	1	1
2	0010	2	STX	DC2	FS	SYN					b	k	s		B	K	S	2	2
3	0011	3	ETX	TM							c	l	t		C	L	T	3	3
4	0100	4	PF	RES	BYP	PN					d	m	u		D	M	U	4	4
5	0101	5	HT	NL	LF	RS					e	n	v		E	N	V	5	5
6	0110	6	LC	BS	ETB	UC					f	o	w		F	O	W	6	6
7	0111	7	DEL	IL	ESC	EOT					g	p	x		G	P	X	7	7
8	1000	8		CAN							h	q	y		H	Q	Y	8	8
9	1001	8-1		EM						`	i	r	z		I	R	Z	9	9
A	1010	8-2	SMM	CC	SM		¢	!	¦ [15]	:									8-2
B	1011	8-3	VT	CU1	CU2	CU3	.	$	,	#									8-3
C	1100	8-4	FF	IFS		DC4	<	*	%	@									8-4
D	1101	8-5	CR	IGS	ENQ	NAK	(	)	_	'									8-5
E	1110	8-6	SO	IRS	ACK		+	;	>	=									8-6
F	1111	8-7	SI	IUS	BEL	SUB	|	¬	?	"								EO	8-7
		Hole Pat.	9	9	9	9									9	9	9	9	
			12				12				12	12		12	12	12	12	12	
				11				11				11	11	11		11	0	11	
					0				0		0		0	0	0			0	

Hole Patterns:

- [1] 9-12-0-8-1
- [2] 9-12-11-8-1
- [3] 9-11-0-8-1
- [4] 9-12-11-0-8-1
- [5] No Pch
- [6] 12
- [7] 11
- [8] 12-11-0
- [9] 12-0
- [10] 11-0
- [11] 0-8-2
- [12] 0
- [13] 0-1
- [14] 9-11-0-1
- [15] 12-11

1	3
2	4

Block	Hole Patterns at:
1	Top and Left
2	Bottom and Left
3	Top and Right
4	Bottom and Right

Fig. 2.28 EBCDIC

Bit Pattern →			A	B	BA
↓	Hole Pattern →		0	11	12
		SP	ƀ [1]	-	& or +
1	1	1	/	J	A
2	2	2	S	K	B
2 1	3	3	T	L	C
4	4	4	U	M	D
4 1	5	5	V	N	E
4 2	6	6	W	O	F
4 2 1	7	7	X	P	G
8	8	8	Y	Q	H
8 1	9	9	Z	R	I
8 2	0	0	‡ [2]	!	?
8 2 1	8-3	# or =	,	$	.
8 4	8-4	@ or '	% or (	*	⌑ or)
8 4 1	8-5	:	γ	]	[
8 4 2	8-6	>	\	;	<
8 4 2 1	8-7	√	⧻	Δ	ǂ

Hole Patterns:

[1] 8–2

[2] 0–8–2

Fig. 2.29 BCDIC

		Lower Case				Upper Case			
Bit Pattern			A	B	BA		A	B	BA
	Hole Pattern		0	11	12		11-0	12-11	12-0
		SP	@ [1]	-	&	SP	¢ [14]	_ [18]	+ [21]
1	1	1	/	j	a	= [3]	? [15]	J	A
2	2	2	s	k	b	⌑ [4]	S	K	B
2 1	3	3	t	l	c	; [5]	T	L	C
4	4	4	u	m	d	: [6]	U	M	D
4 1	5	5	v	n	e	% [7]	V	N	E
4 2	6	6	w	o	f	' [8]	W	O	F
4 2 1	7	7	x	p	g	" [9]	X	P	G
8	8	8	y	q	h	* [10]	Y	Q	H
8 1	9	9	z	r	i	([11]	Z	R	I
8 2	0	0	‡ [2]	<	>	) [12]	‡ [16]	γ [19]	√ [22]
8 2 1	8-3	#	,	$	.	± [13]	, [17]	! [20]	. [23]
8 4	4	PN	BYP	RES	PF	PN	BYP	RES	PF
8 4 1	5	RS	LF	NL	HT	RS	LF	NL	HT
8 4 2	6	UC	EOB	BS	LC	UC	EOB	BS	LC
8 4 2 1	9	EOT	PRE	IL	DEL	EOT	PRE	IL	DEL
	Hole Pattern	9	9-0	9-11	9-12	9	9-0	9-11	9-12

Hole Patterns:

1	8-4	8	8-5	15	12-8-2	22	8-7
2	0-8-2	9	8-1	16	12-8-7	23	12-8-1
3	8-6	10	11-8-4	17	0-8-1		
4	12-8-4	11	12-8-5	18	0-8-6		
5	11-8-6	12	11-8-5	19	0-8-5		
6	8-2	13	0-8-7	20	11-8-2		
7	0-8-4	14	11-8-7	21	12-8-6		

1	3
2	4

Block	Hole Patterns at:
1	Top And Left
2	Bottom and Left
3	Top and Left
4	Bottom and Left

Fig. 2.30 PTTC

Bit pattern	Letter case	Figure case	Bit pattern	Letter case	Figure case
00000	Not used	Not used	10000	E	3
00001	T	5	10001	Z	+ or "
00010	CR	CR	10010	D	(2)
00011	0	9	10011	B	?
00100	SP	SP	10100	S	'
00101	H	(1)	10101	Y	6
00110	N	,	10110	F	(1)
00111	M	.	10111	X	/
01000	LF	LF	11000	A	–
01001	L	)	11001	W	2
01010	R	4	11010	J	Bell
01011	G	(1)	11011	FS	FS
01100	I	8	11100	U	7
01101	P	0	11101	Q	1
01110	C	:	11110	K	(
01111	V	= or ;	11111	(3) LS	LS

(1) For National Use
(1) Used for Answer Back
(3) Also used for Delete

CR Carriage Return
SP Space
LF Line Feed
FS Figure Shift
LS Letter Shift

Fig. 2.31 CCITT #2

Column		0	1	2	3	4	5	6	7
	Bit Pattern b7	0	0	0	0	1	1	1	1
	b6	0	0	1	1	0	0	1	1
	b5	0	1	0	1	0	1	0	1
Row	b4 b3 b2 b1								
0	0 0 0 0	CONTROL (NOT DEFINED)				MS	K	)	0
1	0 0 0 1					UC	L	-	1
2	0 0 1 0					LC	M	+	2
3	0 0 1 1					LF	N	<	3
4	0 1 0 0					CR	O	=	4
5	0 1 0 1					SP	P	>	5
6	0 1 1 0					A	Q	_	6
7	0 1 1 1					B	R	$	7
8	1 0 0 0					C	S	*	8
9	1 0 0 1					D	T	(	9
10	1 0 1 0					E	U	"	'
11	1 0 1 1					F	V	:	;
12	1 1 0 0					G	W	?	/
13	1 1 0 1					H	X	!	.
14	1 1 1 0					I	Y	,	SPEC
15	1 1 1 1					J	Z	STOP	IDLE

Fig. 2.32 FIELDATA

2.26 COMPATABILITY

Compatability between two different codes is not a single, simple aspect. It is a number of aspects:

- *Structural Similarity*. The code table is a compact way to exhibit the relationship between the graphic and control meanings and the associated bit patterns or hole patterns of a coded character set. As can be seen in Figs. 2.28 and 2.29, the 26 alphabetics of EBCDIC and BCDIC are positioned similarly in three contiguous columns of the code tables (although not in the same order of columns). From this columnar positioning is revealed the fact that the low-order four bits of alphabetics are the same in both codes. Equally significant, the noncontiguous alphabetics are noncontiguous in precisely the same way in both codes. Further, the specials in both codes are positioned (mostly) in a 5 by 4 block of the code table. These two codes are said to be structurally similar. By contrast, the alphabetics of ASCII (Fig. 2.26) are positioned in 26 contiguous bit-pattern positions in two columns. EBCDIC and the 7-Bit Code are said to be structurally dissimilar.

- *Collating Sequence.* The collating sequence of the two codes should match. If the codes are of different size, the collating sequence of the smaller code should be embedded in the collating sequence of the larger code (see Chapter 8, The Sequence of EBCDIC, for a full discussion of this embedment).
- *Functional Equivalence.* The codes should be functionally equivalent; that is, they should have the same set of control and graphic meanings, although not necessarily with the same set of bit patterns. A smaller code is said to be functionally equivalent upward to a larger code if the smaller code's set of graphics and control meanings is contained in the set of the larger code. EBCDIC and the Hollerith Card Code are functionally equivalent. ASCII is functionally equivalent upward to EBCDIC.
- *Translation Relationship.* Translation relationships between two codes should be as simple as possible. The translation simplicity is directly related to the structural similarity.

In debates on code compatibility, it often happens that one debater views two codes as incompatible because not all of the four aspects above are present, while the other debater views the two codes as compatible because at least one of the aspects above is present. Certainly, two codes are compatible if all four aspects are present, incompatible if none of the four are present. For codes where some aspects are present and others are not, to determine and agree on which are present and which are not is preferable to arguing about the then indeterminate question of "compatibility."

2.27 GRAPHICS FOR CONTROLS

In some codes, graphic representations are assigned to the control characters. The virtue of this is that when data are listed, particularly in debugging operations, control as well as graphic characters are visible.

In BCDIC for example, graphic representations are assigned to seven control characters:

ƀ	Substitute Blank
Δ	Mode Change
γ	Word Separator
ǂ	Record Mark
⧧	Group Mark
⧺	Segment Mark
√	Tape Mark

Graphic representations have been developed for the 32 control characters, for the Space character, and the Delete character of ASCII.

In Text/360, an IBM programming product for the application of text processing, graphic representations have been assigned to the six control operations (see Chapter 26, Code Extension):

*	Single capitilization
@	Continued capitalization
$	Underscoring
–	Editing
+	Altering
/	Graphic set extension

2.28 COLLAPSE LOGIC

Consider a 256-character, 8-bit code feeding into a 64-character printer. The 64 printing positions of the printing element may be considered to be associated with 64 different 6-bit bit patterns. The hardware logic of the printer will strip off the two high-order bits of 8-bit bit patterns, leaving 6-bit bit patterns. For each different 6-bit bit pattern, there will have been four different 8-bit bit patterns.

Consider Fig. 2.33. The four bit patterns X1, X2, X3, X4 have bit patterns 0010 1010, 0110 1010, 1010 1010, 1110 1010. If the two high-order bits of these 8-bit bit patterns are stripped off, for each of them the same 6-bit bit pattern 101010 will result. Each of these four 8-bit bit patterns then would collapse to the same 6-bit bit pattern; that is, each would go to the same printing position of the printing element. Advantage is taken of the collapse aspect of coded character sets in the design of printing sets.*

---

*Collapse logic varies among printer control units. The examples given here are illustrative only, and do not necessarily reflect any actual printer control unit.

Bits	0, 1 2, 3	00			01			10			11		
	4567	00 01 0 1	10 2	11 3	00 01 4 5	10 6	11 7	00 01 8 9	10 A	11 B	00 01 C D	10	11 F
0 1 2 3 4 5 6 7 8 9	0000												
A	1010		X1			X2			X3			X4	
B C D E F	1111												

Fig. 2.33 Collapse logic

In EBCDIC, the bit patterns of the small letters a, b, c, . . . , z differ from the bit patterns of the corresponding capital letters only in the two high-order bits. On a 64-character printer, therefore, regardless of whether the bit patterns of the small letters or the bit patterns of the capital letters are fed into the hardware logic of the printer control unit, the same alphabetic printing positions on the printing element are reached without any change in logic.

Collapse logic is used in the printing of alphabets other than Latin alphabets. Consider, for example, Fig. 2.34 that shows the assignment in the EBCDIC code table of 31 Cyrillic alphabetics, 10 numerics, and the following 7 specials:

. + ¤ * – / ,

Row	Column	0	1	2	3	4	5	6	7	8	9	A	B	C	D	E	F
	Bit Pat. →	00				01				10				11			
	↓	00	01	10	11	00	01	10	11	00	01	10	11	00	01	10	11
0	0000					SP		–			К						0
1	0001							/		А	Л						1
2	0010									Б	М	Ф					2
3	0011									В	Н	Х					3
4	0100									Г	О	Ц					4
5	0101									Д	П	Ч					5
6	0110									Е	Р	Ш					6
7	0111									Ж	С	Щ					7
8	1000									З	Т	Ы					8
9	1001									И	У	Ь					9
A	1010																
B	1011					.	¤	,					Ю				
C	1100						*			Й		Э	Я				
D	1101																
E	1110					+											
F	1111																

Fig. 2.34 Collapse logic, Cyrillic-48

Row	Column	0	1	2	3	4	5	6	7	8	9	A	B	C	D	E	F
	Bit Pat. →	00				01				10				11			
	↓	00	01	10	11	00	01	10	11	00	01	10	11	00	01	10	11
0	0000					SP	&	–									0
1	0001							/						A	J		1
2	0010													B	K	S	2
3	0011													C	L	T	3
4	0100													D	M	U	4
5	0101													E	N	V	5
6	0110													F	O	W	6
7	0111													G	P	X	7
8	1000													H	Q	Y	8
9	1001													I	R	Z	9
A	1010																
B	1011					.	¤	,	#								
C	1100					<	*	%	@								
D	1101																
E	1110					+											
F	1111																

Fig. 2.35 Collapse logic, Latin-48

Consider also Fig. 2.35 which exhibits a 48-character printing set consisting of 26 Latin alphabetics, 10 numerics, and the following 12 specials:*

+	&	–	/	.	¤
<	*	%	@	,	#

An examination of Figs. 2.34 and 2.35 will show the collapse logic for the 48 printing positions of a 48-character printer as shown in Fig. 2.36.

With the same printer control unit, the collapse logic will automatically provide for a 48-graphic Cyrillic set, or a 48-graphic Latin set, depending on which printing element is mounted by the user.

Cyrillic, 48 graphics Fig. 2.34		Latin, 48 graphics Fig. 2.35	
Hex position	Graphic	Hex position	Graphic
F0 to F9	10 numerics	F0 to F9	10 numerics
81 to 89 91 to 99 A2 to A9	26 Cyrillic alphabetics	C1 to C9 D1 to D9 E2 to E9	26 Latin alphabetics
	5 Cyrillic alphabetics		5 specials
8C		4C	<
90		50	&
AC		6C	%
BB		7B	#
CB		7C	@
	7 specials		7 specials
4B	.	4B	.
4E	+	4E	+
5B	¤	5B	¤
5C	*	5C	*
60	–	60	–
61	/	61	/
6B	,	6B	,
TOTAL	48 graphics	48 graphics	

Fig. 2.36 Cyrillic/Latin collapse

*The special symbol ¤, shown both above and in hex position 5B of Figs. 2.34 and 2.35, is the international "Currency Symbol."

2.29 BOOLEAN EQUATIONS

In some of the cases that are given in this book, the question of the simplicity or complexity of translation relationships from one code to another, or from one representation to another, comes up. Generally, the question is not of absolute simplicity or complexity but of comparative simplicity or complexity. Hardware translation is accomplished by logic circuits. The complete analysis of such circuits and the calculation of hardware costs, estimated or actual, is beyond the scope of this book. However, by making three simplifying assumptions, a reasonably simple procedure can be used that is sufficiently accurate to answer the following question:

> Given two sets of translation relationships, which set would be more complex to implement in circuitry?

Assumption 1. The circuit complexity is equal to implement each of four Boolean operators (to be explained below), AND, Inclusive OR, Exclusive OR, and IDENTITY.

Assumption 2. The circuitry that generates a bit also generates the inverse of the bit with no additional complexity.

Assumption 3. Given two sets of Boolean equations representing two sets of translation relationships, the relative circuit complexity of implementing the relationships is proportional to the number of Boolean operators in the equations.

Example 13

Set 1: $\text{Y1} = A \& Y$ one operator, & (to be explained below).

Set 2: $\text{Y2} = (A \& Y) \mid Z$ two operators, &, | (to be explained below).

Set 2 is more complex than Set 1.

Absolute costs are not determined but relative complexities are; this information is sufficient for making a decision between two sets. The procedure, then, is to derive the Boolean equations, and then count the operators.

There are different notations and conventions used in Boolean Algebra. Some examples are shown below:

$$S = AB + \bar{C}D$$
$$S = A \cdot B + \bar{C} \cdot D$$
$$S = (A \& B) \mid (\neg C \& D)$$
$$S = (A \wedge B) \vee (\bar{C}D)$$

The binary, or two-state, nature of many mechanisms found in computing systems was noted at the beginning of this chapter. For such two-state situations, we might say we have A or we do not have A. Alternatively, we might say we have A or the inverse of A. In Boolean logic, we would say we have "A," or we have "not A." A convention for representing these two possible states is A and $\bar{A}$; that is, $\bar{A}$ represents "not A," or "the inverse of A" or "the negation of A," etc. If we consider A as a binary variable, it can have two values, 0 or 1. By convention, when the variable A has the value of 1, we will represent it by A, and when it has the value of 0, we will represent it by $\bar{A}$.

Example 14

We may represent the three bit positions of a 3-bit register by the Boolean variables A, B, and C. Then the 8 possible states of the 3-bit register can be represented as follows:

State	Representation
000	$\bar{A}\ \bar{B}\ \bar{C}$
001	$\bar{A}\ \bar{B}\ C$
010	$\bar{A}\ B\ \bar{C}$
011	$\bar{A}\ B\ C$
100	$A\ \bar{B}\ \bar{C}$
101	$A\ \bar{B}\ C$
110	$A\ B\ \bar{C}$
111	$A\ B\ C$

Example 15

Another convention is to represent a variable when its value is 1 by the presence of the variable and when its value is 0 by the absence of the variable. This convention is used in a notation based on the decimal equivalents of the powers of 2:

$$2^0 = 1$$
$$2^1 = 2$$
$$2^2 = 4$$
$$2^3 = 8$$

The bit positions of a 4-bit register are represented, from high-order bit position to low-order bit position, by the variables 8, 4, 2, 1. Under the convention of Example 14, 1001 would have been represented as $8\ \bar{4}\ \bar{2}\ 1$, but under the presence/absence convention, 1001 is represented simply as 8 1. Under this convention, the 16 states of a 4-bit register are represented as shown below:

State	Representation
0000	No bits
0001	1
0010	2
0011	2 1
0100	4
0101	4 1
0110	4 2
0111	4 2 1
1000	8
1001	8 1
1010	8 2
1011	8 2 1
1100	8 4
1101	8 4 1
1110	8 4 2
1111	8 4 2 1

Example 16

The 8 states of a 3-bit register, Example 14 under the presence/absence convention, would be represented as shown below:

State	Representation
000	No bits
001	C
101	B
011	BC
100	A
101	AC
110	AB
111	ABC

Comment. The convention of Example 14 yields a *uniform* notation, while the convention of Examples 15 and 16 yields a *compact* notation.

In this book, Boolean equations are used to represent translation relationships. Five Boolean operators (frequently called logical operators) and their representative symbols are shown below:

	Operator	Symbol
1.	AND	$\wedge$
2.	Inclusive OR	$\vee$
3.	Exclusive OR	$\veebar$
4.	IDENTITY	$\equiv$
5.	NOT	$-$

In order to define these operators, we consider two binary input variables, A and B, and one binary output variable, Y, as illustrated below. There are two kinds of operators: (1) dyadic operators; that is, operating on *two* terms or expressions (parts (1–4) above), and (2) monadic operators; that is, operating on one term or expression (part (5) above).

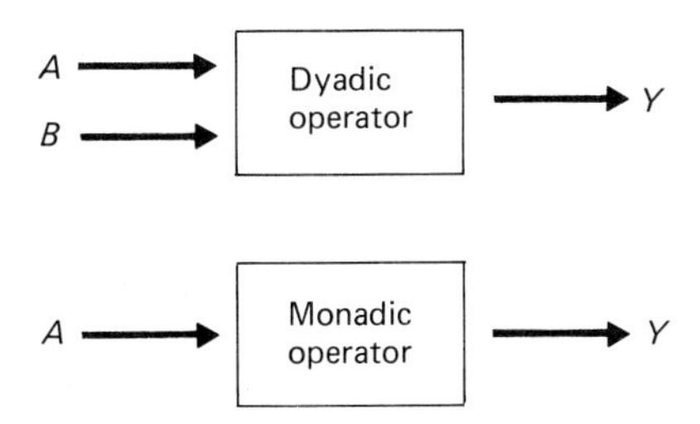

There are two possible states for one variable and four possible states for two variables taken together:

Variable	State
A	1
$\bar{A}$	0
$\bar{A}\bar{B}$	00
$\bar{A}B$	01
$A\bar{B}$	10
AB	11

The operators are defined in the following table:

		NOT		AND	Inclusive OR	Exclusive OR	IDENTITY
A	B	$\bar{A}$	$\bar{B}$	$A \wedge B$	$A \vee B$	$A \veebar B$	$A \equiv B$
0	0	1	1	0	0	0	1
0	1	1	0	0	1	1	0
1	0	0	1	0	1	1	0
1	1	0	0	1	1	0	1

Conceptually, we say

a) AND means both *A and B* are 1.

b) Inclusive OR means either *A or B* is 1, *including* the case when both are 1.

c) Exclusive OR means either *A or B* is 1, *excluding* the case when both are 1.

d) IDENTITY means *A* and *B* are *identical*; that is, both are 0, or both are 1.

3
Early Codes

During the early days of data processing and telecommunications, a number of codes were in use or proposed for use:

a) CCITT #2, a 58-character, shifted 6-bit code, used nationally and internationally on telegraph lines.

b) FIELDATA [3.1, 3.2, 3.3]: a 7-bit code developed by the United States Army for military communications systems.

c) BCDIC [3.4]: a 48-character, 12-row code (initially unnamed) used on computing systems. This code was eventually expanded to be a 64-character, 6-bit code and 12-row card code.

d) The Stretch code: a 120-character, 8-bit code used on the Stretch computer (the IBM 7030) [3.5, 3.6].

e) IPC, Information Processing Code [3.7]: a 128-character, 8-bit code developed by the United States Air Force proposed to be used for information processing and information interchange.

f) A 64-character, 6-bit code proposed by H. S. Bright in 1959 [3.8].

g) A 256-character card code proposed by R. W. Bemer in 1959 [3.9].

h) 4-out-of-8 code: a 70-character, 8-bit data transmission code.

These early codes manifested some of the characteristics of coded character sets described in Chapter 2. Some of these characteristics would be carried forward and incorporated into modern codes. It should not be supposed that these early codes have disappeared from the data processing scene. Products and systems implementing these codes (with the exception of IPC) are still in common use. Figure 3.1 shows the codes and their characteristics.

	CCITT #2	Fiel-data	BCDIC	Stretch	IPC	Bright Proposal	Bemer Proposal	4-out-of-8
Shifted code	yes							
BCD for numerics		yes	yes		yes	yes	yes	yes
Numerics in numeric sequence		yes		yes	yes	yes		
Numerics in contiguous sequence		yes	yes		yes	yes		
Signed numerics			yes				yes	
BCD for alphabetics			yes			yes	yes	yes
Alphabetics in alphabetic sequence		yes		yes	yes	yes	yes	
Alphabetics in contiguous sequence		yes		yes	yes			
Alphabetics in noncontiguous sequence	yes		yes			yes	yes	yes
Alphabetics in interleaved sequence				yes	yes			
Space equals no punches			yes	yes			yes	
Collapse logic			yes	yes	yes			

Fig. 3.1 Characteristics of early codes

3.1 CCITT #2

CCITT #2 was, and is, a 58-character, shifted 6-bit code, standardized as an international telegraph code in 1931 by the Comité Consultatif International Telegraphique et Telephonique (see Fig. 3.2).

Bit pattern	Letter case	Figure case	Bit pattern	Letter case	Figure case
00000	Not used	Not used	10000	E	3
00001	T	5	10001	Z	+ or "
00010	Cr	Cr	10010	D	(2)
00011	O	9	10011	B	?
00100	SP	SP	10100	S	'
00101	H	(1)	10101	Y	6
00110	N	,	10110	F	(1)
00111	M	.	10111	X	/
01000	LF	LF	11000	A	–
01001	L	)	11001	W	2
01010	R	4	11010	J	Bell
01011	G	(1)	11011	FS	FS
01100	I	8	11100	U	7
01101	P	O	11101	Q	1
01110	C	:	11110	K	(
01111	V	= or ;	11111	(3) LS	LS

(1) For National Use
(2) Used for Answer Back
(3) Also used for Delete

CR Carriage Return
SP Space
LF Line Feed
FS Figure Shift
LS Letter Shift

Fig. 3.2 CCITT #2

Figure 3.1 reveals that CCITT #2 manifests few of the characteristics of the other codes, characteristics deemed desirable for data processing codes. The numerics are not BCD, nor contiguous, nor in numeric sequence; the alphabetics are not in alphabetic sequence, and so on. But it should be realized that CCITT #2 was developed as a telegraph code, and characteristics desirable for a data processing code have little importance for a telegraph code.

CCITT #2 did manifest a characteristic that is quite necessary for data processing codes and for telecommunication codes. Three code positions were reserved for "national use." This recognizes a characteristic of certain European languages (German, Danish, Swedish, Finnish, Norwegian, for example) which is that such languages have three letters in addition to the 26 alphabetics of English-speaking languages (see Table 3.1). Such letters are called diacritical letters, or diacritics.

TABLE 3.1 Diacritical Letters

German	Ä	Ö	Ü
Danish/Norwegian	Æ	Ø	Å
Swedish/Finnish	Ä	Ö	Ü

Clearly, telegraph devices operating within national boundaries of countries whose languages require 29 alphabetics would have to have the capability of sending and receiving all 29 letters. The telegraph code, then, must have code positions available for 29 letters and CCITT #2 does.

In English-speaking countries, such code positions could be used to represent other symbols. In the U.S.A., on Western Union telegraph devices, for example, the symbols # $ and & were provided in these three code positions.

3.2 FIELDATA

FIELDATA was a 7-bit plus parity code developed by the United States Army for use on military data communications lines. It became a U.S. Military Standard in 1960 (see Fig. 3.3).

It is to be noted that although there are 128 code positions in the 7-bit code, only 64 were defined, consisting of 9 control functions and 55 graphic characters. The controls are of the kind required by rather simple, typewriter-like devices—Space, Upper Case, Lower Case, Line Feed, Carriage Return, and so on. The 64 undefined code positions were intended to be assigned to the more complex kinds of functions necessary for interconnection and control of data transmission networks.

As it turned out, three different communications systems were developed implementing FIELDATA, and each of these three systems used different control functions in the "not defined" portion of the code table—different in the sense of technical definition and different in the sense of the number of control functions. It was found that because of these different control functions interconnection of these three communication systems, and intercommunication between them, was difficult or impossible.

Row	Column	0	1	2	3	4	5	6	7
	Bit Pattern b7	0	0	0	0	1	1	1	1
	b6	0	0	1	1	0	0	1	1
	b5	0	1	0	1	0	1	0	1
Row	b4 b3 b2 b1								
0	0 0 0 0	CONTROL (NOT DEFINED)				MS	K	)	0
1	0 0 0 1					UC	L	-	1
2	0 0 1 0					LC	M	+	2
3	0 0 1 1					LF	N	<	3
4	0 1 0 0					CR	O	=	4
5	0 1 0 1					SP	P	>	5
6	0 1 1 0					A	Q	_	6
7	0 1 1 1					B	R	$	7
8	1 0 0 0					C	S	*	8
9	1 0 0 1					D	T	(	9
10	1 0 1 0					E	U	"	'
11	1 0 1 1					F	V	:	;
12	1 1 0 0					G	W	?	/
13	1 1 0 1					H	X	!	.
14	1 1 1 0					I	Y	,	SPEC
15	1 1 1 1					J	Z	STOP	IDLE

MS - Master Space
UC - Upper Case
LC - Lower Case
LF - Line Feed
CR - Carriage Return
SP - Space
STOP - Stop
SPEC - Special
IDLE - Idle

Fig. 3.3 FIELDATA

A valuable lesson was learned here. For various reasons, it may be desirable not to complete the assignment of meanings to all code positions of a code table initially. For example, the American National Standard Code for Information Interchange (ASCII), when first standardized in 1963, left some 28 code positions without assigned meanings. And when the extended BCD Interchange Code (EBCDIC) was adopted as an internal standard by IBM in 1964, of the 256 available code positions, only 108 code positions had assigned meanings. Indeed, at this time (almost a decade later) there are still many code positions in EBCDIC with unassigned meanings. However, in the administration of these standards, ASCII and EBCDIC, implementors were advised to provide implementations which did not assign meanings to those code positions without already assigned meaning. These code positions were reserved for future standardization. For FIELDATA, implementors provided implementations with their own local meanings for those code positions not initially assigned. The result was inter-implementation confusion. The disciplined administration of ASCII and EBCDIC prevented such confusion. This point of administrative discipline will be discussed below with IPC, Information Processing Code.

3.3 BCDIC

With modern codes, such as ASCII and EBCDIC, it is common practice to provide implementations which use not the full repertoires of the codes but subsets, subsetted by graphics, or by controls, or by both. By contrast, the code that came to be called the BCD Interchange Code (BCDIC) evolved from a smaller repertoire to a code with a complete repertoire. (The evolution of BCDIC is described in detail in the next two chapters.)

The punched card code devised by Dr. Herman Hollerith at the end of the nineteenth century was a 12-character code consisting of the 10 numerics, 0 through 9, and two control characters in what are now the 12-row and the 11-row of the card. In the statistical applications of the United States Census—for which Dr. Hollerith devised the punched card—these control punches served many purposes. When punched cards came to be used in accounting applications, the 11-punch came to be used to represent a credit balance (mathematicians would call it a negative number).

Somewhere around 1932, the punched card code was expanded to include 26 alphabetics and three special symbols—minus sign, asterisk, and ampersand. The minus sign had replaced the credit symbol, asterisk was used for check protection, and ampersand was used in name-and-address applications (Mr. & Mrs. J. L. Smith, for example). The punched card code for these 39 graphics and space is shown in Fig. 3.4.

During the 1950s, the advent of computers such as the IBM 702, 705, and 1401 saw the expansion of BCDIC into 47 graphics, and also the development of a 6-bit code to represent these graphics. With one

Bit Pattern	Hole Pattern		0	11	12
		SP		–	&
	1	1		J	A
	2	2	S	K	B
	3	3	T	L	C
	4	4	U	M	D
	5	5	V	N	E
	6	6	W	O	F
	7	7	X	P	G
	8	8	Y	Q	H
	9	9	Z	R	I
	0	0			
	8-4			*	

Fig. 3.4 BCDIC, 40-character card code

exception, the 11 special symbols served an obvious purpose in one or another commercial application:

. ' $, # % – & * / ⌑

The exception was the special symbol, ⌑ (lozenge). Because the lozenge appeared on printer chains, it was put to various uses; for example, to indicate, in the margin of a tabulation, final totals as contrasted to subtotals.

The 48-character BCDIC is shown in Fig. 3.5.

3.4 THE STRETCH CODE

In 1961, the IBM 7030 was delivered to the Los Alamos Scientific Laboratory. This computer was developed under "Project Stretch," and this name was popularly used to describe this computer.

Bit Pattern			A	B	BA
	Hole Pattern		0	11	12
		SP		–	&
1	1	1	/	J	A
2	2	2	S	K	B
2 1	3	3	T	L	C
4	4	4	U	M	D
4 1	5	5	V	N	E
4 2	6	6	W	O	F
4 2 1	7	7	X	P	G
8	8	8	Y	Q	H
8 1	9	9	Z	R	I
8 2	0	0			
8 2 1	8–3	#	,	$	.
8 4	8–4	@	%	*	⌑

Fig. 3.5 BCDIC, 48-character code

There were many technological innovations in Stretch. Architecturally, its main innovation was that it had an 8-bit architecture, as contrasted with the 6-bit, or 6-bit oriented, architectures of other computers of the time. With an 8-bit architecture, a 256-character code is possible. In fact, the designers of Stretch chose to provide a 120-character set that, apart from its size (most computer character sets of that day were 48-character sets), had some interesting innovations.

The codes for contemporary computers of that time had evolved from earlier beginnings and compatibility was the primary design criterion. The designers of the Stretch code, E. G. Law, H. J. Smith, Jr., F. A. Williams, W. Buchholz, and R. W. Bemer, did not perceive compatibility with contemporary codes to be a primary criterion. Instead, they themselves set some criteria that they felt were reasonable for a code. The criteria were in regard to the size and structure of the set. The criteria are first stated, and then some of them are discussed.

3.4.1 Size

Criterion 1. The set should contain the contemporary 48-graphic set found on IBM computers:

- Space
- 26 alphabetics (upper case)
- 10 numerics
- 11 specials . ' & % (– / , $ # ¤

Criterion 2. The set should contain the following graphics:

- 26 lower case alphabetics
- The more important punctuation symbols found on office typewriters . , : ; ? ! ' "
- Enough mathematical and logical symbols to satisfy the needs of such programming languages as ALGOL. (The total ALGOL set was well over 100 symbols.)

3.4.2 Structure

Criterion 3. Certain subsets, such as the contemporary 48-character set for high-speed chain printer printing and an 88-graphic set for a typewriter, should be simply derivable.

Criterion 4. The graphics should be blocked contiguously by function; viz., the specials should be in a contiguous block, the alphabetics should be in a contiguous block, the numerics should be in a contiguous block, and so on.

Criterion 5. The binary sequence of the bit patterns representing the graphics should match whatever collating sequence was prescribed for the graphics.

Criterion 6. The 48 graphics of contemporary IBM computer codes should have, in the Stretch code, the same collating sequence, or should be embedded in the same relative collating sequence, as the contemporary collating sequence, namely, Space, then the specials . ¤ & $ * – / , % # @ then the alphabetics, then the numerics.

Criterion 7. The upper and lower case alphabetics should be interleaved.

Criterion 8. There should be unique bit patterns for each unique graphic; that is, duals would not be permitted.

As well as these criteria, there was a constraint on the size of the set. The theoretical constraint was a maximum of 256 characters, since the byte size of Stretch was to be 8 bits. But there was a more pragmatic constraint due to the printer to be used with Stretch, the chain printer. The chain printer, due to its design geometry, had 240 printing positions; so this was clearly the maximum possible set size. However, as a practical consideration, the larger the set size, the lower is the printing speed of the chain printer. The actual choice was 120 characters. This was a matter of judgment; it was decided that this increment over existing sets would be sufficiently large to justify a departure from contemporary codes and would not include many characters of only marginal value. Also, the set size of 120, in terms of the 240 printing positions of the chain printer, meant that each symbol could appear twice on the chain, yielding a not unreasonable printing speed.

The actual character set and the coded representation is shown in Fig. 3.6. It is evident from inspection of the code that not all criteria were met. In fact, the criteria were somewhat mutually conflicting, and some trade-offs were necessary.

<table>
<tr><th rowspan="3">Row</th><th>Column</th><th>0</th><th>1</th><th>2</th><th>3</th><th>4</th><th>5</th><th>6</th><th>7</th><th>8</th><th>9</th><th>A</th><th>B</th><th>C</th><th>D</th><th>E</th><th>F</th></tr>
<tr><th rowspan="2">Bit Pat. →↓</th><th colspan="4">00</th><th colspan="4">01</th><th colspan="4">10</th><th colspan="4">11</th></tr>
<tr><th>00</th><th>01</th><th>10</th><th>11</th><th>00</th><th>01</th><th>10</th><th>11</th><th>00</th><th>01</th><th>10</th><th>11</th><th>00</th><th>01</th><th>10</th><th>11</th></tr>
<tr><td>0</td><td>0000</td><td>SP</td><td>[</td><td>&</td><td>c</td><td>k</td><td>s</td><td>0</td><td>8</td><td></td><td></td><td></td><td></td><td></td><td></td><td></td><td></td></tr>
<tr><td>1</td><td>0001</td><td>±</td><td>⊃</td><td>+</td><td>C</td><td>K</td><td>S</td><td>0</td><td>8</td><td></td><td></td><td></td><td></td><td></td><td></td><td></td><td></td></tr>
<tr><td>2</td><td>0010</td><td>→</td><td>]</td><td>$</td><td>d</td><td>l</td><td>t</td><td>1</td><td>9</td><td></td><td></td><td></td><td></td><td></td><td></td><td></td><td></td></tr>
<tr><td>3</td><td>0011</td><td>≠</td><td>∘</td><td>=</td><td>D</td><td>L</td><td>T</td><td>1</td><td>9</td><td></td><td></td><td></td><td></td><td></td><td></td><td></td><td></td></tr>
<tr><td>4</td><td>0100</td><td>∧</td><td>←</td><td>*</td><td>e</td><td>m</td><td>u</td><td>2</td><td>.</td><td></td><td></td><td></td><td></td><td></td><td></td><td></td><td></td></tr>
<tr><td>5</td><td>0101</td><td>{</td><td>≡</td><td>(</td><td>E</td><td>M</td><td>U</td><td>2</td><td>:</td><td></td><td></td><td></td><td></td><td></td><td></td><td></td><td></td></tr>
<tr><td>6</td><td>0110</td><td>↑</td><td>¬</td><td>/</td><td>f</td><td>n</td><td>v</td><td>3</td><td>-</td><td></td><td></td><td></td><td></td><td></td><td></td><td></td><td></td></tr>
<tr><td>7</td><td>0111</td><td>}</td><td>√</td><td>)</td><td>F</td><td>N</td><td>V</td><td>3</td><td>?</td><td></td><td></td><td></td><td></td><td></td><td></td><td></td><td></td></tr>
<tr><td>8</td><td>1000</td><td>∨</td><td>%</td><td>,</td><td>g</td><td>o</td><td>w</td><td>4</td><td></td><td></td><td></td><td></td><td></td><td></td><td></td><td></td><td></td></tr>
<tr><td>9</td><td>1001</td><td>⊻</td><td>\</td><td>;</td><td>G</td><td>O</td><td>W</td><td>4</td><td></td><td></td><td></td><td></td><td></td><td></td><td></td><td></td><td></td></tr>
<tr><td>A</td><td>1010</td><td>↓</td><td>◇</td><td>'</td><td>h</td><td>p</td><td>x</td><td>5</td><td></td><td></td><td></td><td></td><td></td><td></td><td></td><td></td><td></td></tr>
<tr><td>B</td><td>1011</td><td>||</td><td>|</td><td>"</td><td>H</td><td>P</td><td>X</td><td>5</td><td></td><td></td><td></td><td></td><td></td><td></td><td></td><td></td><td></td></tr>
<tr><td>C</td><td>1100</td><td>></td><td>#</td><td>a</td><td>i</td><td>q</td><td>y</td><td>6</td><td></td><td></td><td></td><td></td><td></td><td></td><td></td><td></td><td></td></tr>
<tr><td>D</td><td>1101</td><td>≥</td><td>!</td><td>A</td><td>I</td><td>Q</td><td>Y</td><td>6</td><td></td><td></td><td></td><td></td><td></td><td></td><td></td><td></td><td></td></tr>
<tr><td>E</td><td>1110</td><td><</td><td>@</td><td>b</td><td>j</td><td>r</td><td>z</td><td>7</td><td></td><td></td><td></td><td></td><td></td><td></td><td></td><td></td><td></td></tr>
<tr><td>F</td><td>1111</td><td>≤</td><td>~</td><td>B</td><td>J</td><td>R</td><td>Z</td><td>7</td><td></td><td></td><td></td><td></td><td></td><td></td><td></td><td></td><td></td></tr>
</table>

Fig. 3.6 Stretch, 120-character set

Comment on Criterion 6

The contemporary collating sequence for the 47 graphics provided on contemporary computers was not achieved. In order to provide an 89-graphic subset and a 49-graphic subset derivable by simple logic (Criterion 3), the specials had to be positioned somewhat arbitrarily (see Figs. 3.7 and 3.8), and this was deemed more advisable than the collating-sequence criterion. Nine of the contemporary specials did collate low to alphabetics and numerics, although even these were not, within themselves, in the contemporary collating sequence. It was felt that the new sequence would be quite usable and that it would be necessary only rarely to resort a file in the transition to the Stretch code. And it is always possible to translate codes to obtain any desired sequence.

Comment on Criterion 3.

As can be seen in Figs. 3.7 and 3.8, both the 49-graphic subset and 89-graphic subset were simply derivable from the 120-graphic code.

	Column	0	1	2	3	4	5	6	7	8	9	A	B	C	D	E	F
	Bit Pat. →	00				01				10				11			
Row	↓	00	01	10	11	00	01	10	11	00	01	10	11	00	01	10	11
0	0000	SP		&				0	8								
1	0001				C	K	S										
2	0010			$				1	9								
3	0011				D	L	T										
4	0100			*				2	.								
5	0101				E	M	U										
6	0110			/				3	-								
7	0111				F	N	V										
8	1000		%	,				4									
9	1001				G	O	W										
A	1010		◇	'				5									
B	1011				H	P	X										
C	1100		#					6									
D	1101			A	I	Q	Y										
E	1110		@					7									
F	1111			B	J	R	Z										

Fig. 3.7 Stretch, 49-character set

Row	Column	0	1	2	3	4	5	6	7	8	9	A	B	C	D	E	F
	Bit Pat. →	00				01				10				11			
	↓	00	01	10	11	00	01	10	11	00	01	10	11	00	01	10	11
0	0000	SP		&	c	k	s	0	8								
1	0001			+	C	K	S	0	8								
2	0010			$	d	l	t	1	9								
3	0011			=	D	L	T	1	9								
4	0100			*	e	m	u	2	.								
5	0101			(	E	M	U	2	:								
6	0110			/	f	n	v	3	-								
7	0111			)	F	N	V	3	?								
8	1000			,	g	o	w	4									
9	1001			;	G	O	W	4									
A	1010			'	h	p	x	5									
B	1011			"	H	P	X	5									
C	1100			a	i	q	y	6									
D	1101			A	I	Q	Y	6									
E	1110			b	j	r	z	7									
F	1111			B	J	R	Z	7									

Fig. 3.8 Stretch, 89-character set

Note that the 49-graphic set included the contemporary 48-graphic set (see Criterion 1) and additionally had the graphic apostrophe or single quote. The provision of a 48-graphic-plus-Space set fitted neatly into the geometry of the 240-printing-position chain printer: $5 \times 48 = 240$. Each graphic was provided in 5 printing positions, yielding very respectable printing speeds.

Note that the 49-graphic set is not entirely a subset of the 89-graphic set. Note also that it was found not practical to retain the upper- and lower-case relationships of punctuation and other special symbols commonly found on typewriter keyboards. (There was no single convention anyway, and typists were accustomed to finding differences in this area.)

Comment of Criterion 7

The benefit of interleaving upper- and lower-case alphabetics is dubious. (For a fuller discussion of this point, see Chapter 25, Contiguous, Noncontiguous, and Interleaved Alphabets.) However, once it is decided to interleave the alphabets, as was done in the Stretch code, a further decision is necessary: Which alphabetic should precede within the pair, the upper-case or the lower-case? The designers of this code had observed that no real precedent existed for the relative position within the code. But the choice had to be made. They chose that lower case should precede upper case within the pair, for reasons not known to the author.

It is interesting to note that had they made the other choice, so that "A" had bit pattern 0010 1100 and "a" had bit pattern 0010 1101, for example, the derivation of the 49-character subset (Fig. 3.7) from the 120-character set (Fig. 3.6) would have been logically simpler. Observe that in Fig. 3.7 the specials chosen alternate in code position with those not chosen and the same is true for the alphabetics and the numerics. However, *two* code positions intervene between the last special and the first alphabetic, and no code position intervenes between the last alphabetic and the first numeric. The logical equations to describe the choice of code positions are somewhat complex because of the double gap and the null gap. Had the opposite choice been made in assigning upper- and lower-case alphabetics, both anomalies would disapppear, and the logical equations would have been quite simple. It should also be noted that this latter choice would not have affected the derivability of the 89-character subset, since the 52 alphabetics would still occupy the same contiguous 52 code positions.

It is interesting that in the design of IPC, Information Processing Code, described below, where the designers also chose to interleave the upper- and lower-case alphabetics, the decision was that upper-case should precede lower-case alphabetics within the pair.

In conjunction with the Stretch bit code, there was a punched card code. The bits of the code were named B0, B1, B2, . . . , B7, from high-order to low-order significance within the byte. A parity bit, odd parity, named Bp was also punched. The card code (see Fig. 3.9) was a binary card code, specified by the following algorithm:

Card Row	Code Bit
12	—
11	—
0	—
1	Bp
2	B0
3	B1
4	B2
5	B3
6	B4
7	B5
8	B6
9	B7

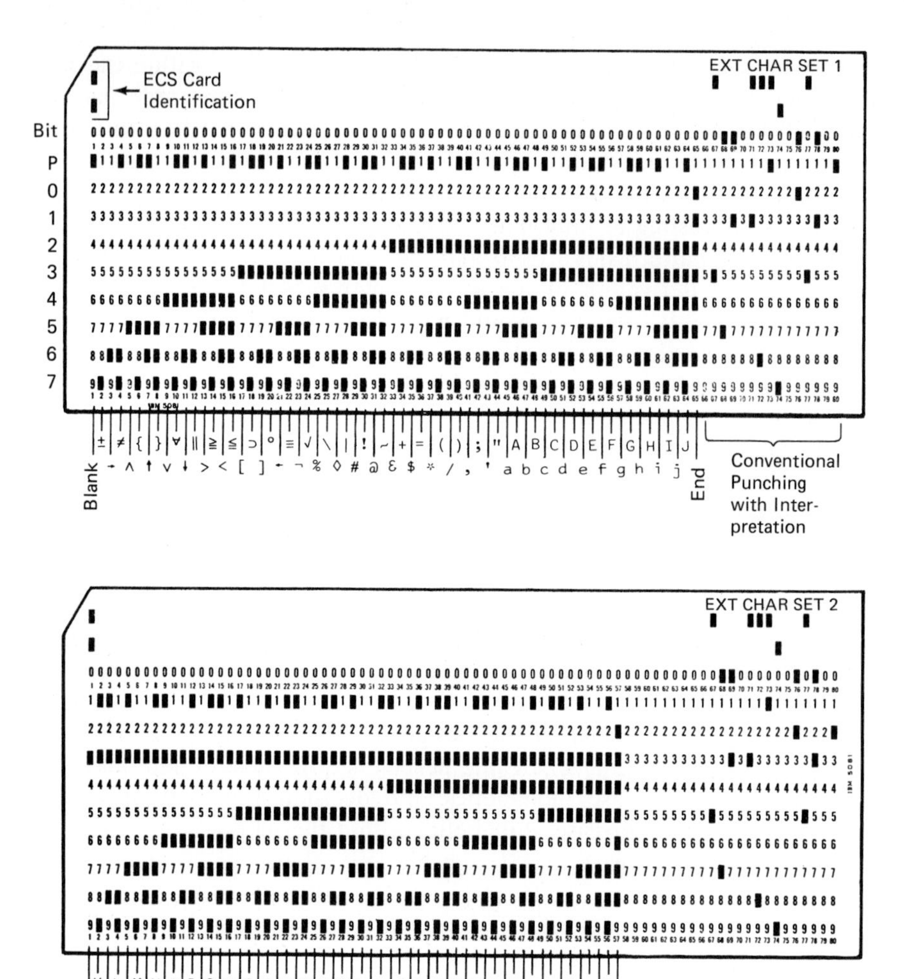

Fig. 3.9 Cards punched with extended character code

In order to distinguish cards with this binary punching from cards punched with the conventional Hollerith card code, binary punched cards had 12-holes and 11-holes punched in column 1. Within an application, conventional Hollerith card code punching could be used in the right end of such cards, as shown in Fig. 3.9. The Space character, having no bits in the code, would nevertheless have a parity bit punched in row 1. However, skipped fields would have no punches, as can be seen in the lower card in Fig. 3.9.

As in discussed in Chapter 16, "Decimal ASCII," the structural strength of a card punched in binary came under serious question particularly if most of the data was numeric (which would lead to one or more rows being laced because of the zone bits in the representation of the numerics). It should be noted that the question of binary card coding in the Decimal ASCII debate was considered in the environment of an individual card, mailed to a human, carried by the human in a pocket in varying conditions of humidity, temperature, and abuse, and subsequently required to be further processed in card equipment. By contrast, the normal environment for a Stretch card was much more protective—generally a deck of cards, handled with reasonable care in a machine room environment. The binary card discussed in Chapter 16 was expected to be subjected to structural stress, the Stretch card was not.

3.5 IPC

IPC, Information Processing Code, was developed by Edward Morenoff, John B. McLean, and Lt. Lawrence Odell in 1964. It was intended as an information manipulation-oriented character set with associated binary code representation. The author does not know if it was actually implemented, but it has some interesting aspects. The design criteria were somewhat similar to those of the Stretch code.

Criterion 1. The set should contain the following graphics:

- Upper- and lower-case alphabetics
- Numerics
- The more important punctuation symbols found on office typewriters
  . , : ; ? ! ' "
- Special symbols peculiar to user operations.

Criterion 2. Certain subsets, 7-bit, 6-bit, 5-bit, 4-bit, should be easily derivable.

Criterion 3. Code positions should be provided that would be dedicated to local interpretation.

IPC was an 8-bit code. However, only 128 characters were specified, and the use of the 8th bit was deliberately left undefined for specification in local environments on the basis of particular applications. For example, the 8th bit might be used as a parity bit to increase the reliability of data transmission. Or it might be used to indicate that some special significance should be attached to a particular character, such as being part of a "keyword," or a part of a highly sensitive piece of information. Since the

	Column	0	1	2	3	4	5	6	7
	Bit Pattern b7 b6 b5	0 0 0	0 0 1	0 1 0	0 1 1	1 0 0	1 0 1	1 1 0	1 1 1
Row	b4 b3 b2 b1								
0	0 0 0 0	0	C	K	S	(	α	Σ	^{3}
1	0 0 0 1	1	c	k	s	!	×	¼	^{2}
2	0 0 1 0	2	D	L	T	?	β	≤	Ⓔ
3	0 0 1 1	3	d	l	t	#	:	½	ⓔ
4	0 1 0 0	4	E	M	U	°	=	≥	Bk_1
5	0 1 0 1	5	e	m	u	/	-	¾	Bk_2
6	0 1 1 0	6	F	N	V		√	∞	Bk_3
7	0 1 1 1	7	f	n	v		∫	↓	Bk_4
8	1 0 0 0	8	G	O	W	*	:	θ	•
9	1 0 0 1	9	g	o	w	)	;	↑	c_1
10	1 0 1 0	SP	H	P	X	.	@	ϕ	c_2
11	1 0 1 1	RES	h	p	x	,	χ	→	c_3
12	1 1 0 0	A	I	Q	Y	π	"	κ	c_4
13	1 1 0 1	a	i	q	y	_	'	←	c_5
14	1 1 1 0	B	J	R	Z	ω	$	[	c_6
15	1 1 1 1	b	j	r	z	+	¢	]	c_7

Fig. 3.10 IPC, 7-bit subset

8th bit is undefined, the code is shown in a 7-bit representation (see Fig. 3.10). The names of the graphics and control characters are given in Table 3.2.

As with the Stretch code, IPC has the upper- and lower-case alphabetics interleaved. And as with the Stretch code, a decision had to

TABLE 3.2. IPC, special graphics and controls

(	Left parenthesis	"	Quotes
!	Exclamation	'	Apostrophe
?	Question	$	Dollars
#	Numbers	¢	Cents
°	Degrees	∑	Summation
/	Slash	1/4	One quarter
*	Asterisk	≤	Equal or less
)	Right parenthesis	1/2	One half
.	Period	≥	Equal or greater
,	Comma	3/4	Three fourths
π	Pi	∞	Infinite
−	Minus	↓	Arrow (down)
ω	Omega	θ	Theta
+	Plus	↑	Arrow (up)
α	Alpha	ϕ	Phi
×	Multiply	→	Arrow (right)
β	Beta	κ	Kappa
÷	Divide	←	Arrow (left)
=	Equals	]	Right bracket
-	Dash	[	Left bracket
√	Square root	^{3}	Cubed
∫	Integral	^{2}	Squared
:	Colon	E (boxed)	Escape code #2
;	Semicolon	e (boxed)	Escape code #1
@	At	Bk_i	Blank key #i
X	Box −x	·	Center dot
		$\dot{C}_i$	Control #i

be made on which should precede within the pair. The IPC designers chose that upper case should precede lower case, so that proper nouns would collate ahead of common nouns. For example, Jack

$$0011110, 0001101, 0010001, 0100001$$

collates ahead of jack

$$0011111, 0001101, 0010001, 0100001.$$

The most interesting aspect of IPC is the design philosphy of Criterion 3—local interpretation. In the design of ASCII, described in later chapters, a set of control characters was defined to include several types of input/output equipments, thus forming a general set, which must of necessity have more characters than the set contained in IPC that is interpreted differently for different equipments.

Example

The seven control characters could be locally interpreted as follows:

C1	backspace	C5	Stop underline
C2	Unformatted tab	C6	Carriage return
C3	Formatted tab	C7	End of message
C4	Start underline		

Row	Column	0	1	2	3	4	5	6	7
	Bit Pattern b7	0	0	0	0	1	1	1	1
	b6	0	0	1	1	0	0	1	1
	b5	0	1	0	1	0	1	0	1
	b4 b3 b2 b1								
0	0 0 0 0	0	C	K	S				
1	0 0 0 1	1	c	k	s				
2	0 0 1 0	2	D	L	T				
3	0 0 1 1	3	d	l	t				
4	0 1 0 0	4	E	M	U				
5	0 1 0 1	5	e	m	u				
6	0 1 1 0	6	F	N	V				
7	0 1 1 1	7	f	n	v				
8	1 0 0 0	8	G	O	W				
9	1 0 0 1	9	g	o	w				
10	1 0 1 0	SP	H	P	X				
11	1 0 1 1	RES	h	p	x				
12	1 1 0 0	A	I	Q	Y				
13	1 1 0 1	a	i	q	y				
14	1 1 1 0	B	J	R	Z				
15	1 1 1 1	b	j	r	z				

Fig. 3.11 IPC, 6-bit subset

Contained within the set were four positions with unassigned meaning and corresponding to two "blank keys" on a keyboard. Thus there are two upper-case and two lower-case characters available for local interpretation.

As stated under Criterion 3, subsets should be simply derivable. By dropping the high-order bit, a 6-bit subset is derived (Fig. 3.11). It contains numerics, upper- and lower-case alphabetics, Space, and the reserved code for local use.

	Column	0	1	2	3	4	5	6	7
	Bit Pattern b7	0	0	0	0	1	1	1	1
	b6	0	0	1	1	0	0	1	1
	b5	0	1	0	1	0	1	0	1
Row	b4 b3 b2 b1								
0	0 0 0 0	UN	C	K	S				
1	0 0 0 1								
2	0 0 1 0	UN	D	L	T				
3	0 0 1 1								
4	0 1 0 0	UN	E	M	U				
5	0 1 0 1								
6	0 1 1 0	UN	F	N	V				
7	0 1 1 1								
8	1 0 0 0	UN	G	O	W				
9	1 0 0 1								
10	1 0 1 0	SP	H	P	X				
11	1 0 1 1								
12	1 1 0 0	A	I	Q	Y				
13	1 1 0 1								
14	1 1 1 0	B	J	R	Z				
15	1 1 1 1								

UN - Unassigned

Fig. 3.12 IPC, 5-bit subset

	Column	0	1	2	3	4	5	6	7
	Bit Pattern b7	0	0	0	0	1	1	1	1
	b6	0	0	1	1	0	0	1	1
	b5	0	1	0	1	0	1	0	1
Row	b4 b3 b2 b1								
0	0 0 0 0	0							
1	0 0 0 1	1							
2	0 0 1 0	2							
3	0 0 1 1	3							
4	0 1 0 0	4							
5	0 1 0 1	5							
6	0 1 1 0	6							
7	0 1 1 1	7							
8	1 0 0 0	8							
9	1 0 0 1	9							
10	1 0 1 0	UN							
11	1 0 1 1	UN							
12	1 1 0 0	UN							
13	1 1 0 1	UN							
14	1 1 1 0	UN							
15	1 1 1 1	UN							

UN - Unassigned

Fig. 3.13 IPC, 4-bit subset

By dropping the highest- and lowest-order bits, a 5-bit subset is derived (Fig. 3.12). It contains upper-case alphabetics, Space and five "unassigned" characters. One of these unassigned characters could be used to indicate either upper- or lower-case representation.

By dropping the three highest-order bits, a 4-bit subset is derived (Fig. 3.13). It contains the numerics and 6 "unassigned" characters for local interpretation.

Bit Pattern	0 0	0 1	1 0	1 1
0 0 0 0	0	+	-	b
0 0 0 1	1	A	J	/
0 0 1 0	2	B	K	S
0 0 1 1	3	C	L	T
0 1 0 0	4	D	M	U
0 1 0 1	5	E	N	V
0 1 1 0	6	F	O	W
0 1 1 1	7	G	P	X
1 0 0 0	8	H	Q	Y
1 0 0 1	9	I	R	Z
1 0 1 0	⊔	e	↓	d
1 0 1 1	=	.	$	,
1 1 0 0	,	)	*	(
1 1 0 1	≡	↑	<	>
1 1 1 0	⌐	]	[	¬
1 1 1 1	;	u	:	n

⊔ - OR
⌐ - AND
e - End of line, end of card, or carriage return
u - Up Shift
b - Space
d - Down Shift
n - Null

Fig. 3.14 Early 64-character proposal

3.6 AN EARLY 64-CHARACTER CODE PROPOSAL

In a Letter to the Editor, Communications of the ACM, 1959 May, H. S. Bright proposed a 64-character, 6-bit code. At that time, most printing and keypunching equipment was limited to 47 or 48 characters. The proposed code is shown in Fig. 3.14.

It is structurally compatible with BCDIC (Fig. 3.5). The sequence of the code table columns containing the alphabetics has been reversed from BCDIC, so that the alphabetics are in relative collating sequence. Oddly enough, Space, which traditionally collates low to numerics, alphabetics, and specials, was not assigned to the bit pattern 000000. This was undoubtedly done so that zero could be given bit pattern 000000, so that the numerics would be in relative collating sequence.

3.7 AN EARLY 256-CHARACTER CARD CODE PROPOSAL

In 1959 September, R. W. Bemer proposed a "Generalized Card Code for 256 Characters." At that time, as stated previously, character sets provided on printers and keypunches were mainly limited to 48.

As described earlier in this chapter, Project Stretch was started in IBM in 1954. It was a project to develop a bigger and faster computer than any then in the field. One decision made was that Stretch would have an 8-bit architecture, in contrast with most computers of that time which had a 6-bit, or 6-bit oriented, architecture.

R. W. Bemer, therefore, foresaw the need for a 256-character card code. The card code he proposed was not, in fact, adopted by Stretch, but it has many ingenious aspects. The card code set had criteria for design.

Criterion 1. The new set must contain the existing 48-character set as a subset, with exactly the same graphic–to–hole-pattern relationship.

Criterion 2. The new set should contain at least 256 combinations and be expansible beyond this number.

Criterion 3. Meanings need not initially be assigned to all hole patterns.

Criterion 4. The hole patterns should be structured, if possible, on existing zone punch/digit punch hole patterns.

Criterion 5. Hole patterns should be constructible and reproducible on existing keypunches (for example, the 024 or 026).

Criterion 6. There should be no duals.

Criterion 7. ALGOL characters should be included.

Criterion 8. Characters not in the current IBM set or ALGOL set, but used by other manufacturers, should be included.

Criterion 9. There should be a simple relationship between upper- and lower-case alphabetic hole patterns.

There are 322 possible combinations of no more than four punches per card column, when no more than two may be zones (12, 11, 0) and no more than two may be digits (1 through 9). Figure 3.15 shows the 256 of these that remain when all combinations with two-digit punches containing a 1-punch and ten other combinations are excluded. The figure also shows assignment of both old and new graphics to hole patterns.

An ingenious aspect of this proposal is that each of the hole patterns may be constructed in a card column by superimposing the hole patterns for two of the alphameric characters in current use. These two characters are chosen for their mnemonic content. Thus [is represented mnemonically by LB (for Left Bracket) and is constructed of the hole pattern 11-3,

Zone Punches			12		11		0		12-11		11-0		0-12	
Digit Punches	G	M	G	M	G	M	G	M	G	M	G	M	G	M
	SP		+	&	—	—	0	0	&	−+	$\bar{0}$	0−	$\overset{+}{0}$	+0
1	1	1	A	A	J	J	/	/	a	−A	j	0J		
2	2	2	B	B	K	K	S	S	b	−B	k	0K	s	+S
3	3	3	C	C	L	L	T	T	c	−C	l	0L	t	+T
4	4	4	D	D	M	M	U	U	d	−D	m	0M	u	+U
5	5	5	E	E	N	N	V	V	e	−E	n	0N	v	+V
6	6	6	F	F	O	O	W	W	f	−F	o	0O	w	+W
7	7	7	G	G	P	P	X	X	g	−G	p	0P	x	+X
8	8	8	H	H	Q	Q	Y	Y	h	−H	q	0Q	y	+Y
9	9	9	I	I	R	R	Z	Z	i	−I	r	0R	z	+Z
2-3									[	LB	:=	LS	;	SC
2-4							_	US						
2-5														
2-6							switch	SW	bool	BO				
2-7			{	BG							stop	SP	□	BX
2-8							‡	SY			√	SQ		
2-9									]	RB	=:	RS		
3-4					∧	LM			comm	CM			~	TD
3-5			¢	CE					⩽	LE	10	TN		
3-6					∨	LO			:	CO	←	LW		
3-7					(	LP			proc	PC			complex	CX
3-8	#	#	.	.	$	$	,	,						
3-9			○	CI	<	LR			cr	CR	△	TR		
4-5									}	ND	≠	UN		
4-6							↑	UW	do	DO			↓	DW
4-7			°	DG	±	PM			dbl pr	DP				
4-8	@	@	⌑	⌑	×	*	%	%	″	DQ	′	QU		
4-9			≡	ID					\	RD				
5-6					¬	NO			∀	EO				
5-7			⩾	GE					!	EP				
5-8			ǂ	EH					=	EQ				
5-9			if ei	IE					return	RE	|	VR		
6-7									go to	GO				
6-8			½	HF										
6-9			?	IF	or if	OR			for	FR	→	RW		
7-8														
7-9					)	RP			>	GR				
8-9					¼	QR					array	RY	∞	IY

G Graphic
M Mnemonic

Fig. 3.15 A 256-character card code proposal

for L, and 12-2, for B. Therefore, the hole pattern chosen to represent Left Bracket is 12-11-2-3. Fig. 3.16 shows the derivation of the mnemonics chosen for new graphics.

Some ALGOL words were arbitrarily assigned to single graphics:

If is assigned to ?
BEGIN is assigned to {
END is assigned to }
INTEGER is assigned to #

Graphic	Mnemonic	Symbolizing	Graphic	Mnemonic	Symbolizing
+	+		,	,	
—	—		.	.	
×	*		;	SC	SemiColon
/	/		:	CO	COlon
\	RD	Reverse Divide	!	EP	Exclamation Point
±	PM	Plus or Minus	'	QU	QUote
√	SQ	SQuare root	"	DQ	Double Quote
=	EQ	EQuals	:=	LS	Left Substitution
≠	UN	UNequals	=:	RS	Right Substitution
>	GR	GreateR	$_{10}$	TN	base TeN
⩾	GE	Greater or Equal	]	RB	Right Bracket
⩽	LE	Lesser or Equal	(	LP	Left Parenthesis
<	LR	LesseR	)	RP	Right Parenthesis
~	TD	TilDe	[	LB	Left Bracket
¬	NO	NOt	↑	UW	Up arroW
∨	LO	Logical Or	↓	DW	Down arroW
∧	LM	Logical Multiply	←	LW	Left arroW
∀	EO	Exclusive Or	→	RW	Right arroW
≡	ID	IDentical to	{	BG	BeGin
			}	ND	eND
¢	CE	CEnts	|	VR	VeRtical
¼	QR	one QuarteR	△	TR	TRiangle
½	HF	one HaLf	□	BX	BoX
cr	CR	CRedit	○	CI	CIrcle
°	DG	DeGree	_	US	UnderScore
∞	IY	InfinitY	procedure	PC	ProCedure
go to	GO	GO to	switch	SW	SWitch
do	DO	DO	array	RY	aRraY
return	RE	REturn	comment	CM	CoMment
stop	SP	StoP	integer	#	
for	FR	FoR	boolean	BO	BOolean
or if	OR	OR if	complex	CX	CompleX
if either	IE	If Either	double pr	DP	Double Precision

Fig. 3.16 Mnemonic derivations for characters

Other words could be assigned to single graphics:

COMMENT could be assigned to "

STOP could be assigned to !

RETURN could be assigned to ←

Record Mark ‡ and Group Mark ⧧ were assigned to existing hole patterns 0-8-2 and 12-8-5, respectively. The mnemonics chosen, SY and EH, are not of course mnemonics for Record Mark and Group Mark, but are mnemonics for the appropriate hole patterns for keypunching:

S, 0-2	Y, 0-8	SY, 0-8-2
E, 12-5	H, 12-8	EH, 12-8-5

Row	Column / Bit Pat.	0	1	2	3	4	5	6	7	8	9	A	B	C	D	E	F
		00				01				10				11			
		00	01	10	11	00	01	10	11	00	01	10	11	00	01	10	11
0	0000																SP
1	0001								A				/		J	1	
2	0010								B				S		K	2	
3	0011				SOR2 ACK2		SOR1 ACK1	C			3	T		L			
4	0100								D				U		M	4	
5	0101				TL		CL	E			5	V		N			
6	0110				<		[	F			6	W		0			
7	0111		7	X		P				G							
8	1000								H				Y		Q	8	
9	1001				IDLE		INQ ERR	I			9	Z		R			
A	1010				\		EOT Y	?			0	‡		!			
B	1011		#	,		$				.							
C	1100				;		TEL]	⌶			@	%		*			
D	1101		√	⧺		Δ				‡							
E	1110		>	ƀ		-				&							
F	1111																

TL - Transmit Leader
CL - Control Leader
SOR1 - Start of Record Odd
ACK1 - Acknowledge Odd
SOR2 - Start of Record Even
ACK2 - Acknowledge Even

INQ - Inquiry
ERR - Error
IDLE - Idle
*TEL - Telephone
*EOT - End of Transmission/Message

* May be sent as valid data characters

Fig. 3.17 4-out-of-8 code

3.8 4-OUT-OF-8 CODE

Another code of the early 1960s had an interesting characteristic. It was used solely for data transmission; it was an 8-bit code. The interesting characteristic was that, of the 8 possible bit positions for any bit pattern of the code, exactly four of the bits would be one-bits. Hence the name, 4-out-of-8 code (see Fig. 3.17). Any single "hit" (the accidental change of a zero-bit to a one-bit, or of a one-bit to a zero-bit) on a bit of a transmitted bit pattern would create an other than 4-out-of-8 bit pattern, and such erroneous bit patterns could be checked by very simple circuitry. Each bit pattern, as received, was fed through a counter. If the count was 4, the bit pattern was accepted as valid, otherwise a data check was raised. Of course, compensating hits (that is to say, hits on a single bit pattern that changed some one-bit to a zero-bit and some zero-bit to a one-bit) would not be detected, but occurrence of such hits was statistically very much less than occurrence of single bit hits.

Mathematically, the code allows exactly 70 valid 4-out-of-8 bit patterns. As can be seen by examination of Fig. 3.17, 64 of these were graphic characters (called "data characters" at that time) and 6 were control characters. Thus this code fittted BCDIC nicely with its 64 characters. As will be described in Chapter 5, 7 of the 64 characters of BCDIC were control characters between various BCDIC CPU's and magnetic tape drives. However, these 7 BCDIC control characters were not 4-out-of-8 control characters; that is to say, they would be transmitted, end to end, without effecting any control actions on the data transmission units.

Some of the 4-out-of-8 control characters did double duty, depending on the data transmission situation. Thus a data transmission unit, sending a data record, would precede it with SOR1, Start of Record Odd. When the transmission unit at the other end received this record, it would send back to the original transmission unit ACKl, Acknowledge Odd (providing no data check had been detected by the receiving unit).

Note from Fig. 3.17 that although the numerics and alphabetics are not contiguous within columns, they are nevertheless BCD, under the definitions in Chapter 2.

CITED REFERENCES

3.1 Military Communication System Technical Standard, MIL-STD-188A, 1958 April 25.

3.2 Military Communication System Technical Standard, MIL-STD-188B, 1964 February 24.

3.3 Military Communication System Technical Standard, MIL-STD-188C, 1969 November 24.

3.4 IBM Corporate Systems Standard, BCD Interchange Code, CSS 3-2-8015-0, 1962 April.

3.5 W. Buchholz, "Planning a Computer System." New York: McGraw-Hill, 1962.

3.6 R. W. Bemer, and W. Buchholz, "An Extended Character Set Standard." IBM Technical Report 00.721 (Rev.), 1960 June 1.

3.7 E. Morenoff; J. B. McLean; and L. Odell, "IPC, A Coded Character Set for Information Processing," Rome Air Development Center Technical Documentary Report No. RADC-TDR-64-426. 1964 October.

3.8 H. S. Bright, "A 64-Character Alphabet Proposal," Letters to the Editor, *Communications of the ACM* **2**:5. 1959 May.

3.9 R. W. Bemer, "A Proposal for a Generalized Card Code for 256 Characters," *Communications of the ACM* **2**:9. 1959 September.

4
The Duals of BCDIC

The code described in the previous chapter as "early BCDIC" will be called BCDIC, Version 1 in this chapter. This coded character set was extended; first by the addition of duals, to BCDIC, Version 2, and then by an expansion to 64 characters, to BCDIC, Version 3.

4.1 BCDIC, VERSION 1

In the late 1950s, the chain printers provided by IBM had a printing repertoire of 48 graphic characters, as follows:

Space		1
Alphabetics: A to Z		26
Numerics: 0 to 9		10
Specials:		
Dollar sign	$	
Slash	/	
Lozenge	⌑	
Asterisk	*	
Percent sign	%	
At sign	@	11
Ampersand	&	
Minus, Hyphen	–	
Number sign	#	
Period	.	
Comma	,	
		48

Hole Pattern → ↓		0	11	12
	SP		–	&
1	1	/	J	A
2	2	S	K	B
3	3	T	L	C
4	4	U	M	D
5	5	V	N	E
6	6	W	O	F
7	7	X	P	G
8	8	Y	Q	H
9	9	Z	R	I
0	0			
8-3	#	,	$	.
8-4	@	%	*	⌑

Fig. 4.1 BCDIC, Version 1

These 48 graphic characters were also keypunchable, interpretable, and verifiable by a single keystroke on the IBM keypunches and verifiers of the day. These 48 characters, which constituted BCDIC, Version 1, are shown in Fig. 4.1

4.2 BCDIC, VERSION 2

Two data processing requirements, European languages and FORTRAN, led to the development of what came to be called "duals."

4.2.1 European Languages Requirements

The languages of some European countries (Germany, Sweden, Denmark, Norway, Finland) require 29 letters—the usual 26 alphabetics of English-speaking countries plus three letters called diacritics. Spanish and Portuguese alphabets have 27 letters. It would be clearly advantageous, from a marketing point of view, to be able to provide these extra alphabetics on printers, keypunches, and verifiers. But how could this be done? The solution that was examined first was to increase the character capability of printers, keypunches, and verifiers from 48 to 51.

In the case of chain printers, this was entirely feasible, since the chain has a possible graphic capability of 240. In fact, on 48-character chains, each of the 48 graphics appears five times on the 240-graphic chain. If there are more than 48 graphics, 51, for example, some of these graphics will *not* appear five times on the chain; in consequence, the printing speed (lines per minute) would be reduced. Since printing speed was (and is) a primary competitive factor for printers, the solution of providing 51 graphics on a chain, with consequent slower printing speeds, was unattractive.

In the case of the keypunch (and verifier), two approaches were examined. Under the first approach, card hole patterns beyond the 48 could be assigned and keypunched by the technique known as multipunching. Under this technique, while a "multipunch" key is held down, other keys may be struck, but the punched card does not advance to the next card column. Accordingly, a number of holes may be punched in a single card column. Clearly, when any of the three diacritic letters is encountered on a data sheet by a keypunch operator, the keypunching mode would have to depart from touch-keying while the operator pays special attention to holding down the multipunch key and to keying such other keys as necessary to generate the appropriate hole pattern. In this approach, then, the keypunching speed would be reduced. As with the line-printer solution discussed above, this approach to keypunching was unattractive.

Under the second approach, either existing keypunches and verifiers could be modified, or new keypunches and verifiers could be designed with additional keys to generate each of the three diacritics with a single keystroke. Presumably (after some training) keypunch operators would be able to touch-key the additional keys, so keypunching speed would be maintained. This approach would result in a relatively costly design and development project, with a product that would have only a small market. The projected additional price for European keypunches and verifiers was unattractive.

A different kind of solution was then proposed. It was observed that three special graphics @ # $ were peculiar in origin and use to English-speaking countries. They were neither needed nor used at that time in continental European countries. The suggestion was to substitute the three diacritics for these three specials, wherever they appeared on the chain. The consequence was that printing speed would not be reduced. Similarly, they could be substituted on the keytops and printing plates of keypunches.

Under this substitution approach, only minor costs would be involved. The solution, then, had the following characteristics:

No reduction in printing speeds.
No reduction in keypunching/verifying speeds.
Small cost.

This approach had the advantages above, and no (known) disadvantages. It was adopted. The approach is still used in current products.

It should be noted, in respect to this approach, that there results a number of graphics—multiple graphics, that is—for three card hole patterns, as shown in Fig. 4.2. However, *within* a country, the graphic set is unique, without duals.

Hole pattern	U.S.A.	Germany	Sweden	Finland	Norway	Denmark
8-3	#	Ä	Ä	Ä	Æ	Æ
8-4	@	ö	ö	ö	ø	ø
11-8-3	$	Ü	Å	Å	Å	Å

Fig. 4.2 Diacritic letters

4.2.2 FORTRAN Requirements

The FORTRAN programming language had, among its other objectives, the objective of a printed listing that would resemble as much as possible the formulae found in mathematical text books. Many of the mathematical symbols found in text books were deemed to be unnecessary for FORTRAN. Some mathematical symbols / − . , were already provided on IBM printers. It was decided that the asterisk * could be used to represent multiplication. But five symbols ( ) + = ′ (not provided on IBM printers) were deemed to be absolutely necessary for FORTRAN. How to provide them?

It was decided that the most economical and efficient solution was to provide them by substitution, as with the European diacritics. The only remaining problem was to choose which five of the specials provided on IBM 48-character printers, keypunches, and verifiers should be replaced by the five mathematical symbols. It was decided to replace % ¤ & # @ by ( ) + = ′ (respectively). This solution resulted in duals within a country. The addition of these five duals led to BCDIC, Version 2, shown in Fig. 4.3.

Hole Pattern → ↓		0	11	12
	SP	[1]	-	& or +
1	1	/	J	A
2	2	S	K	B
3	3	T	L	C
4	4	U	M	D
5	5	V	N	E
6	6	W	O	F
7	7	X	P	G
8	8	Y	Q	H
9	9	Z	R	I
0	0	[2]		
8-3	# or =	,	$	.
8-4	@ or '	% or (	*	⌑ or)
8-5				
8-6				
8-7				

Hole Patterns:
[1] 8-2
[2] 0-8-2

Fig. 4.3 BCDIC, Version 2

Initially, this solution was ideal. With very few exceptions, computing installations in those days were either of a commercial orientation or of a scientific/engineering orientation. In "commercial" installations, such commercial applications as payroll, inventory, premium billing, and utility billing were processed; in such installations, neither scientific nor engineering applications were processed. Similarly, in "scientific" installations, scientific or engineering calculations were processed, and commercial applications were not. (I repeat, there were few exceptions.)

The exceptions that began to be noted were those users who had installations that were commercially oriented, although the company itself was of an engineering or scientific nature. In such companies, there were people who wanted to use the computer for scientific or engineering calculations. It is to be noted that the processes of compiling, debugging, and executing FORTRAN programs could be performed regardless of whether the printers, keypunches, and verifiers had the scientific or commercial graphic sets. However, if the installation had the commercial graphic set, program listings were somewhat bizarre. For example, a FORTRAN statement such as

X = (A + B) * (C − D)/(E + F * G)

would show in the program listing as

X#%A&B¤*%C−D¤/%E&F*G ¤

Such program listings, though bizarre, were unambiguous. To FORTRAN programmers who suffered in the commercial installations of the day, the mental translation of

%	to	(
¤	to	)
&	to	+
#	to	=
@	to	′

became an automatic act.

It should be reemphasized that the scientific symbols seldom (if ever) were needed or used in the listings that were the final results of the executed programs. It was only in the listings of the original FORTRAN programs that programmers had to put up with the graphic substitutions. Programmers were (and, incidentally, still are) notably vocal. If there were something to complain about, they complained vociferously. These complaints gave rise to the question, Could this situation, admittedly infrequent but nonetheless aggravating, be ameliorated?

4.3 BCDIC, VERSION 3

A solution to the "duals problem" was attempted with the IBM 1410. (Another attempt was made in the System/360. See Chapter 9, The Duals of EBCDIC.) The 1410 was to have as its console, a typewriter. The typewriter *could* provide up to 88 graphics. It was decided it *would* provide 63, and Space. (The reasons for a character set size of 64 are detailed in the following chapter, The Size of BCDIC.) The 47 graphics and Space provided on 48-character chain printers are shown in Fig. 4.3. The 63 graphics and space *proposed* to be provided on the 1410 console typewriter are shown in Fig. 4.4; it is called BCDIC, Version 3.

It is to be observed in Fig. 4.4 that four of the five "scientific" graphics ( ) = ′ were to be given unique card hole patterns. Curiously, the

Hole Pattern		0	11	12
	SP	¢ [1]	-	&
1	1	/	J	A
2	2	S	K	B
3	3	T	L	C
4	4	U	M	D
5	5	V	N	E
6	6	W	O	F
7	7	X	P	G
8	8	Y	Q	H
9	9	Z	R	I
0	0	‡ [2]	!	?
8-3	#	,	$	.
8-4	@	%	*	<
8-5	:	=	)	(
8-6	>	'	;	>
8-7	√	"	Δ	⧧

Hole Patterns:
[1] 8-2
[2] 0-8-2

Fig. 4.4 BCDIC, Version 3

fifth scientific graphic + was not to be provided. The author does not know the reason for this curious anomaly.

Beyond the four scientific graphics, 12 new graphics had been added. These were of two kinds:

Kind 1 . ? : ; " < > ¢
Kind 2 ‡ √ △ ⧧

The graphics of Kind 1 were added as a result of market studies for "most-needed graphics" in data processing applications. The graphics of Kind 2 were chosen to meet a criterion which will be described in the next chapter.

This coded character set was announced for the IBM 1410. However, as will be discussed in the next chapter, a review of coded character sets was then undertaken, and this led to the BCD Interchange Code, BCDIC.

5 The Size of BCDIC

5.1 SIZE OF CHARACTER SET

What limits the size of a character set? Is it the number of characters in a character set? The limitation is mathematical, and comes from the binary characteristic of the code that represents the character set. Recall that the binary aspect comes from the nature of the physical medium or hardware that handles the character code. Once the binary aspect of the physical medium is perceived, the binary capacity must next be determined. Some examples follow.

Magnetic tape, seven tracks. One track is for parity, leaving six tracks for storage of characters. The character set size is $2^6 = 64$ characters.

Magnetic tape, nine tracks. One track is for parity, leaving eight tracks for characters. Set size $= 2^8 = 256$ characters.

Paper tape, eight rows. One track is for parity, leaving seven tracks for characters. Set size $= 2^7 = 128$ characters.

Punched cards, twelve rows. Set size $= 2^{12} = 4096$ characters. Most punched card character sets have a set size less than the maximum capacity. For the System/360, for example, the punched card character set size is restricted to 256, in order to match the Nine Track Magnetic Tape character set size of 256 characters.

As described in the previous chapter, IBM character set sizes before the introduction of the 1410 were 48 characters, a limitation imposed by the chain printers and keypunches of the day. The chain-printer limitation of 48 characters was based not on the number of possible different graphic characters on the chain but on marketing considerations having to do with printing speeds.

With the introduction of the 1410, its console typewriter provided a possible character set size of 88 characters. The limitation of printing speed held the chain printer set size to 48 characters, but it was decided to expand the console typewriter set size beyond 48 characters. What should this character set size be?

There were two hardware aspects which limited the set size, happily to the same number. The 1410 architecture was 6 bits, hence maximum set size was 64 characters. Magnetic tape for the 1410 was seven tracks. One track was for parity leaving six tracks for characters. So the magnetic tape also restricted the set size to a maximum of 64.

It was decided to expand the 1410 character set size to 64 characters. Before this time, the 48-character set, BCDIC, Version 2, was as shown in Fig. 5.1.

Bit Pattern →			A	B	BA
↓	Hole Pattern →		0	11	12
		SP		-	& or +
1	1	1	/	J	A
2	2	2	S	K	B
2 1	3	3	T	L	C
4	4	4	U	M	D
4 1	5	5	V	N	E
4 2	6	6	W	O	F
4 2 1	7	7	X	P	G
8	8	8	Y	Q	H
8 1	9	9	Z	R	I
8 2	0	0			
8 2 1	8-3	# or =	,	$	.
8 4	8-4	@ or '	% or (	*	⌑ or)

Fig. 5.1 BCDIC, Version 2

5.2 BCDIC, VERSION 3

The binary coded decimal (BCD) nature of the card-code–to–bit-code relationship pointed to the obvious card-code expansion, to include 8-2, 8-5, 8-6, 8-7 digit punches in conjunction with the zone punches, as shown in Fig. 5.2.

There were two problems to be solved in determining the 64 hole patterns. Since the numeric "0" would clearly retain its card hole pattern 0, what hole patterns would be assigned to code positions in Fig. 5.2 indicated by [1] and [2]? Both of these code positions (following the table column and table row indications) would have the card hole pattern of 0, but three code table positions with the same hole pattern, 0, would be unacceptable.

Bit Pattern →			A	B	BA
↓	Hole Pattern → ↓		0	11	12
		SP	UN. [1]	-	& or +
1	1	1	/	J	A
2	2	2	S	K	B
2 1	3	3	T	L	C
4	4	4	U	M	D
4 1	5	5	V	N	E
4 2	6	6	W	O	F
4 2 1	7	7	X	P	G
8	8	8	Y	Q	H
8 1	9	9	Z	R	I
8 2	0	0 [3]	UN. [2]	UN. [4]	UN. [5]
8 2 1	8-3	# or =	,	$	.
8 4	8-4	@ or '	% or (	*	⌑ or)
8 4 1	8-5	UN.	UN.	UN.	UN.
8 4 2	8-6	UN.	UN.	UN.	UN.
8 4 2 1	8-7	UN.	UN.	UN.	UN.

UN. - Unassigned graphic

Hole Patterns:
[1] 8-2
[2] 0-8-2

Fig. 5.2 Expansion of BCDIC card code to 64

Bit Pattern → / ↓	Hole Pattern → / ↓		A	B	BA
			0	11	12
		SP	¢ [1]	-	&
1	1	1	/	J	A
2	2	2	S	K	B
2 1	3	3	T	L	C
4	4	4	U	M	D
4 1	5	5	V	N	E
4 2	6	6	W	O	F
4 2 1	7	7	X	P	G
8	8	8	Y	Q	H
8 1	9	9	Z	R	I
8 2	0	0	‡ [2]	!	?
8 2 1	8-3	#	,	$	.
8 4	8-4	@	%	*	⌑
8 4 1	8-5	:	=	)	(
8 4 2	8-6	>	'	;	<
8 4 2 1	8-7	√ [4]	"	Δ [5]	‡ [3]

SP - Space

Hole Patterns:
[1] 8-2
[2] 0-8-2

Fig. 5.3 BCDIC, Version 3

Note that the bit pattern for code table position [3] is 82. From the BCD relationship, therefore, a card hole pattern of 8-2 would generate the proper bit pattern. Combining the digit punches 8-2 with a zone punch 0 would therefore generate the correct bit pattern, A82, for code table position [2]. This hole pattern therefore was chosen for this code position.

But what about code table position [1]? Although the numeric "0" occupies code table position [3] and has the hole pattern 0, a proper hole pattern from a BCD relationship point of view would be 8-2.

It should be pointed out that the objective was to determine a set of 64 hole patterns with a BCD relationship. One such set would be the 16 digit combinations, "no-digits", 1, 2, 3, 4, 5, 6, 7, 8, 9, 8-2, 8-3, 8-4, 8-5, 8-6, 8-7 taken with the four zone punches, "no-zone", 0, 12, and 11.

However, this set does not include the hole patterns 12-0 and 11-0, which were widely used in card processing applications. In order to include these in the BCDIC set, two would have to be dropped out of the set of 64 above. The two chosen to be dropped out were 12-8-2 and 11-8-2. (Note: As described in Chapter 10, these *were* included in the 64-character subset for EBCDIC, and 12-0 and 11-0 were *not* included.)

The hole patterns 11-0 and 12-0 fall logically, for the code table of Fig. 5.2, in code positions ^{4} and ^{5}. The only remaining hole pattern from the set of 64 above that has no logical position is 8-2, and the single code table position without an assigned hole pattern is position ^{1}, so that by a process of elimination, the hole pattern 8-2 was assigned to code table position ^{1}.

As described in the previous chapter, 16 graphics had been chosen for the IBM 1410 to expand the character set from 48 to 64. The result, BCDIC, Version 3, is shown in Fig. 5.3.

5.3 BCDIC, VERSION 4

Of these 16 graphics, four had been chosen to eliminate duals and provide ( ) ′ = as unique graphics. Eight had been chosen as a result of market studies for most-wanted additional graphics:

! ? : ; ″ < > ¢

Four had been chosen to meet an interesting criterion:

ǂ ⧣ √ Δ

These four graphics occupied code positions ^{2}, ^{3}, ^{4}, and ^{5}, whose bit patterns had a control function with respect to magnetic-tape devices on one or another of the IBM computing systems.

There is an aspect of human nature which surfaces in data processing. Experience has shown that if graphics are provided on a computing system, they will be used in one way or another by customers, even if they have no intrinsic meaning. The lozenge is an example. It has no intrinsic meaning but customers came to use it to signify things peculiar to their applications—within applications, customers gave the lozenge a meaning. For example, in banking installations, the lozenge was frequently used on tabulation listings to indicate (to the customer) second level totals.

But it would be very undesirable if customers, within an application, used the graphics for code positions ^{2}, ^{3}, ^{4}, and ^{5} so that they would be required to print out on listings. The actual printing of such graphics would not present any hazard, but the data containing the bit patterns representing these graphics, if written on or read from magnetic tape, might cause strange and unwanted results. These bit patterns had a common and interesting characteristic. They were generated or removed automatically by the magnetic-tape hardware. The customer did not have to enter them with his input data.

The obvious criterion for graphics to be assigned to these code positions was that they should cause customers to be disinclined to use them in applications. They should, therefore, be abstract shapes without intrinsic meaning. The graphic shapes finally chosen to meet the criterion are as follows:

2	3	4	5
⧧	√	Δ	⯒

How well the graphics meet the criterion the reader can judge.

As stated in the previous chapter, a reconsideration of BCDIC, Version 3 was undertaken. There were a number of reasons:

1. The plus sign was not provided.
2. The characters

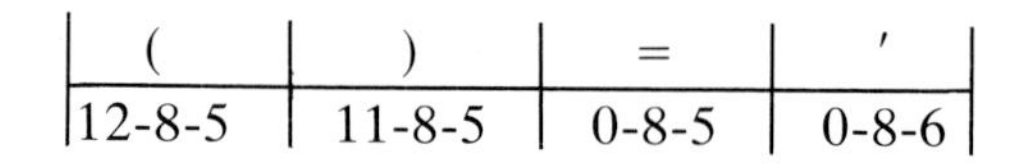

(	)	=	'
12-8-5	11-8-5	0-8-5	0-8-6

   would require multipunching on a keypunch. The speed of keypunching FORTRAN source language programs would be reduced.

3. FORTRAN program decks, keypunched according to the 1410 proposal, BCDIC, Version 3, could not be compiled on any non-1410 computer, because the card hole patterns (and hence the bit patterns) for ( ) = ' had been changed. Similarly, the FORTRAN compiler for the 1410 could not compile any FORTRAN program decks from non-1410 computers.
4. If a 1410 FORTRAN program deck were entered into a 1410, it would not list properly on the chain printer of the 1410. ( ) = ' would not list as ( ) = ' nor indeed even list as % ¤ # @ (the dual graphics). Such a program deck *could* be listed properly on the 1410 console typewriter, but this mode of listing would be excessively slow as compared with listing on a chain printer.

Reason 3 above was crucial. The ability to enter, list, compile, and execute a FORTRAN deck on *any* IBM computing system was a very strong sales point. Therefore, the 1410 coding proposal was changed to remedy the four problems above.

The result of this change became the BCD Interchange Code, BCDIC. The criteria set for BCDIC were as follows:

1. The 48-character code would be extended to 64 characters—63 graphics and Space.
2. Compatibility with the 48 characters of the day—Space, 10 numerics, 26 alphabetics, and 11 specials (including the 5 duals)—would be maintained. That is to say, BCDIC, Version 2 (Fig. 5.1) would be the point of departure.

Bit Pattern ↓ (Bit Pattern →)	Hole Pattern ↓ (Hole Pattern →)		A	B	BA
			0	11	12
		SP	¢ [1]	-	& or +
1	1	1	/	J	A
2	2	2	S	K	B
2 1	3	3	T	L	C
4	4	4	U	M	D
4 1	5	5	V	N	E
4 2	6	6	W	O	F
4 2 1	7	7	X	P	G
8	8	8	Y	Q	H
8 1	9	9	Z	R	I
8 2	0	0	‡ [2]	!	?
8 2 1	8-3	# or =	,	$	.
8 4	8-4	@ or '	% or)	*	⌑ or)
8 4 1	8-5	:	[3]	[5]	[6]
8 4 2	8-6	>	[4]	;	<
8 4 2 1	8-7	√ [7]	" [8]	Δ [9]	⯒ [10]

Hole Patterns:

[1] 8-2

[2] 0-8-2

SP - Space

Fig. 5.4 BCDIC, Version 4

3. As much as possible, compatibility with the announced 1410 set, BCDIC, Version 3 (Fig. 5.3), would be maintained.
4. Graphics for control characters should have no intrinsic meaning.

Initially, these criteria led to the code table of Fig. 5.4.

5.4 BCDIC, FINAL VERSION

Code positions ^{6}, ^{5}, ^{3}, ^{4} that had held ( ) = ′ in the 1410 proposal were left blank, with four new graphics to be chosen.

Code positions ^{2}, ^{7}, ^{9}, ^{10} with graphics ǂ √ Δ ⧧ were deemed to satisfy Criterion 4. But code positions ^{1} and ^{8} had bit patterns that functioned as control characters on one or another IBM computer. Graphics ¢ ″ were clearly a violation of Criterion 4; they were rejected.

This left code positions ^{1}, ^{3}, and ^{8} to be assigned graphics satisfying Criterion 4, and code positions ^{5} and ^{6} to be assigned new graphics.

Code positions ^{6}, ^{5}, and ^{4} had held ( ) ′ under the 1410 proposal. When new graphics [ ] / were suggested to fill these code positions, the suggestion was adopted.

After much debate, ƀ γ ⧻ were chosen for code positions ^{1}, ^{3}, and ^{8} to satisfy Criterion 4. To satisfy Criterion 4, then, eight graphics had been chosen:

ƀ γ ǂ ⧧ ¤ Δ √ ⧻

How well these graphics satisfy the criterion, the reader may judge.

The final result was BCDIC, shown in Fig. 5.5. It was approved as an IBM Corporate Systems Standard in 1962.

Two factors were primary in the development of BCDIC from early BCDIC, Version 1: equipment limitations and compatibility. Equipment limitations led to the introduction of duals both for alphabetic extension and for programming language symbols. Compatibility led to the retention of the duals, even when the 1410 console typewriter removed one equipment limitation. (It may be remarked that the chain of the chain printer, with its capability of 240 graphic positions, did not limit the printing set to 48. Another aspect of the chain printer, printing speed, was responsible for limiting the printing set to 48 graphics.)

Compatibility with existing practice is an important factor in decisions on coded character sets. In summary, the four objections to the 1410 proposal were as follows:

1. Absence of plus sign.

Bit Pattern → ↓			A	B	BA
	Hole Pattern → ↓		0	11	12
		SP	ƀ [1]	-	& or +
1	1	1	/	J	A
2	2	2	S	K	B
2 1	3	3	T	L	C
4	4	4	U	M	D
4 1	5	5	V	N	E
4 2	6	6	W	O	F
4 2 1	7	7	X	P	G
8	8	8	Y	Q	H
8 1	9	9	Z	R	I
8 2	0	0	‡ [2]	!	?
8 2 1	8-3	# or =	,	$	.
8 4	8-4	@ or '	% or (	*	⌑ or)
8 4 1	8-5	:	γ	]	[
8 4 2	8-6	>	\	;	<
8 4 2 1	8-7	√	⧻	Δ	‡

Hole Patterns:
[1] 8-2
[2] 0-8-2

SP - Space

Fig. 5.5 BCDIC, Final version

2. Multipunching required for keypunching,

$$(\quad) \quad = \quad '$$

that is, for keypunching FORTRAN program decks.

3. FORTRAN incompatibility—1410 versus other computing systems.
4. 1410 FORTRAN Programs not listable on 1410 chain printer.

All of these problems were in fact solvable at the time, admittedly at some cost. The incompatibility that would have resulted pre- and post-1410 was unacceptable. The problems were not solved. Duals were assigned into BCDIC.

6
The Size and Structure of PTTC

6.1 INITIAL CONSIDERATIONS

In 1959, engineers had started to design and develop a new communications terminal which came to be the IBM 1050. The keyboard and printing functions were to be provided by an electric typewriter. The typewriter provides a capability of 88 graphics. The question to be decided was what the transmission code should be. Since perforated tape was also envisaged for this terminal, the code came to be named the Perforated Tape and Transmission Code (PTTC).

In today's technology, where hundreds and thousands of electronic circuits can be placed on a small chip, the cost of a bit is negligible. But in the technology of the early 1960s, the cost of a bit was appreciable—6-bit registers cost appreciably more than 5-bit registers, 7-bit registers cost appreciably more than 6-bit registers, and so on.

Another cost factor was implicit in the byte size. On serial data transmission lines, a fixed factor was the number of bits transmittable per second. To transmit, for example, a thousand characters of seven bits per character would take appreciably more time than to transmit a thousand characters of six bits per character. The length of time the data transmission line was in use was a direct factor in determining the amount of money that had to be paid for the use of the data transmission line. In short, data transmission line costs were dependent on the byte size of the transmission code.

These two cost factors, hardware cost and transmission cost, both pointed to the necessity of keeping the byte size of a transmission code as small as possible. In those days, a design engineer built his reputation on his ability to "squeeze the bits."

Before the introduction of the IBM 1050, printing terminals had been limited to single case capability. But the use of an electric typewriter on the 1050 would give the capability of duocase printing—capital letters and small letters. This duocase capability was held to be a very significant marketing factor.

6.2 SIZE OF CHARACTER SET

Recall that the byte size of a code prescribes the number of characters that can be incorporated into the code, by virtue of a simple binary relationship. If the byte size if 5 bits, then $2^5 = 32$, and there are 32 different bit patterns available; that is, a 5-bit code can have 32 characters. If the byte size is 6 bits, then $2^6 = 64$, and there can be 64 characters. Similarly, 7 bits leads to 128 characters, 8 bits leads to 256 characters, and so on.

In designing a coded character set, the first determination must be the number of characters needed to meet the requirements of the applications in which the code will be used. This done, the code size may then be determined by applying the analysis of the preceding paragraph in reverse. For example, if 48 characters are needed, the 32 character positions of a 5-bit code are insufficient, but the 64 character positions of a 6-bit code are (more than) sufficient. A 6-bit code is needed if 48 characters must be provided.

In the case of the 1050, the determination of the number of characters proceeded as follows:

Alphabetics: 26 lower case and upper case

Numerics: 10

Specials: At this time, the character set for most IBM products was 47 and Space. For the console typewriter of the 1410, the set was 63 and Space. From this, it was rationalized that from 11 to 27 specials should be provided. Assume at least 11 would be needed.

Space: 1

Controls: The number of control characters needed was not known in the initial design phase of the 1050. Clearly, characters would be needed to control the typewriter, to control the perforated-tape facility, and to control the data transmission lines. Initially, let the number of control characters be x.

The above tabulates as follows:

Lower case alphabetics	26	73 graphic characters*
Upper case alphabetics	26	
Numerics	10	
Specials	11 at least	
Space	1	
Controls	x	
	$74 + x$	

Therefore, initial analysis showed that (at least) 73 graphic characters, the Space character, and an as yet undetermined number of control characters would be needed for PTTC. This apparently showed that a 6-bit, 64-character set was insufficient; a 7-bit, 128-character set was apparently indicated. But it was pointed out that a particular technique of coding, which involved the use of shift characters, could reduce the size requirement to 6 bits. (A full discussion of this coding technique is found in Chapter 2.)

6.3 PTTC, VERSION 1

Recall from Chapter 2 the formula $2^{x+1} - y$ (where x is the number of bits in the code byte and y is the number of characters wanted to be independent of preceding shift characters). For PTTC it was decided that the Space character and all control characters should be independent of preceding shift characters. At a first analysis, x was taken to be 6.

$$2^{x+1} - y = 2^{6+1} - y$$
$$= 128 - y$$

Thus it was seen that with a byte size of 6 bits, and using the technique of shift characters, $128 - y$ characters could be realized. Also, if $y =$ number of control characters, including Space, then the number of graphic characters is $128 - 2y$. The following possibilities were reviewed.

* It is to be noted that the number of graphic characters needed would be more than 73. This would certainly be realizable on the 88 graphic capability of the electric typewriter. Also, the figure 88 would clearly dictate that the maximum number of specials would be $88 - 62 = 26$.

Number of control characters (y)	Number of graphic characters (128 − 2y)	Number of different characters (128 − y)
17	94	111
18	92	110
19	90	109
20	88	108

At first it was argued that, since the typewriter provides 88 graphics only, the choice should be 20 control characters (including Space) and 88 graphics.

It was counter-argued that extensive analysis of applications suitable for the 1050 showed that 16 control characters and the Space character would be sufficient. Consequently, the choice should be 17 control characters (including Space) and 94 graphic characters. While it was admitted that the typewriter could print 88 graphics only, it was also true that paper tape, punched cards, data transmission lines, and serial printers could certainly implement 94 graphic characters.

At this point, a completely different factor emerged. At this time, standards committees, nationally and internationally, were developing a standard interchange code. All details of this code were not yet decided, but some details *were* decided:

a) The code would be 7 bits.

b) There would be 32 control characters, the Space character, the Delete character, and 94 graphic characters.

It was now proposed that the 1050 should implement the 7-Bit Code, so that it would be compatible with the emerging national and international standards. On the question of 7-bit size for the 1050, two counter-arguments were voiced:

a) A 7-bit 1050 would cost much more than a shifted 6-bit 1050, and low cost was a primary design objective of the 1050.

b) The 1050 development schedule was such that it would certainly be developed and announced before the slowly developing national and international standards were approved. Details of the standards, such as number and choice of control characters and graphic characters, changed from one committee meeting to the next. It was a reasonable certainty, then, that the 1050 character set would disagree, in greater or lesser detail, with the finally approved character set of the standards.

These two factors, particularly the cost factor, were decisive. The earlier decision, to design and develop a shifted 6-bit 1050, was upheld.

However, out of this debate emerged another factor, which was decisive on the earlier question of graphics and controls. A communications system was postulated which would have terminals implementing the 7-bit code, communicating via a computer, with 1050s implementing the shifted 6-bit code. The significant aspect here was that a message, consisting of graphic characters and the Space character, would go from one kind of terminal, through a computer, to the other kind of terminal.

If these different kinds of terminals needed different control characters to send or receive messages, the computer program could accommodate such differences, removing or injecting control characters into the data stream as necessary. But if the terminals had different graphic sets, no computer program could compensate. The number of graphics, and the actual graphics must match.

From this analysis, it was decided that the number of graphics in PTTC and the 7-bit code should be the same, 94. At this stage, the actual graphics could not be matched, since the 7-bit code was not yet finalized. However, after the 7-bit code was finalized, a later model of the 1050 could match the graphics. So the decision was made for the 94 graphic characters. As a consequence, Space and 16 control characters would be independent of shift. The initial code chart for the 1050 looked like Fig. 6.1.

The 16 control characters to be independent of shift were:

PN	Punch On
BYP	Bypass
RES	Restore
PF	Punch Off
RS	Reader Stop
LF	Line Feed
NL	New Line
HT	Horizontal Tab
UC	Upper Case
EOB	End of Block
BS	Back Space
LC	Lower Case
EOT	End of Transmission
PRE	Prefix
IL	Idle
DEL	Delete

Upper Case and Lower Case would be the two required shift characters.

Bit Pattern	Hole Pattern	Lower Case BA	Lower Case B	Lower Case A	Lower Case	Upper Case BA	Upper Case B	Upper Case A	Upper Case
		SP				SP			
1									
2									
2 1									
4									
4 1			47 GRAPHICS				47 GRAPHICS		
4 2									
4 2 1									
8									
8 1									
8 2									
8 2 1									
8 4									
8 4 1			16 CONTROLS				16 CONTROLS		
8 4 2									
8 4 2 1									
	Hole Pattern								

1	3
2	4

Block	Hole Patterns at:
1	Top And Left
2	Bottom and Left
3	Top and Left
4	Bottom and Left

Fig. 6.1 PTTC, Version 1

6.4 PTTC, VERSION 2

Since there was to be a card reader/punch attached to the 1050, a translation would be needed for the bit code of PPTC to/from the card code. In order to minimize the cost of such a translator, it was decided to structure the code, with respect to alphabetics and numerics, so that it

Bit Pattern	Hole Pattern	Lower Case				Upper Case			
			A	B	BA		A	B	BA
		SP				SP			
1		1		j	a			J	A
2		2	s	k	b		S	K	B
2 1		3	t	l	c		T	L	C
4		4	u	m	d		U	M	D
4 1		5	v	n	e		V	N	E
4 2		6	w	o	f		W	O	F
4 2 1		7	x	p	g		X	P	G
8		8	y	q	h		Y	Q	H
8 1		9	z	r	i		Z	R	I
8 2		0	1	1	1		2	2	2
8 2 1									
8 4		PN	BYP	RES	PF	PN	BYP	RES	PF
8 4 1		RS	LF	NL	HT	RS	LF	NL	HT
8 4 2		UC	EOB	BS	LC	UC	EOB	BS	LC
8 4 2 1		EOT	PRE	IL	DEL	EOT	PRE	IL	DEL
	Hole Pattern								

Hole Patterns:

1

2

1	3
2	4

Block	Hole Patterns at:
1	Top And Left
2	Bottom and Left
3	Top and Left
4	Bottom and Left

Fig. 6.2 PTTC, Version 2

resembled BCDIC; that is, the alphabetics would be distributed into three columns of the code table. Clearly, a corresponding upper- and lower-case alphabetic would be on the same 1050 keytop. In order to minimize the logic circuitry between keytops and the generation of the bit patterns of PTTC, lower-case and upper-case alphabetics should occupy corresponding (same bit pattern) locations in the code table. The numerics should be in the lower-case side of the code table, since they are commonly on the lower-case shift of a typewriter. Finally, with these decisions made with respect to numerics, upper-case alphabetics and lower-case alphabetics, it seemed intuitively right that the controls occupy the block of four rows at the bottom of the table. This led to the code table of Fig. 6.2.

6.5 PTTC, VERSION 3

There now remained the assignment of 32 specials to code positions. It is to be noted that, with 94 total graphic positions and a typewriter printing capability of 88 graphics, 6 of these remaining 32 code positions would contain graphics not printable on the typewriter. Clearly, because of the typewriter concept of upper- and lower-case graphics on a key, it would be confusing to an operator if any key had a printable graphic in one case but not in the other case. Also, it would complicate the logic circuitry to realize such an aspect. These considerations led to the conclusion that three of the nonprintable graphics should be in lower case and three in upper case. Also, they should be located in corresponding positions in the code table (to do otherwise would create the undesirable aspect). It was decided that positions [1] and [2] (Fig. 6.2) would be assigned to nonprintable graphics.

Before decisions were made on specific assignment of the 32 specials, some preliminary decisions were made with respect to the associated card code. The reason behind this sequencing of decisions was as follows. Hopefully, card-code assignments could be made on some orderly basis that would optimize the card-code to bit-code relationship, and hence minimize the cost of the hardware translator. If such an assignment of card codes could be worked out, then most of the 32 specials would automatically locate themselves in the code table, because of their already established BCDIC card codes.

The first problem to be solved was with respect to alphabetics. Hitherto, in data processing equipment and applications, only one set of alphabetics was provided. It would be more correct to call these alphabetics "capital letters," rather than "upper-case letters." To refer to them as "upper-case alphabetics" would imply the existence of "lower-case alphabetics," and these latter were not, in general, provided on data processing printers.

The "capital letters" had well-established card codes. Now, however, on the typewriter of the 1050, there were to be both lower- and upper-case letters. The question was, should lower- or upper-case letters be assumed as corresponding to the previous capital letters and hence be assigned their card codes? At first, the answer seems obvious. Upper-case letters should be considered equivalent to the previous capital letters. After all, they would have the same graphic shapes when printed.

There was a counter-argument. Three modes of operation were visualized for the 1050. In the first mode, a communications network would consist of 1050s only, with human operators sending, receiving, and routing messages. In the second mode, the network would consist of 1050s communicating to a computer, and not directly to each other. In this mode, a computer program would do the work on routing or switching messages. In the third mode, the network would be of the same kind as for the second mode, but the 1050s would be considered as data entry points, with the computer executing some data processing application on the data received.

In the first two modes (for which the telegraph network of Western Union might be considered an example), it was assumed that the messages sent and the messages received would use *both* lower- and upper-case letters. There was a human-factor reason for this decision. Human beings are educated to read text in lower- and upper-case letters. Books, magazines, newspapers, etc., display text in both lower- and upper-case letters. It is interesting to read a page of text, printed only in upper case. It is difficult to read; quite possible, of course, but difficult. Interestingly, a page of text in lower-case letters poses very little difficulty in reading. The reason is clear. In a page of text, very few capital letters appear. First word in sentence, people's names, names of cities, towns, countries, etc., are initially capitalized. But all other letters are lower case. A human being is more used to reading lower-case letters. On the Telex telegraphic network, this human factor was recognized, and text on a Telex printer is totally lower-case letters (no capitals). By contrast, a Western Union telegram, printed on Teletype printers, all in capital letters, is more difficult to read.

To repeat, it was assumed that in the first two modes, both lower- and upper-case capability would be used. But in the third mode, remote data entry to a computer, it was assumed that only upper-case letters would be used. This was because the printer of the computer had capital letters only. There would be less confusion if both terminal and computer printers printed letters of the same shape, that is, capital letters. This assumption led to a most interesting conclusion.

The fewer times an operator has to depress the case shift key, the higher the operator productivity. The numerics on the typewriter are in lower case. On the assumption that capital letters would be used, and not small letters, it would be more efficient (in this particular communications mode) if the capital letters were actually reached by the lower-case shift of the printing element. In fact, recognizing this potential efficiency factor, typewriter elements were provided that had capital letters in lower-case shift as well as in upper-case shift.

In the first two communication modes, then, it was assumed that small letters would predominate, with occasional occurrence of capital

Bit Pattern	Hole Pattern	Lower Case				Upper Case			
Bit Pattern →			A	B	BA		A	B	BA
	Hole Pattern →		0	11	12	NYA	NYA	NYA	NYA
		No Pch				No Pch			
1	1	1		11-1	12-1				
2	2	2	0-2	11-2	12-2				
2 1	3	3	0-3	11-3	12-3				
4	4	4	0-4	11-4	12-4				
4 1	5	5	0-5	11-5	12-5		UPPER CASE ALPHABETICS		
4 2	6	6	0-6	11-6	12-6				
4 2 1	7	7	0-7	11-7	12-7				
8	8	8	0-8	11-8	12-8				
8 1	9	9	0-9	11-9	12-9				
8 2	0	0		1	2				
8 2 1	NYA								
8 4	NYA								
8 4 1	NYA		CONTROLS				CONTROLS		
8 4 2	NYA								
8 4 2 1	NYA								
	Hole Pattern →	NYA	NYA	NYA	NYA	NYA	NYA	NYA	NYA

NYA - Not Yet Assigned

1	3
2	4

Block	Hole Patterns at:
1	Top And Left
2	Bottom and Left
3	Top and Left
4	Bottom and Left

Fig. 6.3 PTTC, Version 3

letters. In the third communication mode, it was assumed that capital letters would be used exclusively. At this point, a principle was evolved, as follows:

> In common data processing applications a particular set of card hole patterns is associated with the letters. In such data processing applications, such letters happen to be capital letters. In 1050 communications applications, this same set of card hole patterns should be associated with the set of letters predominantly used in the application. In the first two modes of 1050 communication applications, the predominant letters will be small letters. In the third mode, the predominant (actually, the only) letters will be capital letters. What is significant is that, for all three modes, the predominant letters will appear in the lower-case shift of the typewriter. Therefore, the card hole patterns that have, in data processing applications, been assigned to capital letters, should for PTTC be assigned to the lower-case shift of the code, regardless of whether small or capital letters are implemented in the lower-case shift.

After considerable debate, agreement was reached on this principle. The card code assignment to PTTC then began to take shape. Compare Fig. 6.2, where the assignment of the numerics and lower-case letters is shown, to the preliminary card code for PTTC as shown in Fig. 6.3.

6.6 PTTC, VERSION 4

Some further decisions were now made with respect to card codes:

1. Upper-case alphabetics would have the same digit punches as lower-case alphabetics, but with zones corresponding as shown below:

	Zone punches		
Lower-case alphabetics	0	11	12
Upper-case alphabetics	11-0	12-11	12-0

2. In code positions [1] and [2] in Fig. 6.3, hole patterns of 11-0, and 12-0, respectively, would be assigned.
3. For the sixteen control characters, the digit punches would be 4, 5, 6, and 7, to optimize the bit-code to card-code translation relationship.
4. The control characters would have the zone punches already assigned to the table columns for lower-case alphabetics, and also, for all control characters, an additional zone punch, 9.

These decisions deserve some comment. In choosing the zone punches for the upper-case alphabetics, the reasoning was as follows:

a) There would be no more than two zone punches.

b) Of the two zone punches, one would match that of the corresponding lower-case alphabetic.

		Lower Case				Upper Case			
Bit Pattern →			A	B	BA		A	B	BA
↓	Hole Pattern → ↓		0	11	12	NYA	11-0	12-11	12-0
		No Pch		[3]	[4]				
1	1	1	[5]	11-1	12-1			12-11-1	12-0-1
2	2	2	0-2	11-2	12-2		11-0-2	12-11-2	12-0-2
2 1	3	3	0-3	11-3	12-3		11-0-3	12-11-3	12-0-3
4	4	4	0-4	11-4	12-4		11-0-4	12-11-4	12-0-4
4 1	5	5	0-5	11-5	12-5		11-0-5	12-11-5	12-0-5
4 2	6	6	0-6	11-6	12-6		11-0-6	12-11-6	12-0-6
4 2 1	7	7	0-7	11-7	12-7		11-0-7	12-11-7	12-0-7
8	8	8	0-8	11-8	12-8		11-0-8	12-11-8	12-0-8
8 1	9	9	0-9	11-9	12-9		11-0-9	12-11-9	12-0-9
8 2	0	0		11-0 [1]	12-0 [2]				
8 2 1	NYA	[6]	[7]	[8]	[9]				
8 4	4	9-4	9-0-4	9-11-4	9-12-4	9-4	9-0-4	9-11-4	9-12-4
8 4 1	5	9-5	9-0-5	9-11-5	9-12-5	9-5	9-0-5	9-11-5	9-12-5
8 4 2	6	9-6	9-0-6	9-11-6	9-12-6	9-6	9-0-6	9-11-6	9-12-6
8 4 2 1	7	9-7	9-0-7	9-11-7	9-12-7	9-7	9-0-7	9-11-7	9-12-7
	↑ Hole Pattern →	9	9-0	9-11	9-12	9	9-0	9-11	9-12

NYA - Not Yet Assigned

1	3
2	4

Block	Hole Patterns at:
1	Top And Left
2	Bottom and Left
3	Top and Left
4	Bottom and Left

Fig. 6.4 PTTC, Version 4

In choosing 11-0 and 12-0 for code positions [1] and [2] in Fig. 6.3, the objective was to provide the algebraic sign capability already provided in common practice. That is, the eleven punch over a digit punch in a numeric card field should indicate a negative number for all numerics, 0 through 9. Similarly, the twelve punch over a digit punch in a numeric field should indicate a positive numeric for all numerics, 0 through 9.

The choice of 4, 5, 6, and 7 as digit punches for the bottom four rows of the table would optimize their BCD translation to/from the PTTC bit code.

A zone punch of nine would distinguish all control characters from all graphic characters. Advantage could be taken in the hardware of this distinguishing characteristic.

With these decisions, the card code assignments shown in Fig. 6.3 were increased to those shown in Fig. 6.4.

6.7 PTTC, VERSION 5

There now remained 32 graphic positions in the PTTC code table with unassigned graphics. Of these 32 code positions, 30 had not yet been assigned card hole patterns. The numerics, alphabetics, and Space of BCDIC had been assigned. There remained 27 BCD graphics and hole patterns to be assigned in PTTC. For compatibility reasons, the 27 BCDIC graphics and hole patterns should match the 27 in PTTC. The BCDIC specials were now reviewed:

@	8-4	;	11-8-6
/	0-1	ƀ	8-2
−	11	%(	0-8-4
&+	12	:	8-5
#=	8-3	*	11-8-4
‡	0-8-2	[	12-8-5
,	0-8-3	]	11-8-5
$	11-8-3	⧻	0-8-7
.	12-8-3	⧧	12-8-7
Δ	11-8-7	γ	0-8-5
\	0-8-6	√	8-7
<	12-8-6	!	11-0
>	8-6	?	12-0
⌑	12-8-4		

The card hole patterns 11-0 and 12-0 had been assigned in locations [1] and [2] in Fig. 6.4, so the BCDIC graphics ! and ? would be assigned to these PTTC code positions.

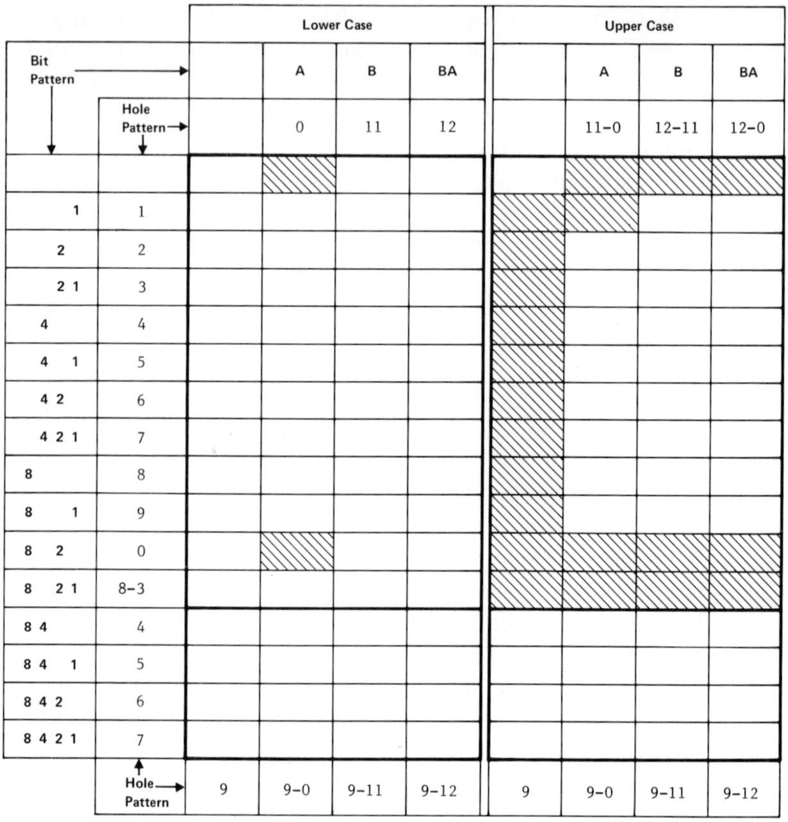

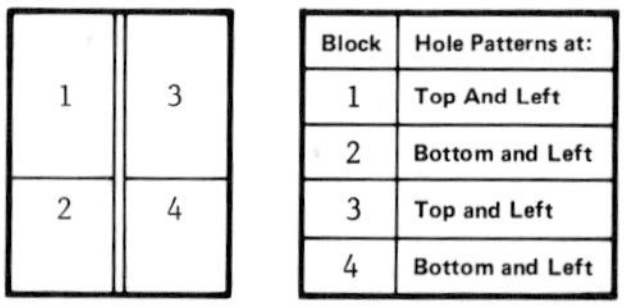

		Lower Case				Upper Case			
Bit Pattern →			A	B	BA		A	B	BA
↓	Hole Pattern → ↓		0	11	12		11-0	12-11	12-0
1	1								
2	2								
2 1	3								
4	4								
4 1	5								
4 2	6								
4 2 1	7								
8	8								
8 1	9								
8 2	0								
8 2 1	8-3								
8 4	4								
8 4 1	5								
8 4 2	6								
8 4 2 1	7								
	↑ Hole → Pattern	9	9-0	9-11	9-12	9	9-0	9-11	9-12

1	3
2	4

Block	Hole Patterns at:
1	Top And Left
2	Bottom and Left
3	Top and Left
4	Bottom and Left

Fig. 6.5 PTTC, Version 5

For translation reasons, the hole pattern 11 should be assigned in position [3], and the hole pattern 12 in position [4] of Fig. 6.4, which would then dictate the assignment of graphics − and & +. For translation purposes, hole patterns 8-3, 0-8-3, 11-8-3, and 12-8-3 should be assigned in positions [6], [7], [8], [9], respectively, which in turn would dictate the location of graphics # = , $. (respectively). For translation purposes, hole pattern 0-1 should be assigned in position [5], which would dictate the location for /.

These decisions resulted in Fig. 6.5.

		Lower Case				Upper Case			
Bit Pattern →			A	B	BA		A	B	BA
↓	Hole Pattern → ↓		0	11	12		11-0	12-11	12-0
		SP	@ or ' [1]	-	& or +	SP	Δ [14]	\ [18]	< [21]
1	1	1	/	j	a	> [3]	? [15]	J	A
2	2	2	s	k	b	⌶ or) [4]	S	K	B
2 1	3	3	t	l	c	; [5]	T	L	C
4	4	4	u	m	d	ƀ [6]	U	M	D
4 1	5	5	v	n	e	% or ([7]	V	N	E
4 2	6	6	w	o	f	: [8]	W	O	F
4 2 1	7	7	x	p	g	" [9]	X	P	G
8	8	8	y	q	h	* [10]	Y	Q	H
8 1	9	9	z	r	i	[[11]	Z	R	I
8 2	0	0	‡ [2]	!	?	] [12]	‡ [16]	Υ [19]	√ [22]
8 2 1	8-3	# or =	,	$	.	⧺ [13]	, [17]	! [20]	. [23]
8 4	4	PN	BYP	RES	PF	PN	BYP	RES	PF
8 4 1	5	RS	LF	NL	HT	RS	LF	NL	HT
8 4 2	6	UC	EOB	BS	LC	UC	EOB	BS	LC
8 4 2 1	7	EOT	PRE	IL	DEL	EOT	PRE	IL	DEL
	↑ Hole Pattern →	9	9-0	9-11	9-12	9	9-0	9-11	9-12

Hole Patterns:

[1] 8-4	[8] 8-5	[15] 12-8-2	[22] 8-7
[2] 0-8-2	[9] 8-1	[16] 12-8-7	[23] 12-8-1
[3] 8-6	[10] 11-8-4	[17] 0-8-1	
[4] 12-8-4	[11] 12-8-5	[18] 0-8-6	
[5] 11-8-6	[12] 11-8-5	[19] 0-8-5	
[6] 8-2	[13] 0-8-7	[20] 11-8-2	
[7] 0-8-4	[14] 11-8-7	[21] 12-8-6	

1	3
2	4

Block	Hole Patterns at:
1	Top And Left
2	Bottom and Left
3	Top and Left
4	Bottom and Left

Fig. 6.6 PTTC, Final Version

6.8 PTTC, FINAL VERSION

There are 23 unassigned code positions (shaded) and 18 unassigned BCDIC graphics. The remaining card hole patterns and remaining PTTC bit patterns were simply not able to be matched to any orderly translation relationship. The assignments were made to optimize the translation relationship as much as possible, while realizing that the relationship could not be very good.

When the 18 BCDIC graphics and hole patterns were assigned in the PTTC code table, there would remain five unassigned code positions. Five graphics and five hole patterns were finally chosen as follows:

Graphic	*Hole pattern*
. (UC)	12-8-1
, (UC)	0-8-1
! (UC)	11-8-2
? (UC)	12-8-2
"	8-1

These five were then assigned into the PTTC table, leading to Fig. 6.6, the final version of PTTC.

These five hole patterns were chosen for the following reason. An examination of the table shows that all combinations of digit punches 1, 2, 3, 4, 5, 6, 7, 8, 9, 8-3, 8-4, 8-5, 8-6, 8-7 with zone punches "no-zones", 0, 11, and 12 (the hole patterns from BCDIC) had been assigned in PTTC. Additionally, for the capital letters, the double-zone combinations 11-0, 12-11, 12-0 had been introduced as previously described. Additionally, the two BCDIC hole patterns 8-2 and 0-8-2 had been assigned. Now five more hole patterns were needed. What should they be?

They could have been some combination of double-zone punches with the double-digit punches 8-3, 8-4, 8-5, 8-6, 8-7, but this would have resulted in hole patterns of four holes. It was thought preferable to choose hole patterns of no more than three holes, and there were six such that suggested themselves; 8-1, 0-8-1, 11-8-1, 12-8-1, 11-8-2, 12-8-2. The 8-1 was first choice, since it was a hole pattern of two holes only. Then four of the five remaining possibles were chosen, 0-8-1, 12-8-1, 11-8-2, and 12-8-2.

7
The Structure of EBCDIC

7.1 INITIAL CONSIDERATION

It is supposed by some people that the requirement that led from computers with a 6-bit architecture to computers with an 8-bit architecture was the requirement for a larger set of characters. It was known that the then current 64-character set of 6-bit computers, while sufficient for most data processing applications, was becoming insufficient for some data processing applications. On the one hand, an insufficient number of graphic code positions had led to the use of duals (Chapter 4). On the other hand, an insufficient number of control code positions had led to the development of PTTC (Chapter 6). The implementation of PTTC on the IBM 1050 (which was based on an electric typewriter) had introduced lower-case as well as upper-case alphabetics to many people in the data processing world. Also, a new data processing application, text processing, had led at least one customer to order a special IBM 1403 print chain and to have special instructions developed for his 1401 computer to allow him to manipulate and process upper- and lower-case alphabetics.

These situations and applications in the data processing field certainly emphasized the needs for a larger coded character set than that of BCDIC. But these needs were very far from sufficient to dictate a *requirement* for an 8-bit computer architecture. There were two other very fundamental aspects of computer architecture that pointed at the requirement for an 8-bit architecture. These aspects led to the development and marketing of the IBM System/360. Once an 8-bit architecture was decided on, with a consequent possible 256 character code positions, the opportunity to enlarge or extend the character set from that of BCDIC was obvious. IBM did indeed take that opportunity; the 8-bit, 256-character EBCDIC was developed and implemented.

The first aspect was the efficiency of representation of numerics in a coded character set. The requirement for 26 (or 29) code positions to represent alphabetics and for 10 code positions to represent numerics together set a requirement for at least 36 code positions. In its turn, the requirement for at least 36 code positions set a requirement for a code byte of at least 6 bits, and BCDIC was (and is) a 6-bit coded character set.

Although 4 bits at most are required to represent the 10 numerics, the 10 numerics of BCDIC are represented by 6 bits, 2 bits more than needed for numeric representation only. That is to say, numerics in BCDIC have an unnecessarily large, and hence inefficient, bit representation.

So numerics are inefficiently represented in BCDIC. Is this significant? Indeed it is. It was variously estimated in the early 1960s that approximately 75 percent of the data used in data processing applications was numeric data. In short, 75 percent of the data was inefficiently represented. Was this fact significant? In previous paragraphs, it has been stated that requirements for larger character sets, although clearly perceived, were not deemed sufficient to increase the bit size of computer architecture. But the inefficiency of numeric data representation affected about 75 percent of the data processed in computers. It hardly needs to be said that efficiency of a computing system was (and is) one of the key elements of any computer marketing strategy. Could the efficiency of numeric representation be improved?

The "packing" of two numeric digits into one 8-bit byte would essentially represent numeric data in 4 bits, the practical minimum. Maximum efficiency of numeric representation would be realized. This was one of the aspects which led to the IBM decision to develop an 8-bit architecture for computers.*

The other aspect had to do with the binary nature of the System/360. In designing the Stretch Computer [7.1], for a number of reasons the organization was chosen to be binary rather than decimal. Similar reasons led to the decision that System/360 would be binary. Not only, of course,

---

* It must be noted that 8 bits, while ideal for representation of packed numerics, is not ideal for the representation of all data, such as alphabetics and special graphics. To represent *all* of numerics, alphabetics, and an adequate number of special graphics, 6 bits is sufficient. So, to represent alphabetics and special graphics by an 8-bit code is *for them*, inefficient. That is an illustration of a design trade-off.

would arithmetic be binary but so also would addressing be binary. For binary addressing of memory words, there is considerable advantage in choosing the number of bits in each word to be a power of 2. The three possibilities looked at were

$$2^5 = 32$$
$$2^6 = 64$$
$$2^7 = 128$$

The choice of 64 bits gives a good compromise between speed and cost of memory, and provides ample space to represent a floating-point number in one memory word.

Since the memory word size of 64 bits was chosen, and since a byte must be an integer submultiple, eight 8-bit bytes was the natural choice.

The decision to go to 8 bits was made, and a coded character set of potentially 256 characters resulted. The 6-bit code had been named the BCD Interchange Code, with BCDIC as the acronym. Since the number of available character positions was to be extended from 64 to 256, the new code came naturally enough to be named the Extended BCD Interchange Code, with EBCDIC as the acronym.

7.2 TECHNICAL DECISIONS

7.2.1 8-Bit Code Table

The first technical decision, then, was that the coded character set would be 8 bits with a potential of 256 characters, although as narrated above, this was more a consequence than a decision. The second decision was how to exhibit it in manuscripts, documents, manuals, and so on. At the time, 6-bit codes were being exhibited in 4-by-16 code tables; 7-bit code tables were being exhibited in 8-by-16 code tables. The natural decision was to exhibit EBCDIC in the form of a 16-by-16 code table.

7.2.2 Bit Numbers

The next step was to decide how to number or name the bits of an 8-bit byte, for reference purposes. The philosophy for BCDIC was bit naming: B, A, 8, 4, 2, 1. The philosophy for ASCII was a combination of bit naming and bit numbering: b7, b6, b5, b4, b3, b2, b1. A common engineering practice was to number from left to right and to associate the order of the numbering with high to low significance; for example, memory addresses in a computer, columns on a punched card, tab stops on a typewriter. It was decided to number the bits of an EBCDIC byte according to this same philosophy (0, 1, 2, 3, 4, 5, 6, 7) from the high-order to the low-order bit of a byte, as shown:

0	1	2	3	4	5	6	7

7.2.3 Hexadecimal Numbers

The next step was to decide how to reference a particular code position. It was decided that the 16 columns (from left to right) and the 16 rows (from top to bottom) would be named 0, 1, 2, 3, 4, 5, 6, 7, 8, 9, A, B, C, D, E, F, as shown in Fig. 7.1.

A particular code position would be referenced by giving its column name followed by its row name: for example, code position A7 in Fig. 7.1. This notation came to be called the hexadecimal notation, or hex notation.

The columns and rows could have been named (numbered) 0, 1, 2, 3, 4, 5, 6, 7, 8, 9, 10, 11, 12, 13, 14, 15, as was done with another 8-bit

	Column	0	1	2	3	4	5	6	7	8	9	A	B	C	D	E	F
	Bit Pat. →	00				01				10				11			
Row	↓	00	01	10	11	00	01	10	11	00	01	10	11	00	01	10	11
0	0000																
1	0001																
2	0010																
3	0011																
4	0100																
5	0101																
6	0110																
7	0111											A7					
8	1000																
9	1001																
A	1010																
B	1011																
C	1100																
D	1101																
E	1110																
F	1111																

Fig. 7.1 Hexadecimal columns and rows

code form (to be discussed in Chapter 20). The hex notation is more compact, and always requires exactly two "typing" spaces for the manuscript representation of a code position; 35, A7, EF, etc. By contrast, the numeric notation requires a separating mark (the slash /) to avoid confusion; 0/9, 3/15, 1/11, etc. Also, if allowed to be a non-uniform notation to gain compactness, as 1/6, 1/11, 11/1, 11/11, the number of "typing" spaces could vary from three to five, while, if uniformity was imposed, as 01/06, 01/11, 11/01, 11/11, the number of "typing" spaces required would be exactly five. Either way, the hex notation, with its always uniform requirement for exactly two "typing" spaces, seems superior.

7.2.4 Quadrants

The final decision, also for purposes of referencing the code table, was to consider the code table to be divided into four equal quadrants, as shown in Fig. 7.2. The quadrants would then be referred to as the first quadrant, the second quadrant, etc., or as quadrant one, quadrant two, etc.

	Column	0	1	2	3	4	5	6	7	8	9	A	B	C	D	E	F
	Bit Pat.	00				01				10				11			
Row		00	01	10	11	00	01	10	11	00	01	10	11	00	01	10	11
0	0000																
1	0001																
2	0010																
3	0011																
4	0100																
5	0101																
6	0110																
7	0111		Quadrant 1				Quadrant 2				Quadrant 3				Quadrant 4		
8	1000																
9	1001																
A	1010																
B	1011																
C	1100																
D	1101																
E	1110																
F	1111																

Fig. 7.2 EBCDIC quadrants

7.2.5 Blocks

The code table would have to be shown with four unequal blocks in order to exhibit the card code (as described in Chapter 2), as shown in Fig. 7.3.

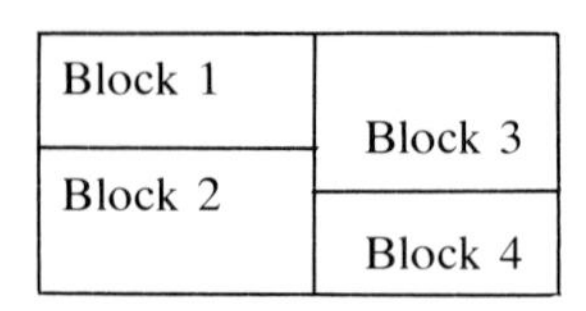

Figure 7.3

7.3 SUMMARY

In summary, then, five decisions were made in order to exhibit and reference the EBCDIC Code Table:

1. The code table would be 8 bits, with a potential of 256 characters.
2. The bits of an EBCDIC byte would be numbered 0, 1, 2, 3, 4, 5, 6, 7, from left to right, that is, from high-order bit to low-order bit of a byte.
3. The 16 columns and 16 rows of the code table would be named according to a hexadecimal notation: 0, 1, 2, 3, 4, 5, 6, 7, 8, 9, A, B, C, D, E, F. A particular code position would be referenced by giving first its column name, then its row name.
4. For purposes of reference the code table would be considered to be divided into four quadrants of four columns each; first quadrant, second quadrant, etc.
5. In order to exhibit the card code, the code table would be shown in four (unequal) blocks.

These decisions having been made (the last four decisions might be considered more of an administrative than of a technical nature), attention was then directed to the technical aspects of EBCDIC. Ten criteria emerged.

7.4 CRITERIA

Criterion 1 (Collatability)

The 64 characters of BCDIC, when embedded in the 256 code positions of EBCDIC, should have the same collating sequence, not necessarily contiguously, as BCDIC.

Criterion 2 (Space collatability)

The Space character should collate low to all EBCDIC graphic characters, those immediately assigned and those to be assigned in the future.

Criterion 3 (Separability)

Control characters should be easily distinguishable, by their bit-patterns, from graphic characters; that is, graphic and control characters should be separable.

Criterion 4 (Duocase capability)

Lower-case alphabetics, as well as upper-case alphabetics, should be assigned.

Criterion 5 (Duocase relationship)

Corresponding upper- and lower-case alphabetics should differ only in high-order, or zone, bits. The bit patterns for corresponding upper- and lower-case alphabetics should have the low-order four bits identical.

Criterion 6 (Sign capability)

The concepts of positive, negative, and absolute numerics, zero through nine, should be incorporated.

Criterion 7 (Card-code compatibility)

The card hole patterns for the 64 BCDIC characters should be the same for BCDIC and EBCDIC.

Criterion 8 (Translation simplicity)

The translation from the 64 6-bit bit patterns of BCDIC to their equivalent 8-bit EBCDIC bit patterns should be as simple as possible.

Criterion 9 (Subsetability)

By dropping the two high-order bits of the 8-bit EBCDIC bit patterns, a compact 64 character subset should emerge. This subset should consist of the 64 BCDIC characters but need not have the same bit patterns.

Criterion 10 (No duals)

The five dual pairs of BCDIC should be eliminated, giving rise to ten unique EBCDIC characters.

It was recognized that Criterion 10 conflicted with Criteria 1, 7, 8, and 9. The resolutions of this conflict led to user dissatisfaction, as described in Chapter 9, The Duals of EBCDIC.

Criteria 1 through 7 are discussed in Chapter 8; Criteria 8 and 9 are discussed in Chapter 10; Criteria 7 and 10 are discussed in Chapter 9.

REFERENCE

7.1 W. Buchholz, "Planning a Computer System." New York: McGraw-Hill, 1962, Chapter 5.

8
The Sequence of EBCDIC

During the late 1950s and early 1960s, the code used on IBM computers was a 64-character, 6-bit code, called BCDIC. It met the requirements of the time well enough. The 64 6-bit bit patterns were sufficient to represent the following:

a) Space, alphabetics, and numerics.
b) The extra diacritic and accent letters needed for the major European Latin alphabets.
c) Special graphics needed for most data processing applications.
d) Special graphics needed for the major programming languages (Assembler, COBOL, FORTRAN, etc.).
e) Control characters needed for control of either data processing devices (mainly tape drives) or formatting of data.

It came to suffer from two defects—duals and collating sequence. (For a discussion of the duals problem, see Chapter 9.)

We learned in Chapter 7 of the decision to go to an 8-bit computer architecture. This led to the potentiality of a 256-character, 8-bit code set and to the establishment of ten criteria. The application of seven of these criteria, beginning with Criterion 1 relating to collatability, are discussed in this chapter.

8.1 BCDIC COLLATING SEQUENCE

The 63 graphics, and Space, of the BCD Interchange Code (BCDIC) are shown in Fig. 8.1, arranged in sequence of bit patterns from low (00,0000)

Bit Pattern		0 0	0 1	1 0	1 1
0 0 0 0		SP	ƀ	-	& or +
0 0 0 1		1	/	J	A
0 0 1 0		2	S	K	B
0 0 1 1		3	T	L	C
0 1 0 0		4	U	M	D
0 1 0 1		5	V	N	E
0 1 1 0		6	W	O	F
0 1 1 1		7	X	P	G
1 0 0 0		8	Y	Q	H
1 0 0 1		9	Z	R	I
1 0 1 0		0	‡	!	?
1 0 1 1		# or =	,	$	.
1 1 0 0		@ or '	% or (	*	⌑ or)
1 1 0 1		:	γ	]	[
1 1 1 0		>	\	;	<
1 1 1 1		√	⧻	Δ	⧧

Fig. 8.1 BCDIC

to high (11,1111). There was, however, an established collating sequence for these 64 characters. Each graphic character, and the Space character, was assigned a collating number, from low (0) to high (63). In Fig. 8.2 are shown the collating numbers assigned to the 64 characters of Fig. 8.1. As can be seen, the bit-pattern sequence of the 64 characters did not correspond in any way to the collating sequence of the 64 characters. The graphic characters, arranged in collating sequence, are shown in Fig. 8.3, with collating numbers running from 0 (low) to 63 (high).

The basic element in any sorting or collating application is a comparison of the magnitude of two quantities. Essentially, the question is asked (by machine instructions in a program):

Is item A greater than, equal to, or less than item B?

Depending on the answer, the item is inserted into an ordered list of items. This comparison (by executing what was generally called a Compare instruction) is generally implemented in hardware by subtracting one

Bit Pattern → ↓					
		0	19	12	6
		55	13	36	26
		56	46	37	27
		57	47	38	28
		58	48	39	29
		59	49	40	30
		60	50	41	31
		61	51	42	32
		62	52	43	33
		63	53	44	34
		54	45	35	25
		20	14	7	1
		21	15	8	2
		22	16	9	3
		23	17	10	4
		24	18	11	5

Fig. 8.2 BCDIC collating numbers

item from the other and inspecting the sign and magnitude of the result (positive, zero, or negative).

In order that the Compare instruction would function correctly on the basis of the established collating sequence, and despite the disordered bit-pattern sequence, one of two approaches has been employed.

8.1.1 Convert/Compare/Reconvert Approach

On the binary machines (704, 709, 7090, etc.) an instruction was provided, generally called a Convert instruction. When executed, this instruction would convert the 6-bit bit patterns to another set of bit patterns. This other set of bit patterns had the characteristic that the bit pattern sequence matched the collating sequence. Thus, when executed, the hardware Compare instruction subsequently would function so that the data would be arranged into the correct collating sequence. After the sorting or collating function was implemented on all the data, that portion

		SP	0	Υ	16	G	32	U	48
		.	1	\	17	H	33	V	49
		⌑ or)	2	⧻	18	I	34	W	50
		[	3	ƀ	19	!	35	X	51
		<	4	# or =	20	J	36	Y	52
		ǂ	5	@ or '	21	K	37	Z	53
		& or +	6	"	22	L	38	0	54
		$	7	>	23	M	39	1	55
		*	8	√	24	N	40	2	56
		]	9	?	25	O	41	3	57
		;	10	A	26	P	42	4	58
		Δ	11	B	27	Q	43	5	59
		–	12	C	28	R	44	6	60
		/	13	D	29	‡	45	7	61
		,	14	E	30	S	46	8	62
		% or)	15	F	31	T	47	9	63

Fig. 8.3 BCDIC graphics in collating sequence

of the data that had been "converted" had to be reconverted back to its correct BCD bit patterns. This reconversion was accomplished by another instruction.

8.1.2 Comparator Approach

On the character machines (1401, 1410, 705, 7080, etc.) special hardware called "comparator" hardware was built in. This hardware, when executing a compare instruction, first performed the equivalent of the Convert instruction described above, then executed the actual comparison of the two items. Thus, the hardware, without actually converting any data (and thus eliminating the need for a subsequent reconversion) allowed the data to be sorted or collated into the correct sequence.

An analysis of the two approaches reveals the following:

- In the Convert/Compare/Reconvert approach, no extra hardware was required, but extra CPU time was required to execute the conversion and reconversion parts of the program.
- In the Comparator approach, no extra CPU time was required, but the Comparator hardware itself increased the cost of the computing system.

There was, therefore, either a performance penalty or a hardware cost penalty.

8.2 EMBEDMENT OF BCDIC COLLATING SEQUENCE

In the design of the new 8-bit CPU code, the Extended BCD Code (EBCDIC), it was postulated that the above penalties could be removed,

	Column	0	1	2	3	4	5	6	7	8	9	A	B	C	D	E	F
	Bit →	00				01				10				11			
Row	Pat. ↓	00	01	10	11	00	01	10	11	00	01	10	11	00	01	10	11
0	0000												I	0	19	12	6
1	0001													55	13	36	26
2	0010													56	46	37	27
3	0011													57	47	38	28
4	0100													58	48	39	29
5	0101												II	59	49	40	30
6	0110													60	50	41	31
7	0111													61	51	42	32
8	1000													62	52	43	33
9	1001													63	53	44	34
A	1010												III	54	55	35	25
B	1011													20	14	7	1
C	1100													21	15	8	2
D	1101												IV	22	16	9	3
E	1110													23	17	10	4
F	1111													24	18	11	5

Fig. 8.4 Blocks in BCDIC

without any deleterious effects on the user. Let us see what actually happened.

In studying Fig. 8.2, it was observed that the code table could be visualized as being in four major blocks, designated I, II, III, and IV in Fig. 8.4. Then if the blocks were rearranged relative to each other, with a view towards coming closer to a correct collating sequence, the result would be as shown in Fig. 8.5. Then, if the two high-order bits of each column were inverted (zero for one, and one for zero) and the columns reordered on the new two high-order bits, the result would be as shown in Fig. 8.6. Finally, given the freedom that columns, or if necessary, partial columns, could be distributed into the 16 column spaces of an 8-bit code table, the results would be as shown in Fig. 8.7.

In Fig. 8.7, observe that the 64 characters are almost (not quite, see character 0 and character 13) in correct collating sequence, albeit not contiguously in bit-pattern sequence. The fact that the BCDIC collating sequence could be embedded in the EBCDIC collating sequence was the primary design factor for EBCDIC.

	Column	0	1	2	3	4	5	6	7	8	9	A	B	C	D	E	F
	Bit Pat. →	00				01				10				11			
Row	↓	00	01	10	11	00	01	10	11	00	01	10	11	00	01	10	11
0	0000													20	15	7	1
1	0001													21	15	8	2
2	0010												IV	22	16	9	3
3	0011													23	17	10	4
4	0100													24	18	11	5
5	0101												I	0	19	12	6
6	0110												III	54	45	35	25
7	0111													55	13	36	26
8	1000													56	46	37	27
9	1001													57	47	38	28
A	1010												II	58	48	39	29
B	1011													59	49	40	30
C	1100													60	50	41	31
D	1101													61	51	42	32
E	1110													62	52	43	33
F	1111													63	53	44	34

Fig. 8.5 BCDIC rearrangement 1

Row	Column	0	1	2	3	4	5	6	7	8	9	A	B	C	D	E	F
	Bit Pat. →	00				01				10				11			
	↓	00	01	10	11	00	01	10	11	00	01	10	11	00	01	10	11
0	0000													1	7	14	20
1	0001													2	8	15	21
2	0010												IV	3	9	16	22
3	0011													4	10	17	23
4	0100													5	11	18	24
5	0101												I	6	12	19	0
6	0110												III	25	35	45	54
7	0111													26	36	13	55
8	1000													27	37	46	56
9	1001													28	38	47	57
A	1010													29	39	48	58
B	1011												II	30	40	49	59
C	1100													31	41	50	60
D	1101													32	42	51	61
E	1110													33	43	52	62
F	1111													34	44	53	63

Fig. 8.6 BCDIC rearrangement 2

Row	Column	0	1	2	3	4	5	6	7	8	9	A	B	C	D	E	F
	Bit Pat. →	00				01				10				11			
	↓	00	01	10	11	00	01	10	11	00	01	10	11	00	01	10	11
0	0000									1	7	14	20				
1	0001									2	8	15	21				
2	0010								IV	3	9	16	22				
3	0011									4	10	17	23				
4	0100									5	11	18	24				
5	0101								I	6	12	19	(0)				
6	0110												III	25	35	45	54
7	0111													26	36	(13)	55
8	1000													27	37	46	56
9	1001													28	38	47	57
A	1010													29	39	48	58
B	1011												II	30	40	49	59
C	1100													31	41	50	60
D	1101													32	42	51	61
E	1110													33	43	52	62
F	1111													34	44	53	63

Fig. 8.7 BCDIC rearrangement 3 (two collating exceptions)

8.3 BCDIC CARD CODE RELATIONSHIP

It was at this point that several other factors were reviewed as design requirements. Following this review, criteria for EBCDIC design were established, and the final EBCDIC was designed. Before looking at the criteria, let us look at the other design factors.

First, in BCDIC, there was a reasonably simple relationship between BCDIC card hole patterns and BCDIC bit patterns (see Fig. 8.8). This relationship, the cornerstone of the binary coded decimal algorithm, results in relatively simple and inexpensive hardware translators in card reader/punch units serving as input/output units to CPU's. It was deemed desirable to maintain this simple bit-pattern–to–hole-pattern relationship in EBCDIC, if possible. The translation relationship, bit patterns to/from hole patterns, reveals itself on examination of Fig. 8.8.

Bit Pattern → ↓		No Zone	A	B	BA
		No Pch	8-2 *	11	12
1		1	0-1	11-1	12-1
2		2	0-2	11-2	12-2
2 1		3	0-3	11-3	12-3
4		4	0-4	11-4	12-4
4 1		5	0-5	11-5	12-5
4 2		6	0-6	11-6	12-6
4 2 1		7	0-7	11-7	12-7
8		8	0-8	11-8	12-8
8 1		9	0-9	11-9	12-9
8 2		0 *	0-8-2	11-0 *	12-0 *
8 2 1		8-3	0-8-3	11-8-3	12-8-3
8 4		8-4	0-8-4	11-8-4	12-8-4
8 4 1		8-5	0-8-5	11-8-5	12-8-5
8 4 2		8-6	0-8-6	11-8-6	12-8-6
8 4 2 1		8-7	0-8-7	11-8-7	12-8-7

* Exception translation

Fig. 8.8 BCDIC–BCD relationship

a) Zone punches—no zone, zero zone, eleven-zone, twelve-zone—translate to/from the two high-order, or zone, bits—No zone, A, B, BA.

b) Digit punches 1, 2, 3, 4, 5, 6, 7 translate to/from their binary equivalents, 1, 2, 21, 4, 41, 42, 421.

c) Eight punch translates to/from its binary equivalent 8. This holds whether or not it is in conjunction with digit punches 1, 2, 3, . . . , 7.

d) Nine punch translates to/from its binary equivalent 8 1.

e) Zero punch translates a little trickily, depending on whether it is a zone punch or a digit punch. It is a zero punch if it is alone, or if it is in conjunction with either zone punch 12 or 11 and then translates to/from its conventional BCD equivalent 8 2. It is a zone punch if it is in conjunction with any other digit punch 1, 2, 3, . . . , 7, 8, 9, and translates to/from the A zone bit.

	Column	0	1	2	3	4	5	6	7	8	9	A	B	C	D	E	F
	Bit Pat. →	00				01				10				11			
Row	↓	00	01	10	11	00	01	10	11	00	01	10	11	00	01	10	11
0	0000									(6)	(12)	(19)	(0)	25	35	45	54
1	0001													26	36	(13)	55
2	0010													27	37	46	56
3	0011													28	38	47	57
4	0100													29	39	48	58
5	0101													30	40	49	59
6	0110													31	41	50	60
7	0111													32	42	51	61
8	1000													33	43	52	62
9	1001													34	44	53	63
A	1010																
B	1011									1	7	14	20				
C	1100									2	8	15	21				
D	1101									3	9	16	22				
E	1110									4	10	17	23				
F	1111									5	11	18	24				

Fig. 8.9 BCDIC rearrangement 4 (five collating exceptions)

In order to maintain this hole-pattern–to–BCD bit-pattern relationship, it is clear that the embedment of the 64 BCDIC characters in the 8-bit code table, as shown in Fig. 8.7, would be wrong. Instead, the embedment shown in Fig. 8.9 would come closer to preserving both the collating sequence and the BCD relationship.

Block I is a little garbled on the collating sequence, and Block III would put the BCD bit patterns 8 2 in the top row. But these are peculiarities which we will study later.

8.4 TECHNICAL DECISIONS

Decision 1

The first decision was with respect to control characters and graphic characters. It was decided (on a purely intuitive basis) that there would be 64 control character code positions and 192 graphic character code positions.

Decision 2

The second decision was with respect to the code table location of the control and graphic characters. It was decided that a quadrant would be devoted to control characters (i.e., control characters should not overlap quadrants) and that the first quadrant would be reserved for control characters. Both Decision 1 and Decision 2 were based on Criterion 3 (see Chapter 7): "Control characters should be easily distinguishable, by their bit patterns, from graphic characters; that is, graphic and control characters should be separable."

The first structuring of EBCDIC began to emerge (Fig. 8.10).

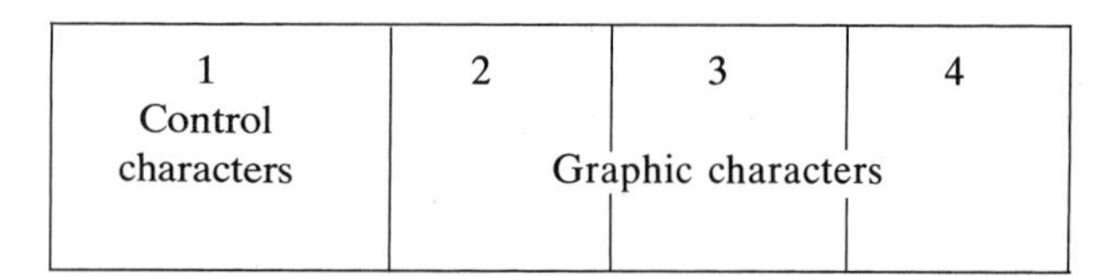

Figure 8.10

Decision 3

It was decided that the special graphics should be contained in one quadrant (mostly) and the alphabetics and numerics in another quadrant, as shown in Fig. 8.9. This decision was based on Criterion 1, the requirement to embed the BCDIC collating sequence in the EBCDIC collating sequence. Letting S stand for special graphics, and AN stand for alphabetics and numerics, this gave rise to three possibilities, as shown in Fig. 8.11.

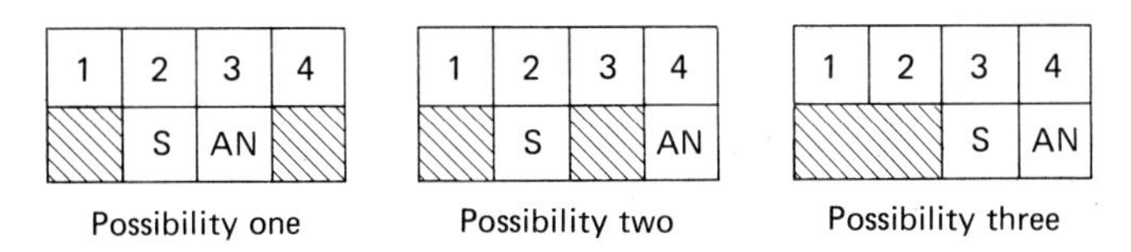

Figure 8.11

Decision 4

Criterion 2 dictated that the Space character should occupy the first code-table position in the Second Quadrant (Fig. 8.12).

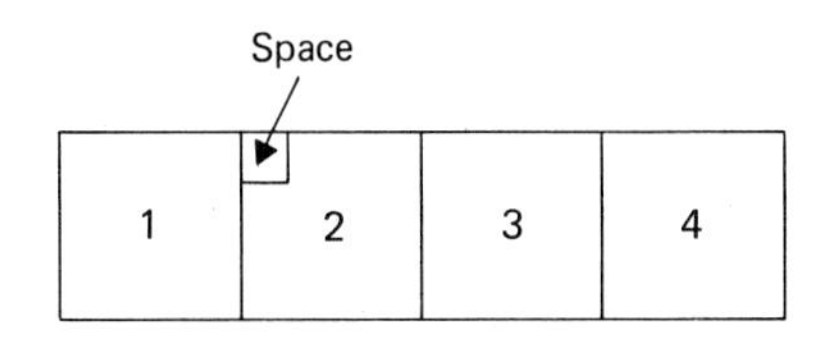

Figure 8.12

Decision 5

The gross collating sequence of BCDIC, and hence of EBCDIC, was specials, alphabetics, numerics. It was decided (intuitively) that specials should collate low to lower-case alphabetics as well as to upper-case alphabetics.

Decision 6

Criterion 4 (inclusion of lower-case alphabetics) and Decision 5 clearly ruled out Possibility 3 of Fig. 8.11.

Decision 7

It was decided (intuitively) that lower-case alphabetics as well as upper-case alphabetics should collate low to numerics.

Decision 8

Decision 7 clearly ruled out Possibility 1 of Fig. 8.11, and left Possibility 2 as the only possible structure for EBCDIC. Decision 2, and Decisions

3, 5, 6, and 7 which led to Decision 8, established the EBCDIC structure as shown in Fig. 8.13.

	Column	0	1	2	3	4	5	6	7	8	9	A	B	C	D	E	F
	Bit Pat.	00				01				10				11			
Row		00	01	10	11	00	01	10	11	00	01	10	11	00	01	10	11
0	0000					SP											
1	0001																N
2	0010																U
3	0011																M
4	0100		CONTROLS								LOWER CASE				UPPER CASE		E
5	0101										ALPHABETICS				ALPHABETICS		R
6	0110																I
7	0111																C
8	1000																S
9	1001																
A	1010																
B	1011																
C	1100																
D	1101						SPECIALS										
E	1110																
F	1111																

Fig. 8.13 EBCDIC gross structure

Decision 9

It was decided that Criterion 1 would be applied absolutely, regardless of other criteria. An examination of Fig. 8.9, therefore, indicated that characters 6, 12, 19, 0, and 13 must be rearranged and Figs. 8.14 and 8.15 show the final result.

It is to be noted that Criterion 6 was also met by Fig. 8.15. The card hole patterns and positive, negative, and absolute numeric equivalents were as shown in Fig. 8.16. Note also that some of the card hole patterns for EBCDIC had now been established, as shown in Fig. 8.17. It was decided at this time that, as regards small letters and capital letters, the capital letters should be assigned to the BCDIC hole patterns for alphabetics, in order to ensure a more reasonable migration from BCDIC to the EBCDIC environments.

Row	Column	0	1	2	3	4	5	6	7	8	9	A	B	C	D	E	F
	Bit Pat.	00				01				10				11			
		00	01	10	11	00	01	10	11	00	01	10	11	00	01	10	11
0	0000				I	0	6	12	19				III	25	35	45	54
1	0001							13						26	36		55
2	0010													27	37	46	56
3	0011													28	38	47	57
4	0100													29	39	48	58
5	0101												II	30	40	49	59
6	0110													31	41	50	60
7	0111													32	42	51	61
8	1000													33	43	52	62
9	1001													34	44	53	63
A	1010																
B	1011					1	7	14	20								
C	1100					2	8	15	21								
D	1101				IV	3	9	16	22								
E	1110					4	10	17	23								
F	1111					5	11	18	24								

Fig. 8.14 BCDIC rearrangement 5 (correct collating sequence)

Row	Column	0	1	2	3	4	5	6	7	8	9	A	B	C	D	E	F
	Bit Pat.	00				01				10				11			
		00	01	10	11	00	01	10	11	00	01	10	11	00	01	10	11
0	0000					SP	&+	-	ƀ					?	!	ǂ	0
1	0001							/						A	J		1
2	0010													B	K	S	2
3	0011													C	L	T	3
4	0100													D	M	U	4
5	0101													E	N	V	5
6	0110													F	O	W	6
7	0111													G	P	X	7
8	1000													H	Q	Y	8
9	1001													I	R	Z	9
A	1010																
B	1011					.	$	,	#=								
C	1100					⌑)	*	% (	@ '								
D	1101					[	]	γ	:								
E	1110					<	;	\	>								
F	1111					ǂ	Δ	⧻	√								

Fig. 8.15 BCDIC graphics in EBCDIC

Graphic	Hole pattern	Numeric equivalent	Graphic	Hole pattern	Numeric equivalent	Graphic	Hole pattern	Numeric equivalent
?	12-0	+0	!	11-0	−0	0	0	0
A	12-1	+1	J	11-1	−1	1	1	1
B	12-2	+2	K	11-2	−2	2	2	2
C	12-3	+3	L	11-3	−3	3	3	3
D	12-4	+4	M	11-4	−4	4	4	4
E	12-5	+5	N	11-5	−5	5	5	5
F	12-6	+6	O	11-6	−6	6	6	6
G	12-7	+7	P	11-7	−7	7	7	7
H	12-8	+8	Q	11-8	−8	8	8	8
I	12-9	+9	R	11-9	−9	9	9	9

Fig. 8.16 EBCDIC–BCD relationship

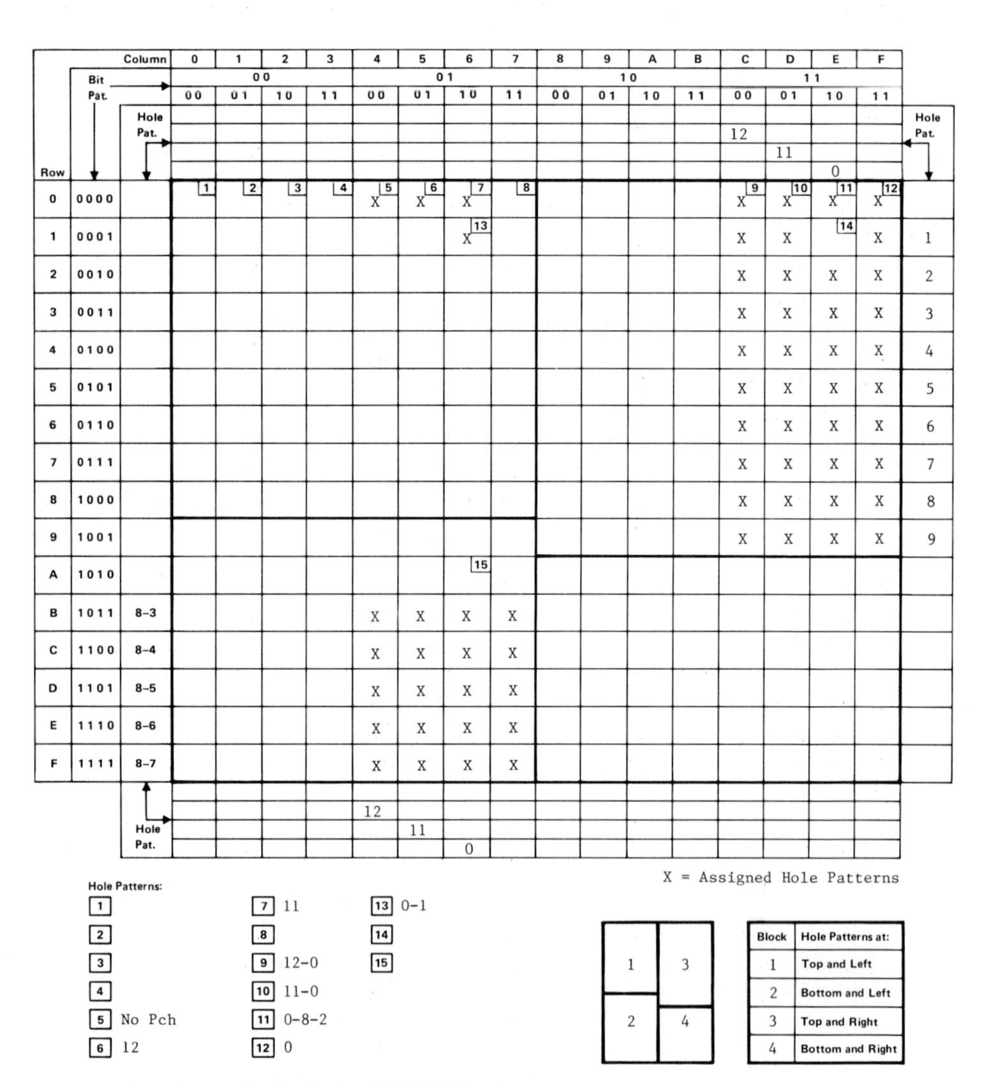

Fig. 8.17 Preliminary EBCDIC hole patterns

9
The Duals of EBCDIC

9.1 A- AND H-DUALS

In Chapters 4 and 5 there is a discussion of the five duals of BCDIC; why they came into being, an attempt to eliminate them, and why they were not eliminated after all. The duals came into existence because of equipment limitations and were retained for reasons of compatibility.

A number of different 48-character chains were provided for the families of 6-bit computers. These chains were designated by letters A, B, C, D, E, F, G, H, I, etc. One of these chains carried the "commerical" graphics and was designated an A chain. Another chain carried the "scientific" graphics and was designated the H chain. In time, the duals came to be designated by these letters; the A-duals and the H-duals.

Hole patterns	A-duals	H-duals
0-8-4	%	(
12-8-4	¤	)
12	&	+
8-3	#	=
8-4	@	'

While EBCDIC was being developed (as described in previous chapters), the question arose again, "Should the duals be eliminated?"

9.2 IMPLICATIONS OF REASSIGNING DUALS

Certainly, the equipment limitations could be removed. While the System/360 was being designed, a new keypunch (which came to be the

IBM 029 Keypunch) was being designed. It would expand from the capability of the 026 Keypunch to key 48 characters by single key-stroke to a capability of 64 characters. New printers were being designed, and it was assumed or hoped that the long-established 48-character printing set could be expanded without sacrificing printing speed. The question of compatibility of card hole patterns with BCDIC would obviously arise and would have to be reviewed. But the full implications of any such incompatibility could not be reviewed in depth until the nature and extent of the incompatibility was known. The first thing to be determined was what the incompatibility might be. There were four possibilities:

Possibility 1. Retain the de facto BCDIC hole patterns for the A-duals, and assign new hole patterns to the H-duals.

Possibility 2. Retain the de facto BCDIC hole patterns for the H-duals, and assign new hole patterns for the A-duals.

Possibility 3. Retain the de facto BCDIC hole patterns for some of the A-duals and for some of the H-duals, and assign new hole patterns to the other A-duals and to the other H-duals.

Possibility 4. Assign new hole patterns to the A-duals *and* to the H-duals.

It was clear that, whatever the implications of Possibilities 1 and 2, these must be determined first, after which the implications of Possibilities 3 and 4 could be determined easily. So Possibilities 1 and 2 were looked at first.

Three data processing customer situations were reviewed:

Situation 1. Customer now, or in the future, will take a successfully performing application on a BCDIC computer and convert it to run on an EBCDIC computer.

Situation 2. An application will be organized so that it is processed partially on a BCDIC computer and partially on an EBCDIC computer.

Situation 3. An application will be processed completely on an EBCDIC computer.

With respect to Possibilities 1, 2, 3, and 4, Situation 3 seemed to display no implications, so it was disregarded in further review.

Two assumptions were now made:

Assumption 1. A-duals will appear mainly in data. That is, they will be required to be input to the system, will exist in data during various stages of processing, and may be required in output listings or other output data.

Assumption 2. H-duals will appear mainly in programming source language statements. That is, they will require to be input to the system, and will be required for source language program listings, and will be required during compile processes, but will not then be required in further stages of processing.

Some implications now emerged:

Implication 1. Possibility 1 posed no adverse implications under Assumption 1 for any of Situations 1, 2, or 3, but it posed adverse implications under Assumption 2.

Implication 2. Possibility 2 posed no adverse implications under Assumption 2 for any of Situations 1, 2, or 3, but it posed adverse implications under Assumption 1.

Implication 3. Possibilities 3 and 4 posed adverse implications for all of Situations 1, 2, and 3 under both Assumptions 1 and 2.

Before we consider adverse implications, let us look at another assumption that was made.

Assumption 3. Under Possibilities 1, 2, 3, and 4, the "new" hole patterns would nevertheless be contained within the set of 64 BCDIC hole patterns. That is to say, the "new" hole patterns* could still be input to BCDIC computing systems, even though their graphic meanings had been changed.

Now, let us examine the adverse implications in detail. First we need some terminology to cover the four Possibilities precisely.

If the old and therefore compatible hole patterns are retained for the A-duals, the data containing these duals will be called "compatible BCDIC A-data," or "compatible EBCDIC A-data," depending on which code is used. Similarly, if old hole patterns are retained for the H-duals, reference will be made to "compatible BCDIC H-data," or to "compatible EBCDIC H-data."

If new and therefore incompatible hole patterns are assigned to the A-duals, reference will be made to "incompatible BCDIC A-data" or to

---

*An intriguing aspect of "new" hole patterns emerged in EBCDIC. A 64-character subset of the 256 EBCDIC hole patterns was the set that was single-stroke keypunchable on the 029 Keypunch. But the EBCDIC set of 64 hole patterns did not match the BCDIC set of 64 hole patterns. EBCDIC subset contained 12-8-2 and 11-8-2, but not 12-0 and 11-0 (12-0 and 11-0 were, of course, contained in the total set of 256 EBCDIC hole patterns). BCDIC set contained 12-0 and 11-0, but did not contain 12-8-2 and 11-8-2. This anomaly is fully discussed in Chapter 10.

"incompatible EBCDIC A-data." Similarly, if new hole patterns are assigned to the H-duals, reference will be made to "incompatible BCDIC H-data" or to "incompatible EBCDIC H-data."

9.2.1 Situation 1 Consequences

Consider Situation 1 under each of the four Possibilities:

Situation 1/Possibility 1. There will be no problem with A-data, but all programs will have to be either reprogrammed or rekeypunched for the incompatible EBCDIC H-data, then recompiled and redebugged.

Situation 1/Possibility 2. Data containing compatible BCDIC A-data will have to be converted to incompatible EBCDIC A-data. Programs will have to be either reprogrammed or recompiled and redebugged (but not rekeypunched).

Situation 1/Possibility 3. The actual situation here would depend on which A- and H-duals were, or were not, changed. However, for those applications with A-data whose A-duals had been changed, data would have to be converted. Programs would have to be either reprogrammed or rekeypunched, then recompiled and redebugged.

Situation 1/Possibility 4. Data containing compatible BCDIC A-data would have to be converted. Programs would have to be rekeypunched, recompiled, and redebugged.

9.2.2 Situation 2 Consequences

Now, Situation 2 had to be defined in greater depth. There are three considerations:

- BCDIC computer does, or does not, process H-data.
- BCDIC computer does, or does not, pass A-data to EBCDIC computer.
- BCDIC computer does, or does not, receive A-data from EBCDIC computer.

The various possible situations are shown in the left column of Fig. 9.1. For these various situations, the table indicates whether implications are unsatisfactory (U) or satisfactory (S). For the various Situations under the four Possibilities, Situation/Possibilities were unsatisfactory in 12 instances because of change of H-duals, unsatisfactory in 18 instances because of change of A-duals. Assuming all Situation/Possibilities were

BCDIC Computer	Possibility 1	Possibility 2	Possibility 3	Possibility 4
Processes H, yes	U	S	U	U
Passes A, yes	S	U	U	U
Receives A, yes	S	U	U	U
Processes H, yes	U	S	U	U
Passes A, yes	S	U	U	U
Receives A, no	S	S	S	S
Processes H, yes	U	S	U	U
Passes A, no	S	S	S	S
Receives A, yes	S	U	U	U
Processes H, yes	U	S	U	U
Passes A, no	S	S	S	S
Receives A, no	S	S	S	S
Processes H, no	S	S	S	S
Passes A, yes	S	U	U	U
Receives A, yes	S	U	U	U
Processes H, no	S	S	S	S
Passes A, yes	S	U	U	U
Receives A, no	S	S	S	S
Processes H, no	S	S	S	S
Passes A, no	S	S	S	S
Receives A, yes	S	U	U	U
Processes H, no	S	S	S	S
Passes A, no	S	S	S	S
Receives A, no	S	S	S	S

Figure 9.1

equally likely to occur, the table shows more unsatisfactory implications for A-dual changes than for H-dual changes.

There was another consideration. There are vastly more tapes containing application data (i.e., containing A-duals) than there are source language program tapes (i.e., tapes containing H-duals). In general, it was reasoned that the costs of converting data (if A-duals were changed) would be vastly greater than the costs of converting programs (if H-duals were changed). Possibility 1, therefore, seemed to pose very much less of a cost implication for users than Possibility 2. Possibilities 3 and 4 seemed to pose more cost implications for users than either Possibilities 1 or 2. In short, Possibility 1 seemed to be the least onerous choice.

9.3 FIRST DECISION

The first decision was made. If one of the four Possibilities were chosen, it would be Possibility 1—to retain the BCDIC hole patterns for the A-duals and to change the hole patterns for the H-duals. This Possibility would be taken together with Assumption 3—to retain the 64 BCDIC hole patterns.

The next step was to decide which five BCDIC graphics would be replaced by the EBCDIC H-duals:

( ) + = '

9.4 FURTHER DECISIONS

Some further decisions were made:

1. Space, numerics, alphabets would not be changed.
2. @ # % & ¤ would not be changed.
3. . , * / $ – would not be changed.

This left the following BCDIC graphics for consideration:

: ; ! ? < > [ ]
/ ƀ γ ǂ ⧧ ⧻ Δ √

This problem was being considered in the same time frame as the design and development of the System/360. It had already been decided that none of the control functions provided by the seven BCDIC control characters would be provided as functions on the System/360, and the seven graphics would not be provided on the System/360. Therefore, the seven graphics would not be provided in EBCDIC, and the seven corresponding code positions were available for assignments of the H-duals or of new graphics as seemed appropriate.

It was decided that the five H-duals would not be assigned to any of these seven code positions. The reasoning went as follows. The five H-duals were graphics used in programming languages. It was entirely possible that source language programs intended for execution on System/360 might first pass through a BCDIC computer, for one or another reason. But such programs would have to have the "new" codes for the H-duals, whatever they might be. If the H-duals were assigned to the hole patterns of BCDIC control characters, then, when such programs were entered into a BCDIC computer, the H-duals would have bit patterns of control characters. And if such programs were then recorded on seven-track magnetic tape, during the recording or subsequent reading of such tapes on BCDIC computers, the control bit patterns might cause unexpected and undesirable effects. Might not, of course, if care was taken, but the feeling was, it was better to be safe than sorry. The H-duals should not be assigned to the hole patterns of the BCDIC control characters.

This left the following set of BCDIC graphics, of which five were to be replaced by H-duals:

: ; ! < > \ [ ]

Intuitively, it was decided to replace

[ ] by ( )

leaving

: ; ! < > \

three of which were to be replaced by

+ = ′

As has been mentioned before, it had already been decided not to provide on the 029 keypunch the hole patterns 12-0 and 11-0 as single-stroke keypunchable characters. In consequence, the hole patterns of the BCDIC graphics ? ! were not available to be replaced by any of the H-duals. (The reason for this aspect of the design of the 029 Keypunch is discussed more fully in Chapter 10.) This left BCDIC graphics

: ; < > \

The ; was a required graphic in COBOL, so it could not be replaced. Both < and > were also COBOL graphics, but the COBOL standard stipulated that they could be represented by two-character representations; GT (Greater Than) for > and LT (Less Than) for <. It was decided that < and > were the two BCDIC graphics to be replaced, and + and = were chosen to replace them, respectively. BCDIC ; and EBCDIC ; would be matched. This left BCDIC graphic either : or \ to be replaced by ′; on not much more than a toss of a coin basis, it was decided to replace

: by ′

The situation was now as follows:

BCDIC		EBCDIC
Space	match	Space
0–9	match	0–9
A–Z	match	A–Z
. , * / $ – ;	match	. , * / $ – ;
@ # % & ¤	match	@ # % & ¤
: > [<]	replaced by	' = (+)
? !	hole patterns not to be assigned EBCDIC graphics	
\	undecided	
ƀ γ ǂ ⧧ ⧻ Δ √	not to be assigned in EBCDIC	

At this stage then, five BCDIC graphics < > [ ] : were to be replaced, the card hole patterns of two BCDIC graphics ? ! were not to be on the 029 Keypunch, and no decision has been made with respect to the BCDIC graphic \ .

The next question was whether any of the seven BCDIC graphics

< > [ ] : ! ?

should be reassigned to BCDIC hole patterns to be vacated by

ǂ ⧧ ⧻ ƀ Δ √ γ

9.5 PL/I CONSIDERATIONS

While this question was being considered, a new factor came on the scene. A new higher-level programming language, PL/I was being developed. PL/I itself has some character set requirements. The Space character would be needed and so would the following 59 graphics:

10 numerics — 0 to 9
26 alphabetics — A to Z
3 alphabetic extenders — # $ @
20 syntactics* — + = – / * ( ) < > _ . , : ; ? | ¬ ' % &

---

*A syntactic is a character that has some specific meaning within the syntax of a programming language.

Actually, the PL/I designers had wanted more graphics, in particular [and], but the requirement to implement the set on a 60-character chain made it impossible to provide the brackets to PL/I.

It was decided that these 59 graphics must definitely be assigned in EBCDIC and to hole patterns that are single-stroke keypunchable on the 029 Keypunch. Many of them had already been assigned, under the discussion above. The Space character, numerics, alphabetics, and alphabetic extenders had been assigned. Of the syntactics, 13 had been assigned:

( ) + = ' . , ; % & / * −

Seven syntactics remained to be assigned:

< > | ¬ _ ? :

Also, graphics for three lower-case alphabetic extenders needed to be assigned. Ten BCDIC graphics had not yet been replaced:

? ! \ ƀ ‡ ⧧ ⧺ Δ √

Of these ten, as mentioned previously, the hole patterns 12-0 and 11-0 for ? and ! were not to be available on the 029 Keypunch. Compensating for this, two new hole patterns would be available, 12-8-2 and 11-8-2.

It seemed like a fortuitous match—seven syntactics and three lower-case alphabetic extenders needed to be assigned, and eight BCDIC hole patterns and two new hole patterns were available. This fortuity quickly disappeared, for the following reasons.

9.6 "88 − 26 = 62"

The console typewriter for the System/360 would provide 88 graphics and the Space character. Of these 88, 26 are lower-case alphabetics, leaving 62 graphics. The 029 Keypunch can provide 63 graphics and the Space character, but if it does so, one of those 63 graphics cannot be typed on the console typewriter. The system would be out of balance. To resolve this system imbalance, the 029 Keypunch must be allowed to provide only 62 graphics, and the Space character. The 029 would have the physical capability of providing a 63rd graphic, but it must not do so. This reasoning was accepted. (A fuller discussion is given in Chapter 10.) The 029 Keypunch was designed to have a key that will generate the 0-8-2 hole pattern, but no graphic is interpreted on the punched card. Since the 0-8-2 hole pattern was selected, no EBCDIC graphic would be assigned to replace the BCDIC ‡.

9.7 ASCII CONSIDERATIONS

The consequence of this decision was that there were 9 hole patterns available and 10 graphics to be assigned. This dilemma was resolved by consideration of another factor. It would be helpful in the long run if EBCDIC provided the same set of graphics as ASCII. A corollary of this was that EBCDIC should not have graphics that were not in ASCII. This focused attention on three EBCDIC graphics:

| ¬ ¤

The first part of the solution involved | and ¬ . As described in Chapter 24, this problem was solved when the standards committees decided that the ASCII graphics ! and ^ could be stylized as (that is, substituted by) | and ¬.

This left the graphic ¤ to be resolved. Attempts to persuade the standards committees to assign this graphic in ASCII were unavailing. Eventually, it was decided not to assign ¤ in EBCDIC. This decision, as it turned out, was not subsequently accepted by many customers, who requested that it be provided on printers for the System/360. It was provided to these customers, although it ostensibly did not exist in EBCDIC.

9.8 BCDIC CONTROL CHARACTERS

This brought the counts back to match—EBCDIC graphics for BCDIC hole patterns. The question that now arose concerned the fact that six of these BCDIC hole patterns represented BCDIC control characters. As stated above, BCDIC hole patterns that represented BCDIC control characters were avoided in reassigning the H-duals. Shouldn't they also be avoided in assigning the rest of the PL/I syntactics?

It would not be possible to avoid them, however, if Assumption 3 above was to be valid. So the question was not how to avoid assigning PL/I graphics to BCDIC control characters, but rather what the implications of such an assignment might be. The reasons for avoiding BCDIC control characters for H-duals were reviewed:

- H-duals were used in FORTRAN and COBOL source language programs.
- Such programs, intended for execution on a System/360, might nevertheless be processed in some way on a BCDIC computer before arriving at the System/360.
- During the processing on a BCDIC computer, the source language program might be stored on magnetic tape.

- The control bit patterns might cause unpredictable and unwanted results.

Since PL/I, as a programming language, was not being developed for use on a BCDIC computer, it seemed unlikely that any PL/I source language programs intended for execution on a System/360 would be entered into a BCDIC computer for any reason. Therefore, it seemed that assigning PL/I syntactics to BCDIC control bit patterns was unlikely to lead to trouble. Two of these syntactics < and > were also COBOL syntactics, so it was decided not to assign < and > to BCDIC control characters. There were just two BCDIC noncontrol characters remaining unassigned, ⌑ (freed up as described above) and \. These two hole patterns were assigned to < and > (respectively).

The five remaining PL/I syntactics | ¬ _ : ? were assigned to the hole patterns previously assigned to BCDIC graphics ǂ γ Δ ƀ ⧻ respectively.

9.9 LOWER-CASE ALPHABETIC EXTENDERS

The sole remaining problem, then, was assignments for the three lower-case alphabetic extenders. While this development work on EBCDIC was going on, a new PTTC was being developed for the System/360 (see Chapter 12). The criterion developed for lower-case alphabetic extenders for the new PTTC was as follows:

> U.S.A. graphics for the three lower-case alphabetic extender code positions must be such that they will not be required or wanted in any European country with a Latin alphabet. That is, in such countries, the U.S.A graphics can be "throwaways."

The three graphics ¢ ! " were chosen to meet this criterion. (These graphics also met the requirement that they be ASCII graphics, although ¢ disappeared from ASCII before ASCII was finally approved as an American National Standard.) And so ¢ ! were assigned to the two new hole patterns, 12-8-2 and 11-8-2, and " was assigned to the sole remaining BCDIC graphic √ with its hole pattern of 8-7.

It should be pointed out that because of their card hole patterns 12-8-2 and 11-8-2, the EBCDIC ¢ and ! came in time to be associated with the ASCII graphics [and] associated with those hole patterns. When this association became firm (when the American National Standard Hollerith Punched Card Code was approved), it was suggested that EBCDIC be changed, replacing ¢ and ! with [and] (respectively). This suggestion was reviewed, but not adopted, for the following reasons.

1. The cost to replace 029 Keypunch printing plates and keytops, printer chains and trains, typewriter printing elements, graphic display character generators, etc., would be considerable.
2. Graphics [and] were in ASCII code positions which corresponded to National Use positions in the ISO 7-Bit Code. ISO 7-Bit Code National Use graphics, like EBCDIC alphabetic extenders, were expected to be replaced in those European countries with Latin alphabets of more than 26 letters; that is, the graphics [and] would not, in fact, appear in Europe.
3. In FORTRAN and PL/I, there had long been an unfulfilled requirement for a second pair of "parentheses." The [and] would certainly serve that purpose. If the brackets were put on the 029 Keypunch, that would make them available for just such a second level of parentheses.
4. But such a compiler would not serve in Europe, where the brackets would be replaced by letters.
5. To avoid such a potential dichotomy for programming languages between Europe and the U.S.A., graphics [and] were not put on the 029 Keypunch.
6. A small glitch between ASCII and EBCDIC—[and] corresponding respectively to ¢ and !—seemed preferable to the potential programming language dichotomy of reason 5 above.

9.10 FINAL ASSIGNMENT OF SPECIALS

Figure 9.2 shows the final assignment of specials into EBCDIC in 1970, as a result of reassigning the H-duals. Figure 9.3 shows, for comparison, the graphics that would have been assigned in EBCDIC if the BCDIC specials, complete with A/H-duals, had been assigned according to their BCDIC card hole patterns. Of the 27 BCDIC specials, only 11 ended up with unchanged code positions in EBCDIC.

9.11 CONSEQUENCES OF REASSIGNMENT

A question that arose was whether the collating sequence had been affected by these changes. The primary criterion in the development of EBCDIC was that the collating sequence of BCDIC should be embedded in the EBCDIC collating sequence (see Chapter 8). In a very real sense, this criterion had not been aborted, even though many BCDIC graphics ended up with EBCDIC card hole patterns different than their BCDIC card hole patterns.

Row	Column	0	1	2	3	4	5	6	7	8	9	A	B	C	D	E	F
	Bit Pat.	00				01				10				11			
		00	01	10	11	00	01	10	11	00	01	10	11	00	01	10	11
0	0000					SP	&	-									
1	0001							/									
2	0010																
3	0011																
4	0100																
5	0101																
6	0110																
7	0111																
8	1000																
9	1001																
A	1010					¢	!		:								
B	1011					.	$	,	#								
C	1100					<	*	%	@								
D	1101					(	)	_	'								
E	1110					+	;	>	=								
F	1111					|	¬	?	"								

Fig. 9.2 EBCDIC specials

Row	Column	0	1	2	3	4	5	6	7	8	9	A	B	C	D	E	F
	Bit Pat.	00				01				10				11			
		00	01	10	11	00	01	10	11	00	01	10	11	00	01	10	11
0	0000					SP	&+	-						?	!	‡	
1	0001							/									
2	0010																
3	0011																
4	0100																
5	0101																
6	0110																
7	0111																
8	1000																
9	1001																
A	1010								ƀ								
B	1011					.	$	,	#=								
C	1100					⌑)	*	% (	@'								
D	1101					[	]	γ	:								
E	1110					<	;	\	>								
F	1111					ǂ	Δ	⧻	√								

Fig. 9.3 BCDIC specials in EBCDIC

Example

The field on which records are sorted or collated is called a keyword. Keypunch a set of records, and keywords, on an 026 (BCDIC) Keypunch. Enter the data into a BCDIC computer. Sort the records in sequence of keywords.

Take the same card deck, and enter it into an EBCDIC computer. Sort the records in sequence of keywords.

The sequence of records in the BCDIC computer and the sequence of records in the EBCDIC computer will be identical.

The sequence of records will be identical, but will anything be different? List the keywords and records on the printer of the BCDIC computer. List the keywords and records on the EBCDIC computer. Compare the listings. If all graphics in the keywords and records are in the following set, the listings will be identical:

Space	
Numerics	0 to 9
Alphabetics	A to Z
Specials	. , / * $ – ; & % # @

If graphics are used in keywords or records beyond the set above, the listings will look different, the differences corresponding to the differences between Fig. 9.2 and 9.3. But it must be reemphasized that the sequence of records will be identical.

Were there any adverse effects of the reassignment of the H-duals? Yes, indeed! The first effect showed up for programmers who were developing various programs for the System/360. Engineering models of the System/360 were available for the use of programmers, but 029 Keypunches were not. Programmers could not get their programs keypunched according to the EBCDIC card hole patterns. If programs could not be keypunched, they could not be entered and debugged. The solution to this impasse was to modify several 026 keypunches to generate the EBCDIC hole patterns for ( ) + = ′ . Then the programs could be keypunched, entered, and debugged.

The second effect was on customers who had received a System/360. Of course, old BCDIC machine language programs would not work on the System/360, but, to the extent that customers had retained source language program decks or program tapes for COBOL or FORTRAN, the programs could be recompiled, a task which was a far less onerous proposition than reprogramming. Unfortunately, such program decks or tapes would have the old BCDIC H-dual hole patterns or bit patterns for ( ) + = ′ and the System/360 compilers for COBOL and FORTRAN had

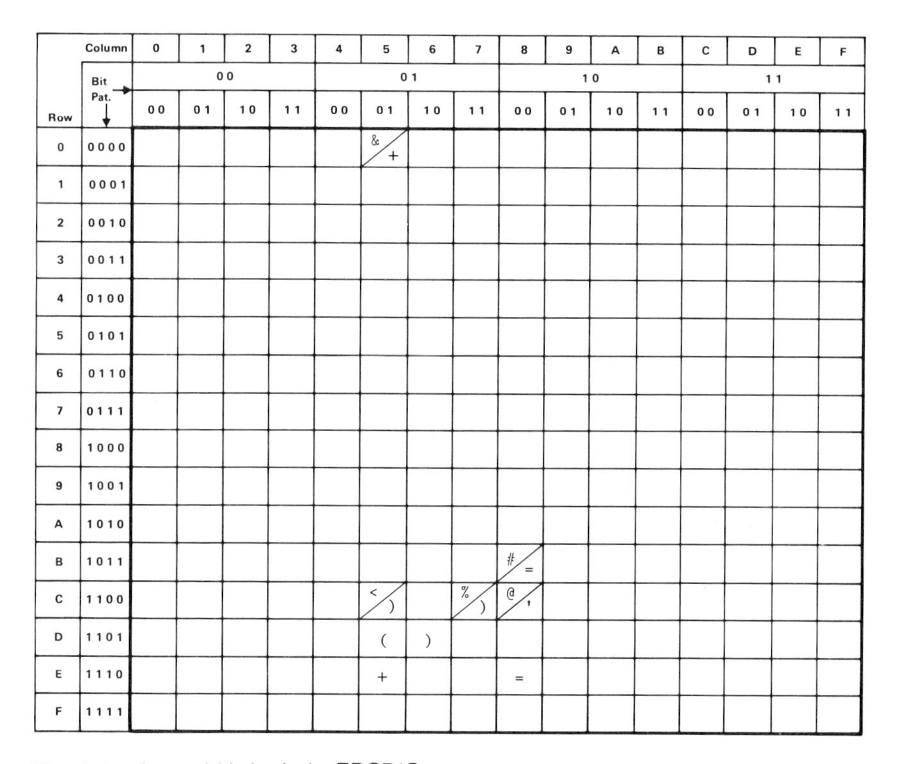

Fig. 9.4 A- and H-duals in EBCDIC

been written assuming the new EBCDIC patterns for these graphics. Could this dilemma be resolved?

It could, and was, with the aid of some IBM customers. Consider Fig. 9.4. EBCDIC hex positions 4D, 5D, 4E, 7E, and 7D were the assigned positions for the bit patterns of ( ) + = '. EBCDIC hex positions 6C, 4C, 50, 7B, and 7C were where these graphics would have been assigned according to their old BCDIC hole patterns or bit patterns. Three things were done.

1. The logic in the control unit of the chain and train printers was modified, as shown in Fig. 9.4, so that
   - either hex position 4D or 6C printed (
   - either hex position 5D or 4C printed)
   - either hex position 4E or 50 printed +
   - either hex position 7E or 7B printed =
   - either hex position 7D or 7C printed '

2. The scan portion of the FORTRAN compiler was modified so that either of the equivalent pairs of bit patterns would be accepted for ( ) + = '.
3. The scan portion of the COBOL compiler was similarly modified.

By these actions, old FORTRAN or old COBOL program decks or tapes could be read into a System/360, listed for debug purposes, compiled, and executed.

Clearly, if these actions had been taken during the development cycle of System/360 programs, the first adverse effect above would not have occurred, and unmodified 026 Keypunches could have been used. Hindsight is easily come by.

With the reassignment of H-duals in EBCDIC, and with the assignment or reassignment of the remaining PL/I syntactics and of the lowercase alphabetic extenders, the 88 graphics of EBCDIC were set in place. Attention now centered on completing the 256 card-hole-patterns–to–bit-patterns assignments. This will be discussed in Chapter 11.

10
The Graphic Subsets of EBCDIC

The 256-character code EBCDIC was designed as the CPU code for the System/360. As described in Chapter 8, a decision was made to reserve 64 code positions for control meanings and 192 code positions for graphic positions. The physical capability of chain/train printers of providing up to 240 different graphics did not limit the total numbers of graphics to be assigned in EBCDIC. Other factors did set limits and gave rise to graphic subsets of EBCDIC.

10.1 88-GRAPHIC SETS

The console printer for the System/360 was based on an electric typewriter, duocase, with 44 keys, and a capability of printing the following 88 graphics:

10 numerics — 0 to 9
26 lower-case alphabetics — a to z
3 lower-case alphabetic extenders — ¢ ! "
26 upper-case alphabetics — A to Z
3 upper-case alphabetic extenders — # $ @
20 specials — / * + = − & % | ¬ _ . , : ; ? < > ( ) '

As described in Chapter 8, these 88 graphics were assigned to code positions as shown in Fig. 10.1.

Row	Column	0	1	2	3	4	5	6	7	8	9	A	B	C	D	E	F
	Bit Pat.	00				01				10				11			
		00	01	10	11	00	01	10	11	00	01	10	11	00	01	10	11
0	0000					SP	&	-									0
1	0001							/		a	j			A	J		1
2	0010									b	k	s		B	K	S	2
3	0011									c	l	t		C	L	T	3
4	0100									d	m	u		D	M	U	4
5	0101									e	n	v		E	N	V	5
6	0110									f	o	w		F	O	W	6
7	0111									g	p	x		G	P	X	7
8	1000									h	q	y		H	Q	Y	8
9	1001									i	r	z		I	R	Z	9
A	1010					¢	!		:								
B	1011					.	$	,	#								
C	1100					<	*	%	@								
D	1101					(	)	_	'								
E	1110					+	;	>	=								
F	1111					|	¬	?	"								

Fig. 10.1 EBCDIC 88-graphic set

10.2 62-GRAPHIC SUBSET

From the duocase set of 88 graphics emerged a monocase set. The IBM 029 Keypunch was being designed at the same time as the System/360, and it had been decided to provide 64 hole patterns on the 029. One of these hole patterns would be the "no-holes" hole pattern for the Space character, leaving 63 hole patterns to be assigned. It was decided that the 029 Keypunch would provide a monocase set, and that the hole patterns for the monocase alphabetics would be those already assigned to the upper case alphabetics of EBCDIC. A keypunch keyboard is represented in Fig. 10.2.

By the decision to assign the hole patterns of the EBCDIC upper-case alphabetics to the keypunch monocase alphabetics, the EBCDIC lower-case alphabetics were excluded from the keypunch—excluded in the sense of being single-stroke punchable. That is to say, of the 88

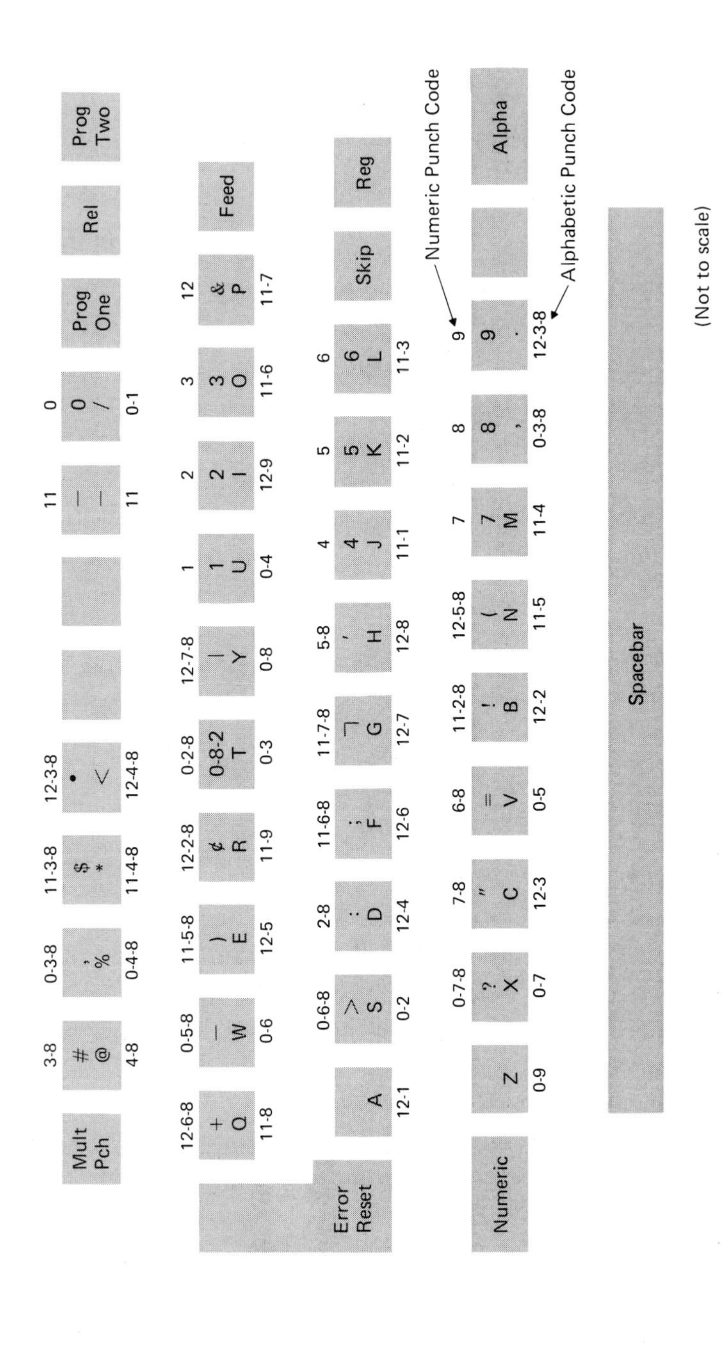

Fig. 10.2 029 keyboard

Hole Pattern	12	11	0	
	&	-	0	SP
1	A	J	/	1
2	B	K	S	2
3	C	L	T	3
4	D	M	U	4
5	E	N	V	5
6	F	O	W	6
7	G	P	X	7
8	H	Q	Y	8
9	I	R	Z	9
0	¢	!		:
8-3	.	$	,	#
8-4	<	*	%	@
8-5	(	)	_	'
8-6	+	;	>	=
8-7	|	¬	?	"

Fig. 10.3 EBCDIC 64-graphic set

EBCDIC graphics, 88 − 26 = 62 could be provided. But the keypunch could provide 63 graphics. There were, then, two possible choices:

1. Assign 63 graphics on the keypunch, and add a graphic to EBCDIC, making a total of 89.
2. Assign 62 graphics on the keypunch, and thus leave one of the 63 hole patterns unassigned.

If choice (1) were made, the 89th graphic could then not be printed on the 88-graphic console typewriter. An imbalance in the system would be created. For this reason, choice (1) was rejected.

Under choice (2), the keypunch could punch and interpret 62 characters and the Space character. The hole pattern 0-8-2 has no graphic

	Hole Pattern → ↓	12	11	0	
		& or +	-	0	SP
	1	A	J	/	1
	2	B	K	S	2
	3	C	L	T	3
	4	D	M	U	4
	5	E	N	V	5
	6	F	)	W	6
	7	G	P	X	7
	8	H	Q	Y	8
	9	I	R	Z	9
	0	? [1]	! [2]	‡	ƀ
	8-3	.	$	,	# or =
	8-4	⌑ or)	*	% or (	@ or '
	8-5	[	]	γ	:
	8-6	<	;	\	>
	8-7	ǂ	Δ	⧻	√

Hole Patterns:
[1] 12-0
[2] 11-0

Fig. 10.4 BCDIC 64-graphic set

assigned. As can be seen from Fig. 10.2, 0-8-2 is engraved on a keytop. When this key is depressed, the hole pattern 0-8-2 is punched in the card, but no graphic is interpreted on the card. (As described later in Section 10.3, a graphic was assigned some years later to the hole pattern 0-8-2, but it is not interpreted on the 029 Keypunch.)

The 64 characters of this EBCDIC subset are shown in Fig. 10.3. Figure 10.4 shows the 64-character set of BCDIC. It is to be noted that the two sets of 64 hole patterns are not quite the same. EBCDIC-64 has hole patterns 12-8-2 and 11-8-2, and does not have 12-0 and 11-0. BCDIC has hole patterns 12-0 and 11-0, and does not have 12-8-2 and

11-8-2. The hole patterns 12-8-2 and 11-8-2 were chosen instead of 12-0 and 11-0 for the 029 Keypunch because of a mechanical problem.*

It is of interest that these 64 hole patterns of the 029 Keypunch are the hole patterns assigned to the 64 graphics and Space in columns 2, 3, 4, and 5 of the 7-Bit Code (Fig. 2.26). In that code, the graphic \ is assigned to the hole pattern 0-8-2, and the graphics [and] are assigned to the hole patterns 12-8-2 and 11-8-2, respectively, as contrasted to the EBCDIC graphics ¢ and ! . Further, in the 7-Bit Code, graphics ! and ∧ are assigned to hole patterns 12-8-7 and 11-8-7, respectively, as contrasted to the stylistically similar EBCDIC graphics | and ¬ . This 64-graphic set is shown in Fig. 10.5.

Another 64-character set emerged during the design of the IBM System/3. It was decided to provide a printing set of 63 graphics and Space. Of these 63 graphics, it was quickly decided that 62 would be those of EBCDIC previously described. But what should the 63rd graphic be? It will be recalled that for the System/360, a console typewriter of 88-graphic capacity limited the EBCDIC monocase set to 62 graphics. But for the System/3, a 63 monocase printer would be provided for the console, so the system imbalance limitation did not appear.

In the System/3, as with other BCD computers, the BCD relationship for alphabetics would be provided. That is, as discussed in Chapter 2, hole patterns 12-1, 12-2, . . . , 12-9 would mean A, B, . . . , I as alphabetics, but would mean +1, +2, . . . , +9 as signed numerics; hole patterns 11-1, 11-2, . . . , 11-9 would mean J, K, . . . , R as alphabetics, but would mean −1, −2, . . . , −9 as signed numerics.

When signed numerics are printed out in final listings, the sign − is separated from the units position of a numeric field and printed separately. But during debugging runs, the sign is generally not printed out. That is to say, −1, −2, . . . , −9 will print as J, K, . . . , R. While this may look peculiar, it is quite unambiguous to the programmer, and is acceptable. Similarly, +1, +2, . . . , +9 will print as A, B, . . . , I.

The problem is, what will print for −0 and for +0 ? The problem with +0 is not so pressing, since input data for debugging generally has absolute (unsigned) numbers instead of positive (signed) numbers. But the

* Without going into details on this mechanical problem, let it suffice that to interpret from hole patterns 12-0 and 11-0 would be quite difficult, while to interpret from hole patterns 12-8-2 and 11-8-2 was quite easy, so the latter pair were chosen. The hole patterns 12-0 and 11-0 are included in the total set of 256 hole patterns of the EBCDIC card code, but they are not in the 64-character set of the 029 Keypunch.

Hole Pattern→ ↓	12	11	0	
	&	-	0	SP
1	A	J	/	1
2	B	K	S	2
3	C	L	T	3
4	D	M	U	4
5	E	M	V	5
6	F	O	W	6
7	G	P	X	7
8	H	Q	Y	8
9	I	R	Z	9
0	[	]	\	:
8-3	.	$	,	#
8-4	<	*	%	@
8-5	(	)	_	'
8-6	+	;	>	=
8-7	!	^	?	"

Fig. 10.5 7-Bit code 64-graphic set

problem for −0 remains. It was decided that there must be an actual graphic to represent −0. The bit pattern for −0 is in hex-position D0. As explained later in this chapter, the graphic } had been assigned to this EBCDIC code position. Therefore, it was chosen to represent −0 in the System/3.

It seemed strange to provide, in a printing set, } and not to provide {. However, with the addition of } to the 62 graphics and Space, all positions of the 64-character set were filled. If { were to be provided, then one of the 62 graphics would not be provided, and this possibility was rejected by the System/3 designers. The 64-character set of the System/3 is shown in Fig. 10.6.

	Column	0	1	2	3	4	5	6	7	8	9	A	B	C	D	E	F
	Bit Pat. →	00				01				10				11			
Row	↓	00	01	10	11	00	01	10	11	00	01	10	11	00	01	10	11
0	0000					SP	&	-							}		0
1	0001							/						A	J		1
2	0010													B	K	S	2
3	0011													C	L	T	3
4	0100													D	M	U	4
5	0101													E	N	V	5
6	0110													F	O	W	6
7	0111													G	P	X	7
8	1000													H	Q	Y	8
9	1001													I	R	Z	9
A	1010					¢	!		:								
B	1011					.	$	,	#								
C	1100					<	*	%	@								
D	1101					(	)	_	'								
E	1110					+	;	>	=								
F	1111					|	¬	?	"								

Fig. 10.6 System/3 64-graphic set

10.3 94-GRAPHIC SUBSETS

ASCII, the U.S.A. version of the ISO 7-Bit Code, has 94 graphics. When the card code for ASCII was approved (to be discussed in Chapter 17), it was possible to match the graphics of EBCDIC with the graphics of ASCII, through their associated card hole patterns. At that time, the four anomalies previously described were revealed:

Hole pattern	ASCII	EBCDIC
12-8-2	[	¢
11-8-2	]	!
12-8-7	!	|
11-8-7	∧	¬

(A fuller discussion of the respective matching of ! and ∧ with | and ¬ is found in Chapter 24.)

In addition to these four anomalies, the 94-graphic set of ASCII contained 6 more graphics than the 88-graphic set of EBCDIC. Since these 6 graphics had associated hole patterns, and since the hole patterns had associated code positions in EBCDIC, it was possible to determine where to locate them in EBCDIC, as follows:

<table>
<tr><th>Graphic</th><th></th><th>Hole pattern</th><th>Hexadecimal position</th></tr>
<tr><td>Back slash</td><td>\</td><td>0-8-2</td><td>E0</td></tr>
<tr><td>Grave accent</td><td>`</td><td>8-1</td><td>79</td></tr>
<tr><td>Opening brace</td><td>{</td><td>12-0</td><td>C0</td></tr>
<tr><td>Vertical line</td><td>|</td><td>12-11</td><td>6A</td></tr>
<tr><td>Closing brace</td><td>}</td><td>11-0</td><td>D0</td></tr>
<tr><td>Tilde</td><td>~</td><td>12-11-0-1</td><td>A1</td></tr>
</table>

These six graphics were assigned in EBCDIC, as shown in Fig. 10.7.

<table>
<tr><td></td><td>Column</td><td>0</td><td>1</td><td>2</td><td>3</td><td>4</td><td>5</td><td>6</td><td>7</td><td>8</td><td>9</td><td>A</td><td>B</td><td>C</td><td>D</td><td>E</td><td>F</td></tr>
<tr><td></td><td>Bit Pat.</td><td colspan="4">00</td><td colspan="4">01</td><td colspan="4">10</td><td colspan="4">11</td></tr>
<tr><td>Row</td><td></td><td>00</td><td>01</td><td>10</td><td>11</td><td>00</td><td>01</td><td>10</td><td>11</td><td>00</td><td>01</td><td>10</td><td>11</td><td>00</td><td>01</td><td>10</td><td>11</td></tr>
<tr><td>0</td><td>0000</td><td></td><td></td><td></td><td></td><td>SP</td><td>&</td><td>-</td><td></td><td></td><td></td><td></td><td></td><td>{</td><td>}</td><td>\</td><td>0</td></tr>
<tr><td>1</td><td>0001</td><td></td><td></td><td></td><td></td><td></td><td></td><td>/</td><td></td><td>a</td><td>j</td><td>~</td><td></td><td>A</td><td>J</td><td></td><td>1</td></tr>
<tr><td>2</td><td>0010</td><td></td><td></td><td></td><td></td><td></td><td></td><td></td><td></td><td>b</td><td>k</td><td>s</td><td></td><td>B</td><td>K</td><td>S</td><td>2</td></tr>
<tr><td>3</td><td>0011</td><td></td><td></td><td></td><td></td><td></td><td></td><td></td><td></td><td>c</td><td>l</td><td>t</td><td></td><td>C</td><td>L</td><td>T</td><td>3</td></tr>
<tr><td>4</td><td>0100</td><td></td><td></td><td></td><td></td><td></td><td></td><td></td><td></td><td>d</td><td>m</td><td>u</td><td></td><td>D</td><td>M</td><td>U</td><td>4</td></tr>
<tr><td>5</td><td>0101</td><td></td><td></td><td></td><td></td><td></td><td></td><td></td><td></td><td>e</td><td>n</td><td>v</td><td></td><td>E</td><td>N</td><td>V</td><td>5</td></tr>
<tr><td>6</td><td>0110</td><td></td><td></td><td></td><td></td><td></td><td></td><td></td><td></td><td>f</td><td>o</td><td>w</td><td></td><td>F</td><td>O</td><td>W</td><td>6</td></tr>
<tr><td>7</td><td>0111</td><td></td><td></td><td></td><td></td><td></td><td></td><td></td><td></td><td>g</td><td>p</td><td>x</td><td></td><td>G</td><td>P</td><td>X</td><td>7</td></tr>
<tr><td>8</td><td>1000</td><td></td><td></td><td></td><td></td><td></td><td></td><td></td><td></td><td>h</td><td>q</td><td>y</td><td></td><td>H</td><td>Q</td><td>Y</td><td>8</td></tr>
<tr><td>9</td><td>1001</td><td></td><td></td><td></td><td></td><td></td><td></td><td></td><td>`</td><td>i</td><td>r</td><td>z</td><td></td><td>I</td><td>R</td><td>Z</td><td>9</td></tr>
<tr><td>A</td><td>1010</td><td></td><td></td><td></td><td></td><td>¢</td><td>!</td><td>¦</td><td>:</td><td></td><td></td><td></td><td></td><td></td><td></td><td></td><td></td></tr>
<tr><td>B</td><td>1011</td><td></td><td></td><td></td><td></td><td>.</td><td>$</td><td>,</td><td>#</td><td></td><td></td><td></td><td></td><td></td><td></td><td></td><td></td></tr>
<tr><td>C</td><td>1100</td><td></td><td></td><td></td><td></td><td><</td><td>*</td><td>%</td><td>@</td><td></td><td></td><td></td><td></td><td></td><td></td><td></td><td></td></tr>
<tr><td>D</td><td>1101</td><td></td><td></td><td></td><td></td><td>(</td><td>)</td><td>_</td><td>'</td><td></td><td></td><td></td><td></td><td></td><td></td><td></td><td></td></tr>
<tr><td>E</td><td>1110</td><td></td><td></td><td></td><td></td><td>+</td><td>;</td><td>></td><td>=</td><td></td><td></td><td></td><td></td><td></td><td></td><td></td><td></td></tr>
<tr><td>F</td><td>1111</td><td></td><td></td><td></td><td></td><td>|</td><td>¬</td><td>?</td><td>"</td><td></td><td></td><td></td><td></td><td></td><td></td><td></td><td></td></tr>
</table>

Fig. 10.7 EBDIC 94-graphic set

10.4 CHAIN/TRAIN PRINTER SETS

It is necessary to understand the fundamental principles of chain/train printers in order to see the rationale for printer sets of graphics.

Chains and trains are similar in concept. They are loops of printing slugs which are continuously circulated in a plane normal to the plane of the paper on which printing is to take place (Fig. 10.8).

One principle of a chain/train is significant: the more times a graphic is repeated around the chain/train, the more frequently it will pass a printing position. It is common practice to repeat sets of graphics around the chain/train. Thus a 48-graphic set can be repeated 5 times ($5 \times 48 = 240$), a 60-graphic set can be repeated 4 times ($4 \times 60 = 240$), and so on. The chain/train does not move more rapidly, but individual graphics pass a given printing position more frequently. The following table presents comparative information. Nominal printing speed is given in number of lines printed per minute (LPM).

Number of graphics	Repeated sets	Nominal printing speed
40	6	1250 LPM
48	5	1100 LPM
60	4	950 LPM
120	2	570 LPM
240	1	300 LPM

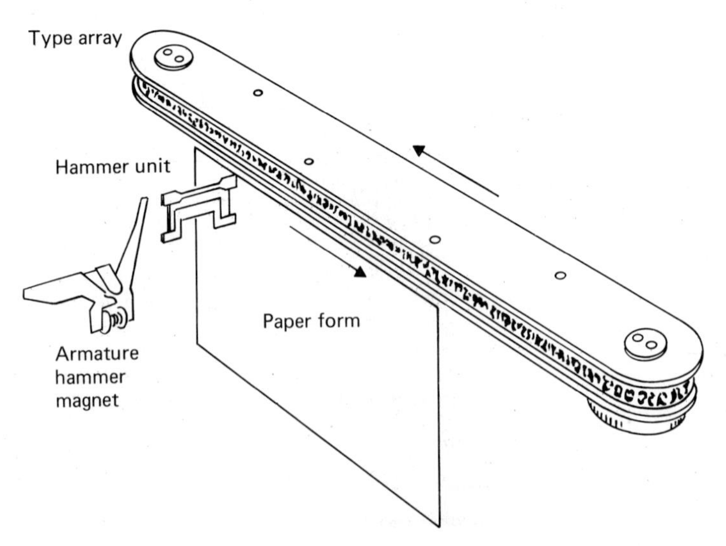

Fig. 10.8 Schematic representation of chain/train printer

10.5 "PREFERRED" GRAPHICS

A more subtle method is to repeat more frequently used graphics more often than less frequently used graphics. The sets of more frequently used graphics are called "preferred" graphics. Of course, the principle is still the same—the more times a graphic is repeated around the chain/train, the more frequently it passes a given printing position.

Consider a 60-graphic set, which could be repeated 4 times around the 240 position chain/train, with nominal printing speed of 950 LPM. But it is also possible to repeat 45 of the graphics 5 times and 15 graphics just once:

$$(45 \times 5) + (15 \times 1) = 240$$

Then, if all the data being printed on a line contain graphics *only* in the set of 45, the nominal printing speed will be 1100 LPM. If the data of a line contains one or more graphics in the set of 15, the printing speed of those lines will be 300 LPM. If the data consists *mostly* of graphics in the set of 45, printing speeds will approach 1100 LPM, as compared with 950 LPM for a chain/train with 60 graphics repeated 4 times.

Some examples of chain/train sets with preferred sets are given, with both 48- and 60-character chains for comparison:

Chain/train sets	Repeat pattern	Nominal printing speed
48		1100 LPM
60, with 45 preferred	45 × 5 = 225 15 × 1 = 15 60 240	950 LPM
52, with 47 preferred	47 × 5 = 235 5 × 1 = 5 52 240	950 LPM
42, with 39 preferred	39 × 6 = 234 3 × 2 = 6 42 240	1250 LPM
84, with 78 preferred	78 × 3 = 234 6 × 1 = 6 84 240	770 LPM
120	120 × 2 = 240	570 LPM

10.6 48-GRAPHIC SETS

We know that 48-character sets are very popular. They strike a good balance between reasonably fast printing speeds and adequate graphic capability. Two well-known sets emerged in the days of BCDIC (Chapter 4) called the A-set and the H-Set, and were perpetuated into EBCDIC (Chapter 9). Some care must be taken with the terminology. A 48-graphic set for BCDCIC consisted of 47 graphics and Space, while a set for EBCDIC consisted of 48 graphics and Space. The ⌑ of BCDIC was replaced by the < of EBCDIC.

			11 specials
BCDIC A-set	Space 0 to 9	A to Z	. , / * – $ % ⌑ # @ &
H-set	Space 0 to 9	A to Z	. , / * – $ ( ) = ' +
			12 specials
EBCDIC A-set	Space 0 to 9	A to Z	. , / * – $ & + % < # @
H-set	Space 0 to 9	A to Z	. , / * – $ & + ( ) = '

10.7 PL/I SUBSETS

The 60-character set for the programming language PL/I consists of 59 graphics and Space:

1 Space
10 numerics — 0 to 9
26 alphabetics — A to Z
3 alphabetic extenders — # $ @
20 specials — / * + = – | ¬ _ & %
( ) < > ' . , : ; ?

In addition, four 2-character operators are recognized by PL/I:

\>= Greater than or equal to
<= Less than or equal to
¬= Not equal to
|| Concatenation

A 48-graphic subset of PL/I consists of 48 single-graphic representations and some 2- and 3-graphic representations:

1 Space
10 numerics

26 alphabetics
12 specials . , ' $ * /
+ – ( ) = &

<table>
<tr><th>Operator</th><th>Representation</th><th>Meaning</th></tr>
<tr><td>:</td><td>..</td><td>Colon</td></tr>
<tr><td>;</td><td>.,</td><td>Semicolon</td></tr>
<tr><td>%</td><td>//</td><td>Percent</td></tr>
<tr><td>></td><td>GT</td><td>Greater than</td></tr>
<tr><td><</td><td>LT</td><td>Less than</td></tr>
<tr><td>>=</td><td>GE</td><td>Greater than or equal to</td></tr>
<tr><td><=</td><td>LE</td><td>Less than or equal to</td></tr>
<tr><td>¬=</td><td>NE</td><td>Not equal to</td></tr>
<tr><td>¬</td><td>NOT</td><td>Logical NOT</td></tr>
<tr><td>|</td><td>OR</td><td>Logical OR</td></tr>
<tr><td>&</td><td>AND</td><td>Logical AND</td></tr>
<tr><td>||</td><td>CAT</td><td>Concatenation</td></tr>
</table>

10.8 KATAKANA SUBSETS

The Japanese written language, like the Chinese written language on which it is based, consists of ideographs—one ideograph per word. Kanji, as it is called, consists of many thousands of ideographs. For normal data processing printers, with limited graphic repertoires, the printing of Kanji is quite impossible.

Another alphabet, invented by the Japanese and called Katakana, is more amenable to data processing printer technology. Katakana is a phonetic alphabet; each Katakana character consists of a vowel, or of a consonant and a vowel, as shown in Fig. 10.9. Thus, Japanese spoken words can be phonetically approximated by a written or printed alphabetic.

As originally assigned in EBCDIC, Katakana consisted of 47 graphics assigned to bit patterns as shown in Fig. 10.10. From this assignment, two Katakana sets were available.

64-character
1 Space
10 numerics
47 Katakana graphics
6 specials – / ¥ . , *

Shape	Name	Shape	Name
ア	A	ハ	HA
イ	I	ヒ	HI
ウ	U	フ	FU
エ	E	ヘ	HE
オ	O	ホ	HO
カ	KA	マ	MA
キ	KI	ミ	MI
ク	KU	ム	MU
ケ	KE	メ	ME
コ	KO	モ	MO
サ	SA	ヤ	YA
シ	SHI		
ス	SU	ユ	YU
セ	SE		
ソ	SO	ヨ	YO
タ	TA	ラ	RA
チ	CHI	リ	RI
ツ	TSU	ル	RU
テ	TE	レ	RE
ト	TO	ロ	RO
ナ	NA	ワ	WA
ニ	NI	ン	N
ヌ	NU		
ネ	NE	゛	Voiced Sound Symbol
ノ	NO	゜	Semi-voiced Sound Symbol

Fig. 10.9 Katakana-47, phonetics

	Column	0	1	2	3	4	5	6	7	8	9	A	B	C	D	E	F
	Bit Pat. →	00				01				10				11			
Row	↓	00	01	10	11	00	01	10	11	00	01	10	11	00	01	10	11
0	0000					SP		–			ソ						0
1	0001							/		ア	タ			A	J		1
2	0010									イ	チ	ヘ		B	K	S	2
3	0011									ウ	ツ	ホ		C	L	T	3
4	0100									エ	テ	マ		D	M	U	4
5	0101									オ	ト	ミ		E	N	V	5
6	0110									カ	ナ	ム		F	O	W	6
7	0111									キ	ニ	メ		G	P	X	7
8	1000									ク	ヌ	モ		H	Q	Y	8
9	1001									ケ	ネ	ヤ		I	R	Z	9
A	1010									コ	ノ	ユ	レ				
B	1011					.	¥	,					ロ				
C	1100						*			サ		ヨ	ワ				
D	1101									シ	ハ	ラ	ン				
E	1110									ス	ヒ	リ	゛				
F	1111									セ	フ	ル	゜				

Fig. 10.10 Katakana 89-graphic set

This set, outlined by heavy lines in Fig. 10.11 is provided by collapse logic (as described in Chapter 2).

	89-character
1	space
10	numerics
26	Latin alphabetics
47	Katakana alphabetics
5	Specials . , * – /

The 64-character set was sufficient for most normal data processing applications. The 89-character set was provided on 44-key electric typewriters. The 89-graphic set is shown in Fig. 10.10.

We shall learn in Chapter 18 that the assignment of Katakana in EBCDIC created complications.

	Column	0	1	2	3	4	5	6	7	8	9	A	B	C	D	E	F
	Bit Pat. →	00				01				10				11			
Row	↓	00	01	10	11	00	01	10	11	00	01	10	11	00	01	10	11
0	0000					SP		–			ソ						0
1	0001							/		ア	タ						1
2	0010									イ	チ						2
3	0011									ウ	ツ	ホ					3
4	0100									エ	テ	マ					4
5	0101									オ	ト	ミ					5
6	0110									カ	ナ	ム					6
7	0111									キ	ニ	メ					7
8	1000									ク	ヌ	モ					8
9	1001									ケ	ネ	ヤ					9
A	1010									コ	ノ	ヘ	レ				
B	1011					.	¥	,					ロ				
C	1100						*			サ		ヨ	ワ				
D	1101									シ	ハ	ラ	ン				
E	1110									ス	ヒ	リ	゛				
F	1111									セ	フ	ル	゜				

Fig. 10.11 Katakana 64-graphic set

11
The Card Code of EBCDIC

As described in Chapters 8 and 9, some 63 graphic and card hole-pattern and bit-pattern assignments had been made in EBCDIC. In Fig. 11.1, the code positions designated X indentify the hole patterns assigned in EBCDIC.

11.1 PTTC CONSIDERATIONS

In Chapter 6, it was noted that the de facto monocase card hole patterns 12-1, . . . , 12-9, 11-1, . . . , 11-9, 0-2, 0-3, . . . , 0-9 were assigned to lower-case alphabetics A, B, . . . , I, J, K, . . . , R, S, T, . . . , Z, and that new card hole patterns 12-0-1, 12-0-2, . . . , 12-0-9, 12-11-1, 12-11-2, . . . , 12-11-9, 11-0-2, 11-0-3, . . . , 11-0-9 had been assigned to upper-case alphabetics. However, as will be described in Chapter 12, a new version of the IBM 1050 terminal was being designed for the System/360, and with it, a new PTTC emerged, which reversed the assignments of lower-case and upper-case alphabetics noted above. In Chapter 8, it had been decided to locate the lower-case alphabetics in hex-columns 8, 9, and A. The card hole-pattern–to–bit-pattern assignments for EBCDIC were thus extended from those of Fig. 11.1 to those of Fig. 11.2. In Chapter 9, two hole patterns, 12-8-2 and 11-8-2, were noted and assigned to graphics ¢ and ! in hex-positions 4A and 5A, respectively.

Figure 11.2, then, shows the hole patterns assigned at this point. Where a graphic is shown in the code table, the corresponding hole pattern was assigned.

Row	Bit Pat.	Hole Pat.	0	1	2	3	4	5	6	7	8	9	A	B	C	D	E	F	Hole Pat.
	Bit Pat.		00				01				10				11				
			00	01	10	11	00	01	10	11	00	01	10	11	00	01	10	11	
		Hole Pat.														12	11	0	
0	0000		[1]	[2]	[3]	[4]	[5] X	[6] X	[7] X	[8]					[9] X	[10] X	[11] X	[12] X	
1	0001								[13] X						X	X	[14]	X	1
2	0010														X	X	X	X	2
3	0011														X	X	X	X	3
4	0100														X	X	X	X	4
5	0101														X	X	X	X	5
6	0110														X	X	X	X	6
7	0111														X	X	X	X	7
8	1000														X	X	X	X	8
9	1001														X	X	X	X	9
A	1010						X	X	[15]	X									
B	1011	8–3					X	X	X	X									
C	1100	8–4					X	X	X	X									
D	1101	8–5					X	X	X	X									
E	1110	8–6					X	X	X	X									
F	1111	8–7					X	X	X	X									
		Hole Pat.					12	11	0										

X = Assigned Hole Patterns

Hole Patterns:

[1]	[7] 11	[13] 0–1
[2]	[8]	[14]
[3]	[9] 12–0	[15]
[4]	[10] 11–0	
[5] No Pch	[11] 0–8–2	
[6] 12	[12] 0	

1	3
2	4

Block	Hole Patterns at:
1	Top and Left
2	Bottom and Left
3	Top and Right
4	Bottom and Right

Fig. 11.1 EBCDIC card code, Version 1

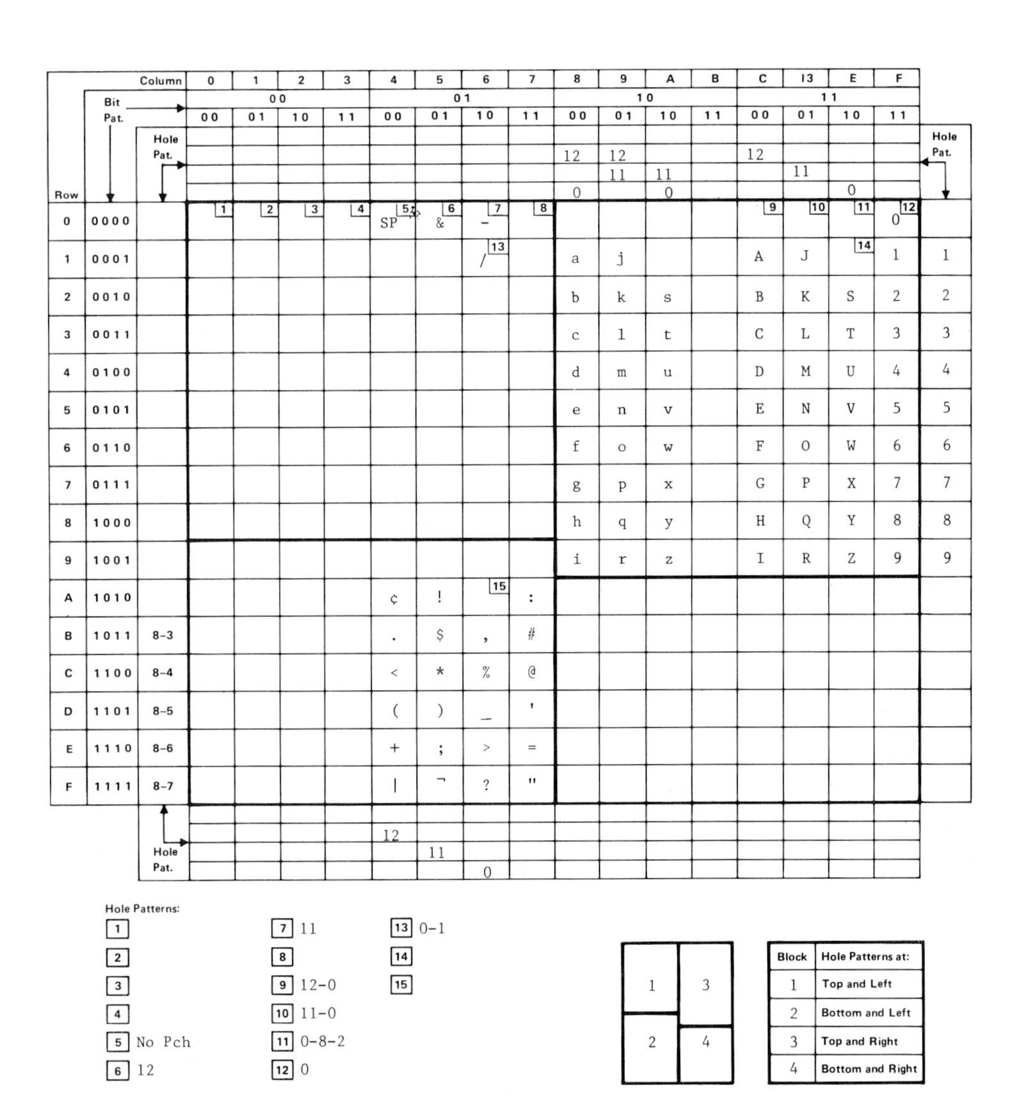

Row	Bit Pat.	Hole Pat.	0	1	2	3	4	5	6	7	8	9	A	B	C	13	E	F	Hole Pat.
		(bits)	00 00	00 01	00 10	00 11	01 00	01 01	01 10	01 11	10 00	10 01	10 10	10 11	11 00	11 01	11 10	11 11	
		(top hole pat.)									12, 0	12, 11	11, 0		12	11	0		
0	0000		[1]	[2]	[3]	[4]	[5] SP	[6] &	[7] -	[8]					[9]	[10]	[11]	[12] 0	
1	0001								[13] /		a	j			A	J	[14]	1	1
2	0010										b	k	s		B	K	S	2	2
3	0011										c	l	t		C	L	T	3	3
4	0100										d	m	u		D	M	U	4	4
5	0101										e	n	v		E	N	V	5	5
6	0110										f	o	w		F	O	W	6	6
7	0111										g	p	x		G	P	X	7	7
8	1000										h	q	y		H	Q	Y	8	8
9	1001										i	r	z		I	R	Z	9	9
A	1010						¢	!	[15]	:									
B	1011	8-3					.	$	,	#									
C	1100	8-4					<	*	%	@									
D	1101	8-5					(	)	_	'									
E	1110	8-6					+	;	>	=									
F	1111	8-7					|	¬	?	"									
		Hole Pat. (bottom)					12	11	0										

Hole Patterns:

- [1]
- [2]
- [3]
- [4]
- [5] No Pch
- [6] 12
- [7] 11
- [8]
- [9] 12-0
- [10] 11-0
- [11] 0-8-2
- [12] 0
- [13] 0-1
- [14]
- [15]

Block	Hole Patterns at:
1	Top and Left
2	Bottom and Left
3	Top and Right
4	Bottom and Right

Fig. 11.2 EBCDIC card code, Version 2

Hole Pat.																
	9	9	9	9	9	9	9	9								
	12	12	12	12					12	12	12	12				
	11	11			11	11			11	11			11	11		
	0		0		0		0		0		0		0		0	
1																
2																
3																
4																
5																
6																
7																
8																
8–1																
8–2																
8–3																
8–4																
8–5																
8–6																
8–7																

Fig. 11.3 256 hole patterns

In order to arrive at the total EBCDIC set of 256 different hole patterns, two decisions were made:

Decision 1 All 32 possible combinations of the zone punches 9, 12, 11, 0, 8 (including "no-zones") would be used.

Decision 2 With each of the 32 possible zone-punch combinations, one of the digit punches 1, 2, 3, 4, 5, 6, 7 (including "no-digits") would be used.

The logical set of 256 hole patterns is shown in Fig. 11.3.

In BCDIC, 0 had served both as a zone punch and as a digit punch for the numeric 0. Thus, in 0, 12-0, and 11-0, the 0 is regarded as a digit punch rather than a zone punch. In a sense 8 also served as both a zone punch and a digit punch. With the decision for EBCDIC that 9 would serve as a zone punch, 9 would also serve both as a zone punch and as a digit punch for the graphics 9, I, R, Z, i, r, and z.

As described in Chapter 6, in PTTC 16 hole patterns had been assigned to control characters, as shown in Fig. 11.4. It was decided to carry these assignments forward into EBCDIC. The control characters might, probably would, not be needed for EBCDIC as a CPU code, but it was sensible to preempt these hole patterns in EBCDIC, so that they could not subsequently be assigned to EBCDIC control characters that would conflict with the PTTC control characters. Besides, with the decision to attach the IBM 1050 (implementing PTTC) to System/360, it was clear that PTTC data would enter the System/360. It would be necessary to have EBCDIC bit patterns into which all PTTC bit patterns, controls, and graphics could be translated.

In Chapter 8, it was decided that the first quadrant of EBCDIC would be reserved for control characters. In consequence of this decision, the PTTC control characters would be located in the first quadrant. Therefore, zone punches 9-12, 9-11, 9-0, and 9 would be assigned to Quadrant 1.

Zone punches / Digit punches	9	9-0	9-11	9-12
4	PN	BYP	RES	PF
5	RS	LF	NL	HT
6	UC	ETB	BS	LC
7	EOT	ESC	IL	DEL

Fig. 11.4 PTTC hole patterns for control characters

11.2 TRANSLATION CONSIDERATIONS

From Fig. 11.2, it was noted that zone patterns 12, 11, 0, and "No-zone" would appear for the bottom six rows of Quadrant 2 and for the top ten rows of Quadrant 4. It was decided for purposes of reducing translation complexity (bit patterns to/from hole patterns) that the zone patterns for the *top* ten rows of Quadrant 2 should also be the zone patterns for the bottom six rows of Quadrant 4. (This decision was later slightly amended, but the spirit of it was maintained.) Fig. 11.5 represents decisions up to this point.

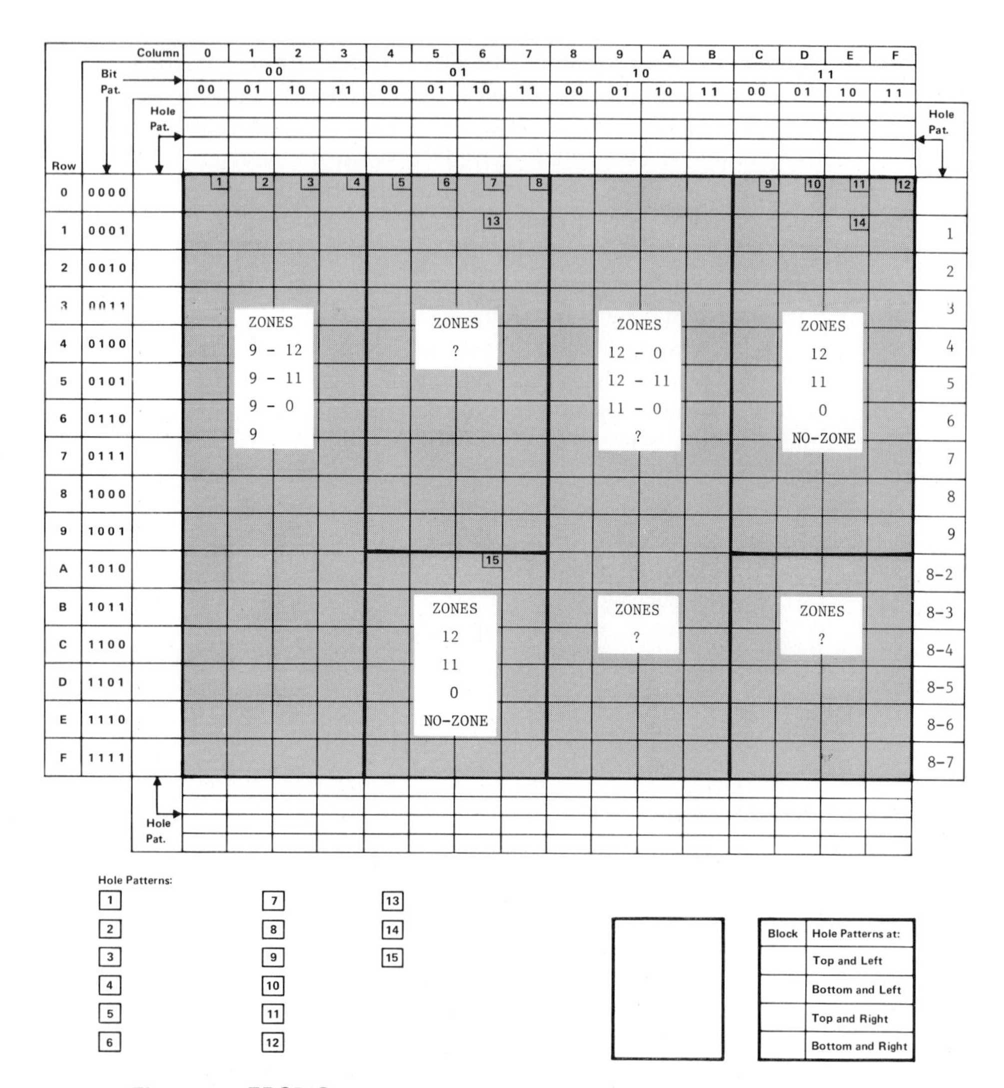

Fig. 11.5 EBCDIC card code, Version 3

This left zone patterns 12-11-0, 9-12-0, 9-12-11, 9-11-0, 9-12-11-0 unassigned. It seemed intuitive that the fourth zone pattern for Quadrant 3 should be one of these five without a 9-zone, that is, 12-11-0.

To meet the criterion above for the top ten rows of Quadrant 2 and the bottom six rows of Quadrant 4, the zone patterns 12-0, 12-11, 11-0, 12-11-0 clearly could not be assigned, because they had already been assigned to the top ten rows of Quadrant 3. Also since zone patterns

9-12, 9-11, 9-0, 9 were to be assigned to Quadrant 1 (not yet decided if to the top ten rows, the bottom six rows, or to both the top ten and the bottom six rows), they could not be assigned to the top ten rows of Quadrant 2 and the bottom six rows of Quadrant 4. This left only one choice; zones 9-12-0, 9-12-11, 9-11-0, 9-12-11-0 for the top ten rows of Quadrant 2 and for the bottom six rows of Quadrant 4. We now had Fig. 11.6.

This now left two choices:

Choice 1

- 9-12, 9-11, 9-0, 9 for the top ten rows of Quadrant 3.
- 12-0, 12-11, 11-0, 12-11-0 for the top ten rows of Quadrant 3 and the bottom six rows of Quadrant 1.

Choice 2

- 9-12, 9-11, 9-0, 9 for both the top ten and the bottom six rows of Quadrant 1.

Row	Column → Bit Pat. ↓	0	1	2	3	4	5	6	7	8	9	A	B	C	D	E	F
		00				01				10				11			
		00	01	10	11	00	01	10	11	00	01	10	11	00	01	10	11
0	0000																
1	0001																
2	0010																
3	0011																
4	0100						9-12- 0 9-12-11 9-11- 0 9-12-11- 0				12- 0 12-11 11- 0 12-11- 0				12 11 0 NO ZONE		
5	0101																
6	0110		9-12 9-11 9- 0 9														
7	0111																
8	1000																
9	1001																
A	1010																
B	1011																
C	1100						12 11 0 NO ZONE				?				9-12- 0 9-12-11 9-11- 0 9-12-11- 0		
D	1101																
E	1110																
F	1111																

Fig. 11.6 EBCDIC card code, Version 4

- 12-0, 12-11, 11-0, 12-11-0 for both the top ten and the bottom six rows of Quadrant 3.

Choice 2 posed a less complex translation relationship (hole patterns to/from bit patterns) and Choice 2 was decided. This led to Fig. 11.7.

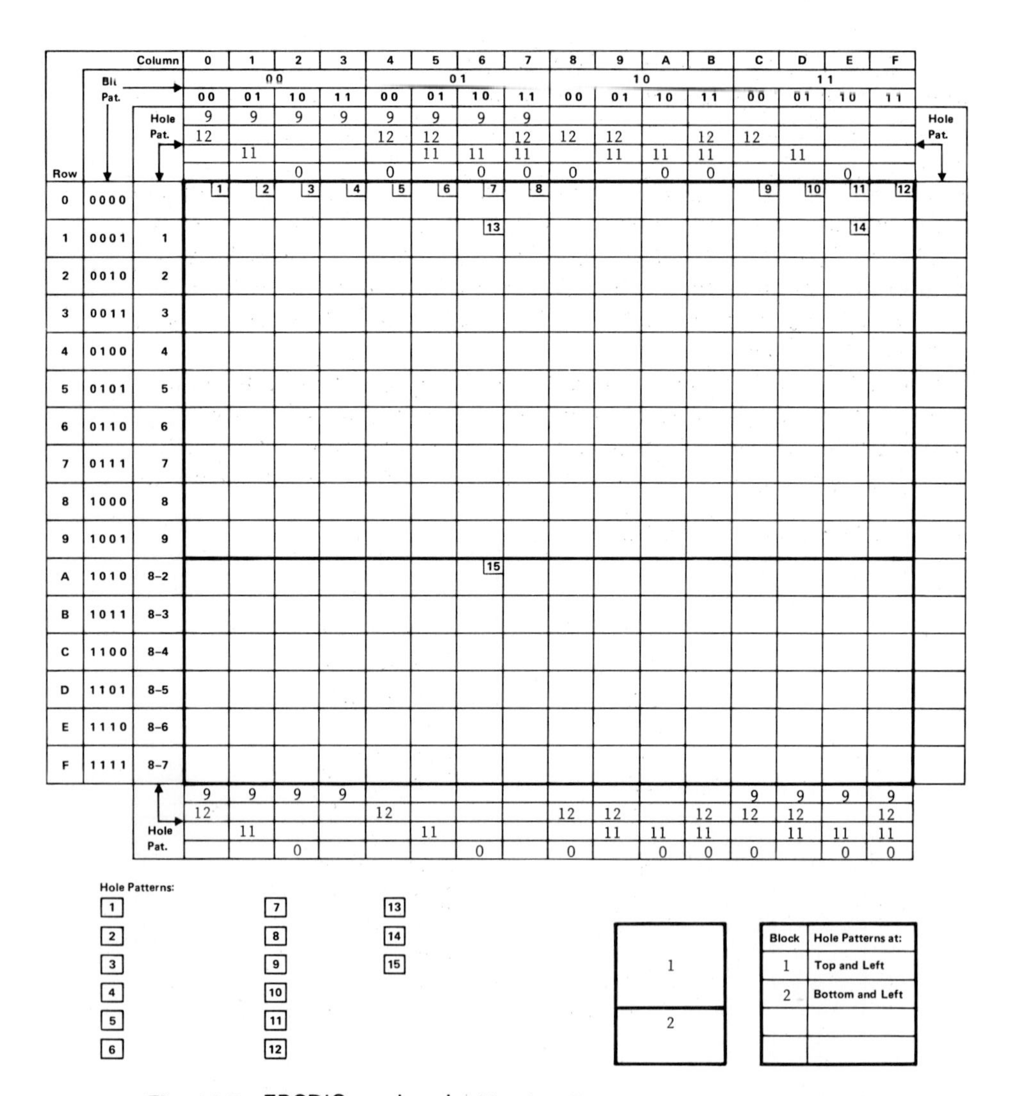

Fig. 11.7 EBCDIC card code, Version 5

11.3 8-1 VERSUS 9

It was now noted (Fig. 11.7) that certain hole patterns appeared twice: for example, 9 in hex F9 and in hex 30, 9-12 in hex C9 and hex 00. Further, missing from the set of hole patterns were zone punches combined with

Row	Bit Pat.	Hole Pat.	0	1	2	3	4	5	6	7	8	9	A	B	C	D	E	F
Column			00				01				10				11			
			00	01	10	11	00	01	10	11	00	01	10	11	00	01	10	11
		Hole Pat.	9	9	9	9	9	9	9	9								
			12				12	12		12	12	12		12	12			
				11				11	11	11		11	11	11		11		
					0		0		0	0	0		0	0			0	
0	0000		1	2	3	4	5	6	7	8					9	10	11	12
1	0001	1							13								14	
2	0010	2																
3	0011	3																
4	0100	4																
5	0101	5																
6	0110	6																
7	0111	7																
8	1000	8																
9	1001	8-1																
A	1010	8-2							15									
B	1011	8-3																
C	1100	8-4																
D	1101	8-5																
E	1110	8-6																
F	1111	8-7																
		Hole Pat.	9	9	9	9									9	9	9	9
			12				12				12	12		12	12	12		12
				11				11				11	11	11		11	11	11
					0				0		0		0	0	0		0	0

Hole Patterns:

1, 2, 3, 4, 5, 6, 7, 8, 9, 10, 11, 12, 13, 14, 15

1
2

Block	Hole Patterns at:
1	Top and Left
2	Bottom and Left

Fig. 11.8 EBCDIC card code, Version 6

8-1 hole patterns. This glitch could be fixed by applying the digit–punch combinations 8-1, rather than the digit 9, to hex row 9. The result was Fig. 11.8.

While the card code of Fig. 11.8 would lead to a translation (bit code to/from card code) of not unreasonable complexity, it was not acceptable.

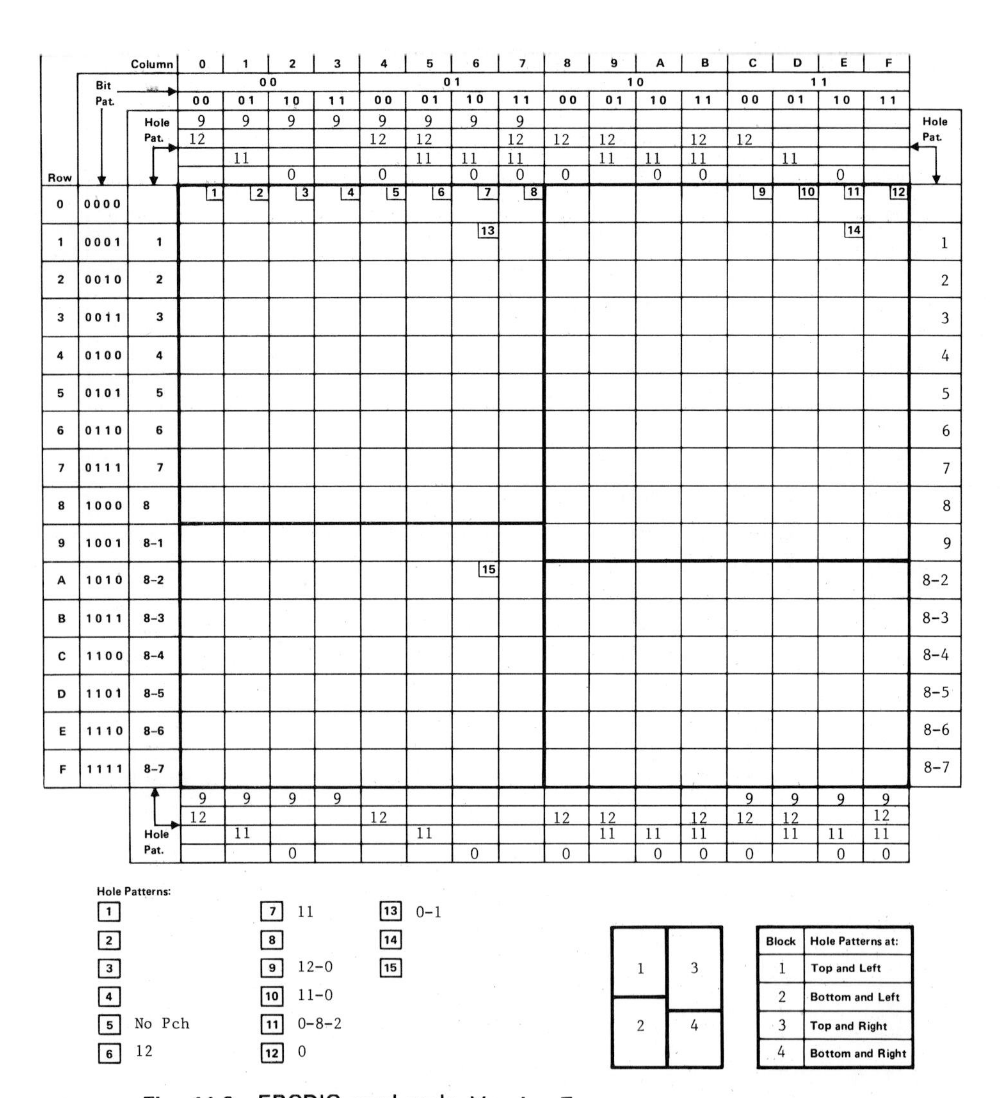

Fig. 11.9 EBCDIC card code, Version 7

The digit–punch combination 8-1 could not be assigned to hex row 9 of Quadrants 3 and 4, because i, r, z, I, R, Z, and 9 (all of which had the digit punch 9) were already assigned to that row.

But if hole patterns 9-12-0, 9-12-11, 9-11-0, 9-12-11-0, 9-12, 9-11, 9-0, and 9 are assigned to hex row 9 of Quadrants 3 and 4, then hole patterns 9-12-0-8-1, 9-12-11-8-1, 9-11-0-8-1, 9-12-11-0-8-1, 9-12-8-1, 9-11-8-1, 9-0-8-1, and 9-8-1 must be displaced. Since the hole pattern 8-1 translates in BCD the same as 9, these displaced hole patterns were assigned intuitively to hex row 9, Quadrants 1 and 2, as shown in Fig. 11.9. Note that the horizontal line is now staggered as it crosses between hex columns 7 and 8.

11.4 EXCEPTION TRANSLATIONS

As shown in Fig. 11.2, there were eight code positions with exception hole patterns. These are also noted in Fig. 11.9. These eight exception hole patterns would, of course, displace eight more hole patterns, as shown in Fig. 11.10. These exception hole patterns, if they had occupied their "theoretical" code positions in Fig. 11.9, would have occupied positions as shown as shown in Fig. 11.11.

Thus there were twelve code positions affected directly or indirectly by the exception hole patterns:

40, 50, 60, 61, 6A, 80, 90, C0, D0, E0, E1, F0

Code-table location	Exception hole patterns	Displaced hole patterns
40	No punches	9-12-0
50	12	9-12-11
60	11	9-11-0
61	0-1	9-11-0-1
C0	12-0	12
D0	11-0	11
E6	0-8-2	0
F0	0	No punches

Fig. 11.10 Exception and displaced hole patterns

Exception hole patterns	Theoretical code-table location
No punches	F0
12	C0
11	D0
0-1	E1
12-0	80
11-0	90
0-8-2	6A
0	E0

Fig. 11.11 Theoretical code-table locations

In the accommodation of the displaced hole patterns, even more hole-pattern exceptions were generated, giving rise to a total of 15, as shown in Fig. 11.12.

The card code shown in Fig. 11.12 became the EBCDIC card code. It was incorporated into IBM's Corporate System Standard CSS 2-8015-002 [11.1], later designated CSS 3-3220-002 [11.2]. The EBCDIC code chart of that time (1964 October) was completed with the assignment of the 16 control characters of PTTC (from Fig. 11.4).

11.5 A DIFFERENT BLOCKING

It was subsequently discovered that if the blocking into four blocks was done in a slightly different way, and if the four zone patterns above block 1 were amended as shown in Fig. 11.13, four of the exception translations (hole pattern to/from bit patterns) would disappear, namely those in hex positions 00, 10, 20, and 30. It is to be emphasized that while the tableau of Fig. 11.13 is different than that of Fig. 11.12, the actual translation relationship (hole patterns to/from bit patterns) is, in fact, identical for both tableaux. For both tableaux, the hole patterns for hex positions 00, 10, 20 and 30 are 9-12-0-8-1, 9-12-11-8-1, 9-11-0-8-1, 8-12-11-0-8-1, respectively.

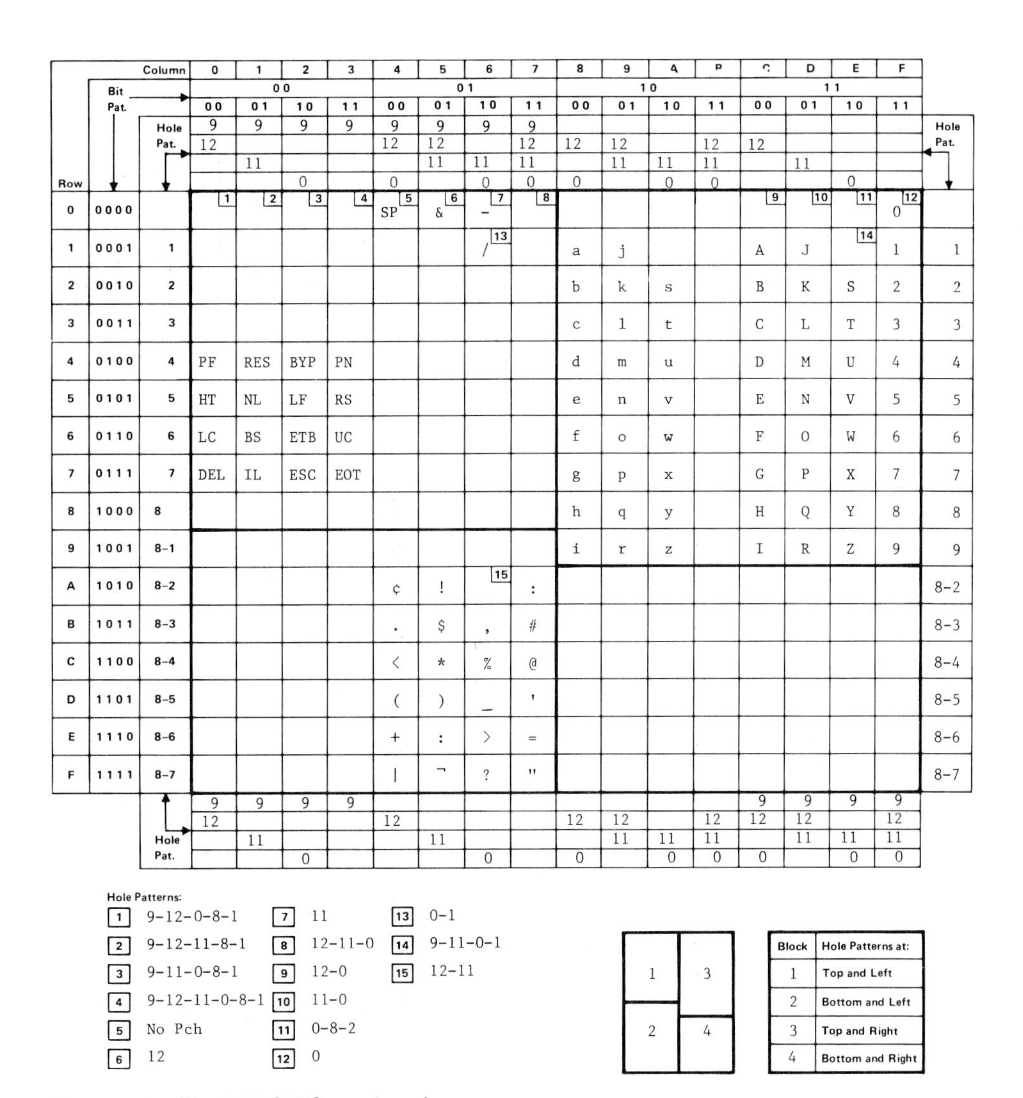

Row	Bit Pat.	Hole Pat.	0	1	2	3	4	5	6	7	8	9	A	B	C	D	E	F	Hole Pat.
	Bit Pat. (first two bits)		00				01				10				11				
	Bit Pat. (second two bits)		00	01	10	11	00	01	10	11	00	01	10	11	00	01	10	11	
		Hole Pat. (top)	9	9	9	9	9	9	9	9									
			12				12	12		12	12	12		12	12				
				11				11	11	11		11	11	11		11			
					0		0		0	0	0		0	0			0		
0	0000		[1]	[2]	[3]	[4]	SP [5]	& [6]	- [7]	[8]					[9]	[10]	[11]	0 [12]	
1	0001	1							/ [13]		a	j			A	J	[14]	1	1
2	0010	2									b	k	s		B	K	S	2	2
3	0011	3									c	l	t		C	L	T	3	3
4	0100	4	PF	RES	BYP	PN					d	m	u		D	M	U	4	4
5	0101	5	HT	NL	LF	RS					e	n	v		E	N	V	5	5
6	0110	6	LC	BS	ETB	UC					f	o	w		F	O	W	6	6
7	0111	7	DEL	IL	ESC	EOT					g	p	x		G	P	X	7	7
8	1000	8									h	q	y		H	Q	Y	8	8
9	1001	8-1									i	r	z		I	R	Z	9	9
A	1010	8-2					¢	!	[15]	:									8-2
B	1011	8-3					.	$	,	#									8-3
C	1100	8-4					<	*	%	@									8-4
D	1101	8-5					(	)	_	'									8-5
E	1110	8-6					+	;	>	=									8-6
F	1111	8-7					|	¬	?	"									8-7
		Hole Pat. (bottom)	9	9	9	9									9	9	9	9	
			12				12				12	12		12	12	12		12	
				11				11				11	11	11		11	11	11	
					0				0		0		0	0	0		0	0	

Hole Patterns:

[1]	9-12-0-8-1	[7]	11	[13]	0-1
[2]	9-12-11-8-1	[8]	12-11-0	[14]	9-11-0-1
[3]	9-11-0-8-1	[9]	12-0	[15]	12-11
[4]	9-12-11-0-8-1	[10]	11-0		
[5]	No Pch	[11]	0-8-2		
[6]	12	[12]	0		

Block	Hole Patterns at:
1	Top and Left
2	Bottom and Left
3	Top and Right
4	Bottom and Right

Fig. 11.12 Final EBCDIC card code

Row	Bit Pat.	Hole Pat.	0	1	2	3	4	5	6	7	8	9	A	B	C	D	E	F	Hole Pat.
	Bit Pat.		00				01				10				11				
			00	01	10	11	00	01	10	11	00	01	10	11	00	01	10	11	
		Hole Pat.	9	9	9	9	9	9	9	9									Hole Pat.
			12	12		12	12	12		12	12	12		12	12				
				11	11	11		11	11	11		11	11	11		11			
			0		0	0	0		0	0	0		0	0			0		
0	0000	8-1					5	6	7	8					9	10	11	12	8-1
1	0001	1							13								14		1
2	0010	2																	2
3	0011	3																	3
4	0100	4																	4
5	0101	5																	5
6	0110	6																	6
7	0111	7																	7
8	1000	8																	8
9	1001	8-1																	9
A	1010	8-2							15										8-2
B	1011	8-3																	8-3
C	1100	8-4																	8-4
D	1101	8-5																	8-5
E	1110	8-6																	8-6
F	1111	8-7																	8-7
		Hole Pat.	9	9	9	9									9	9	9	9	
			12				12				12	12		12	12	12		12	
				11				11				11	11	11		11	11	11	
					0				0		0		0	0	0		0	0	

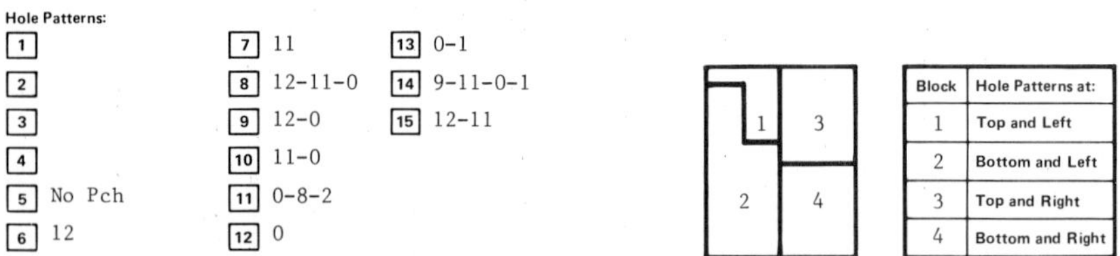

Block	Hole Patterns at:
1	Top and Left
2	Bottom and Left
3	Top and Right
4	Bottom and Right

Fig. 11.13 EBCDIC card code, modified tableau

REFERENCES

11.1 IBM Corporate Systems Standard, CSS 2-8015-002, "Extended BCD Interchange Code," 1964 October.

11.2 IBM Corporate Systems Standard, CSS 3-3220-002, "Extended BCD Interchange Code," 1968 November.

12 The New PTTC

In Chapter 6, the development of a shifted 6-bit code for paper tape and for transmission was described. In Chapter 9, how the graphic assignment of some graphics to hole patterns was changed in order to eliminate duals was described. The IBM 1050, a terminal implementing PTTC, had been designed for use with BCDIC computers. The 1050 had an associated punched card code.

12.1 A NEW 1050

A new model of the 1050 was being designed for use with the then-being-designed System/360. This new model would also implement PTTC and have an associated card code. Since some graphic–to–hole-pattern assignments had been changed between BCDIC and EBCDIC and since the new 1050 would be used with the System/360, an EBCDIC computer, it was clear that some corresponding changes would have to be made in the PTTC card code. There would have to be a new PTTC.

Since the old and new PTTC would be different, it was decided that they should be distinguished by different names. The old PTTC was designed for use in the environment of 6-bit, BCDIC computers. The new PTTC would be used in the environment of 8-bit, EBCDIC computers. Initially, then, the codes were named

PTTC/6, the PTTC for 6-Bit Environments, and
PTTC/8, the PTTC for 8-Bit Environments.

When these two names were published an unexpected confusion arose.

Some people interpreted PTTC/6 to mean a 6-bit code and PTTC/8 to mean an 8-bit code. The latter interpretation, of course, was incorrect. This confusion became manifest to such an extent that it was decided the names must be changed to eliminate the source of confusion. Eventually, the two codes were renamed:

PTTC/BCD, the PTTC for BCDIC Environments, and
PTTC/EBCD, the PTTC for EBCDIC Environments.

12.2 CRITERIA

Some criteria were established for the design of PTTC/EBCD:

Criterion 1. PTTC structure

PTTC/EBCD should have the same structure as PTTC/BCD; that is, be a shifted 6-bit code, with Space, 16 shift-independent control positions, and 94 graphic positions.

Criterion 2. PTTC/BCD compatibility

PTTC/EBCD should be as compatible as possible with PTTC/BCD.

Criterion 3. EBCDIC compatibility

The graphic–to–card-hole-pattern assignments for PTTC/EBCDIC should match those of EBCDIC.

Criterion 4. Monocase/Duocase*

There should be a monocase alphabet set of the 62 graphics of the 029 Keypunch, and a duocase alphabet set of the 88 graphics of EBCDIC.

Criterion 5. Basic/extended card code

There should be a card-code subset of 64 hole patterns that is a subset of the full set of 111 hole patterns.

---

* The Monocase Alphabet Set of PTTC/EBCD is so entitled because it contains only the capital-letter representations of the alphabet independent of whether the case shift is upper or lower. The Duocase Alphabet Set contains both capital-letter representations and small-letter representations of the alphabet. A clear distinction must be kept between the concept on the one hand of small and capital letters and the concept on the other hand of lower-case shift and upper-case shift on a typewriter-like device. Normally, small letters *are* implemented on the lower-case shift, and capital letters *are* implemented on the upper-case shift. This was the way the 1050 implemented the duocase alphabet for PTTC/BCD. But for PTTC/EBCD, the Monocase Alphabet Set was implemented on the 1050 with capital letters in *both* upper- and lower-case shift (as will be described).

As with the 1050 implementing PTTC/BCD, the 1050 implementing PTTC/EBCD used an electric typewriter as the keyboard and printer. The typewriter forced its arithmetic on the decision for the code.

12.3 TYPEWRITER ARITHMETIC

1. 44 keys, lower-case shift
2. 44 keys, upper-case shift
3. 26 keys, alphabetic in both shifts
4. 10 keys, numeric in lower-case shift, specials in upper-case shift
5. 8 keys, specials in both shifts

The structure of the code also forced its arithmetic on decisions for the code.

12.4 PTTC/EBCD ARITHMETIC

1. 47 graphic positions in lower-case shift, 3 of which would be non-printing
2. 47 graphic positions in upper-case shift, 3 of which would be non-printing
3. 1 Space position, shift independent
4. 16 control positions, shift independent
5. 64 lower-case shift positions, 17 of them shift independent
6. 64 upper-case shift positions, 17 of them shift independent
7. 111 different characters (94 shift-dependent graphic characters plus 1 shift-independent Space character plus 16 shift-independent control characters)

This structure is illustrated in Fig. 12.1.

12.5 MONOCASE AND DUOCASE SETS

In the design of the 1050, there would be two variables that were essentially independent. The first variable would be the graphic set, Monocase and Duocase (Criterion 4). The particular set in use at any particular time would be determined by which printing element the customer mounted on the 1050. The printing elements for both would have 88 printing positions. In the case of the Duocase Alphabet Set element, all 88 graphics would be different, and small and capital letters

Bit Pattern	Hole Pattern	Lower Case				Upper Case			
		SP				SP			
1									
2									
2 1									
4									
4 1			47 GRAPHICS				47 GRAPHICS		
4 2									
4 2 1									
8									
8 1									
8 2									
8 2 1									
8 4									
8 4 1			16 CONTROLS				16 CONTROLS		
8 4 2									
8 4 2 1									
	Hole Pattern								

1	3
2	4

Block	Hole Patterns at:
1	Top And Left
2	Bottom and Left
3	Top and Left
4	Bottom and Left

Fig. 12.1 PTTC structure

would be provided. In the case of the Monocase Alphabet Set, 26 printing positions would provide capital letters, and 26 *other* printing positions would *also* provide capital letters, so that there would, in fact, be $88-26=62$ different graphics.

12.6 BASIC SET AND EXTENDED SETS

The other variable would be the card-code set, which came to be called the Basic Set and the Extended Set (Criterion 5). The Extended Set would consist of 111 different hole patterns; the Basic Set would consist of 64 different hole patterns. The particular card-code set in use on the 1050 would depend on which of the two card-code features the customer had ordered.

12.7 INITIAL DECISIONS

The code-structure arithmetic spoke quickly to Criterion 5. The 64-character, Basic Card-Code Set would be assigned to both sets of 64-character upper- and lower-shift code positions.

1	space character
47	graphic characters
17	control characters
64	characters

Although it would be possible to use the Basic Card-Code Set with the Ducocase Alphabet Set (as will be described), it is more reasonable to discuss the Basic Card-Code Set in the context of the Monocase Alphabet Set.

It was decided that PTTC/EBCD should have not only the same structure as PTTC/BCD (Criterion 1) but also the same set of control characters (Criterion 2). Since the positioning of the alphabetics and numerics was implicit in the structure, the Monocase Alphabet Set would start as shown in Fig. 12.2.

As described in Chapter 6, it was decided to assign the BCDIC hole patterns for alphabetics to the lower-case shift, regardless. of whether these were small or capital letters. This decision was reviewed for PTTC/EBCD in the context of the Duocase Alphabet Set (as will be described), but in the context of the Monocase Alphabet Set and the Basic Card-Code Set, it seemed obvious that these alphabetic hole patterns should be assigned to the capital letters in both shifts.

Further, it was observed in PTTC/BCD (see Fig. 6.6) that the 8 printable graphics

@ / , . — $ &

in the lower-case shift had not changed card hole patterns between BCDIC and EBCDIC. Therefore, it was decided, in view of Criteria 2 and 3, that these specials should have the same bit patterns in

Bit Pattern	Hole Pattern	Lower Case				Upper Case			
			A	B	BA		A	B	BA
		SP				SP			
1		1		J	A			J	A
2		2	S	K	B		S	K	B
2 1		3	T	L	C		T	L	C
4		4	U	M	D		U	M	D
4 1		5	V	N	E		V	N	E
4 2		6	W	O	F		W	O	F
4 2 1		7	X	P	G		X	P	G
8		8	Y	Q	H		Y	Q	H
8 1		9	Z	R	I		Z	R	I
8 2		0	N.P.	N.P.	N.P.		N.P.	N.P.	N.P.
8 2 1									
8 4		PN	BYP	RES	PF	PN	BYP	RES	PF
8 4 1		RS	LF	NL	HT	RS	LF	NL	HT
8 4 2		UC	EOB	BS	LC	UC	EOB	BS	LC
8 4 2 1		EOT	PRE	IL	DEL	EOT	PRE	IL	DEL
	Hole Pattern								

N.P. - Non-Printing Positions

1	3
2	4

Block	Hole Patterns at:
1	Top And Left
2	Bottom and Left
3	Top and Left
4	Bottom and Left

Fig. 12.2 PTTC/EBCD Monocase Alphabet Set, Version 1

PTTC/EBCD as in PTTC/BCD and the same hole patterns as in BCDIC, EBCDIC, and PTTC/BCD.

Further, in view of Criterion 2, it was decided that the 16 control characters should have both the same bit patterns and the same hole patterns in PTTC/EBCD as in PTTC/BCD. (This decision led to the decision, as described in Chapter 11, that these 16 control characters

		Lower Case				Upper Case			
Bit Pattern →			A	B	BA		A	B	BA
	Hole Pattern →		0	11	12		0	11	12
		SP	@ [1]	-	&	SP	[1]		
1	1	1	/	J	A			J	A
2	2	2	S	K	B		S	K	B
2 1	3	3	T	L	C		T	L	C
4	4	4	U	M	D		U	M	D
4 1	5	5	V	N	E		V	N	E
4 2	6	6	W	O	F		W	O	F
4 2 1	7	7	X	P	G		X	P	G
8	8	8	Y	Q	H		Y	Q	H
8 1	9	9	Z	R	I		Z	R	I
8 2	0	0	[2]	[3]	[4]		[2]	[3]	[4]
8 2 1	8-3	#	,	$	.				
8 4	4	PN	BYP	RES	PF	PN	BYP	RES	PF
8 4 1	5	RS	LF	NL	HT	RS	LF	NL	HT
8 4 2	6	UC	EOB	BS	LC	UC	EOB	BS	LC
8 4 2 1	7	EOT	PRE	IL	DEL	EOT	PRE	IL	DEL
	↑ Hole Pattern →	9	9-0	9-11	9-12	9	9-0	9-11	9-12

Hole Patterns:

[1] 8-4
[2] 0-8-2
[3] 11-0
[4] 12-0

1	3
2	4

Block	Hole Patterns at:
1	Top And Left
2	Bottom and Left
3	Top and Left
4	Bottom and Left

Fig. 12.3 PTTC/EBCD, Version 2, Monocase Alphabet Set, Basic Card-Code Set

would be assigned in EBCDIC, and with the PTTC/EBCD hole patterns. This decision, therefore, also satisfied Criterion 3.)

Finally, for the Basic Card-Code Set, it was decided, for the nonprinting graphic code positions (in both shifts), to maintain card-code compatibility with PTTC/BCD (Criterion 2).

With these decisions, the Monocase Alphabet Set and Basic Card-Code Set shaped up as in Fig. 12.3. Blank spaces in the code table are for as yet unassigned graphics. The 64-character, Basic Card-Code Set was complete.

In the development of PTTC/BCD, as described in Chapter 6, it was decided for various reasons to assign the BCDIC card hole patterns for alphabetics to the lower-case shift, regardless of whether small or capital letters were assigned to that shift. That decision was now reviewed for PTTC/EBCD.

For the Monocase Alphabet Set, capital letters would be assigned to both lower- and upper-case shift code positions. For the Duocase Alphabet Set, small letters would be assigned to lower-case shift and capital letters to upper-case shift. (The same decision had been made for PTTC/BCD.) In assigning card hole patterns for PTTC/BCD, it had been decided at that time to assign the BCDIC hole patterns for alphabetics to *small letters* and another (related) set of hole patterns to capital letters. Criterion 2 should dictate the same decision for PTTC/EBCD. But for EBCDIC (Chapter 8) exactly the opposite had been decided. Criterion 3 should dictate the same decision for PTTC/EBCD.

12.8 FURTHER DECISIONS

Since PTTC/EBCD was being designed for a 1050 to operate with the System/360, an EBCDIC computer, it was decided that Criterion 3, EBCDIC compatibility, outweighed Criterion 2, PTTC/BCD compatibility.

It was also decided for the Duocase Alphabet Set and Extended Card-Code Set, that the eight specials

@ / . , – $ &

should have the hole patterns previously decided for the Basic Card Set, in order to ensure compatibility with EBCDIC (Criterion 3).

Since no reason could be found not to do so, it was decided to carry forward from positions ^{2}, ^{3}, and ^{4} in Fig. 12.3 the hole patterns 0-8-2, 11-0, and 12-0 for lower-case shift. This would be in accord with Criterion 2, PTTC/BCD compatibility.

With these decisions, a beginning was made on the Duocase Alphabet Set and Extended Card-Code Set for PTTC/EBCD, as shown in Fig. 12.4. Blank spaces on the code table are for as yet unassigned graphics or hole patterns.

Bit Pattern	Hole Pattern	Lower Case				Upper Case			
Bit Pattern →			A	B	BA		A	B	BA
	Hole Pattern →		11–0	12–11	12–0		0	11	12
		SP	@ [1]	– [5]	& [8]	SP			
1	1	1	/ [2]	j	a			J	A
2	2	2	s	k	b		S	K	B
2 1	3	3	t	l	c		T	L	C
4	4	4	u	m	d		U	M	D
4 1	5	5	v	n	e		V	N	E
4 2	6	6	w	o	f		W	O	F
4 2 1	7	7	x	p	g		X	P	G
8	8	8	y	q	h		Y	Q	H
8 1	9	9	z	r	i		Z	R	I
8 2	0	0	[3]	[6]	[9]				
8 2 1	8–3	# [32]	, [4]	$ [7]	. [10]				
8 4	4	PN	BYP	RES	PF	PN	BYP	RES	PF
8 4 1	5	RS	LF	NL	HT	RS	LF	NL	HT
8 4 2	6	UC	EOB	BS	LC	UC	EOB	BS	LC
8 4 2 1	7	EOT	PRE	IL	DEL	EOT	PRE	IL	DEL
	Hole Pattern	9	9–0	9–11	9–12	9	9–0	9–11	9–12

Hole Patterns:

[1]	8–4	[8]	12
[2]	0–1	[9]	12–0
[3]	0–8–2	[10]	12–8–3
[4]	0–8–3	[32]	8–3
[5]	11		
[6]	11–0		
[7]	11–8–3		

1	3
2	4

Block	Hole Patterns at:
1	Top And Left
2	Bottom and Left
3	Top and Left
4	Bottom and Left

Fig. 12.4 PTTC/EBCD, Version 1, Duocase Alphabet Set, Extended Card-Code Set

There now remained the question of 18 printable graphic positions and the 3 nonprinting graphic positions in upper-case shift. In PTTC/BCD, for the 3 nonprinting graphic positions in upper-case shift, hole patterns 12-8-7, 0-8-5, and 8-7 had been assigned. At the same time, hole patterns 8-1, 0-8-1, and 12-8-1 has been assigned to graphics in printable positions. The typewriter arithmetic referred to earlier would yield 88 printable graphic positions (hence the 6 nonprintable graphic positions in the code structure's 94 graphic positions). Of these 88 graphics, 26 would be small letters, leaving 62 graphic positions for numerics, specials, and capital letters.

		Lower Case				Upper Case			
Bit Pattern →			A	B	BA		A	B	BA
↓	Hole Pattern →		11–0	12–11	12–0		0	11	12
		X	X [1]	X [5]	X [8]	X	[22]	[26]	[29]
1	1	X	X [2]	X	X	[11]	[23]	X	X
2	2	X	X	X	X	[12]	X	X	X
2 1	3	X	X	X	X	[13]	X	X	X
4	4	X	X	X	X	[14]	X	X	X
4 1	5	X	X	X	X	[15]	X	X	X
4 2	6	X	X	X	X	[16]	X	X	X
4 2 1	7	X	X	X	X	[17]	X	X	X
8	8	X	X	X	X	[18]	X	X	X
8 1	9	X	X	X	X	[19]	X	X	X
8 2	0	X	X [3]	X [6]	X [9]	[20]	X [24]	X [27]	X [30]
8 2 1	8–3	X [32]	X [4]	X [7]	X [10]	[21]	[25]	[28]	[31]
8 4	4	X	X	X	X	X	X	X	X
8 4 1	5	X	X	X	X	X	X	X	X
8 4 2	6	X	X	X	X	X	X	X	X
8 4 2 1	7	X	X	X	X	X	X	X	X
	↑ Hole Pattern →	9	9–0	9–11	9–12	9	9–0	9–11	9–12

X - Assigned Hole Patterns

Hole Patterns:

- [1] 8–4
- [2] 0–1
- [3] 0–8–2
- [4] 0–8–3
- [5] 11
- [6] 11–0
- [7] 11–8–3
- [8] 12
- [9] 12–0
- [10] 12–8–3
- [24] 0–8–1
- [27] 12–8–1
- [30] 8–1
- [32] 8–3

1	3
2	4

Block	Hole Patterns at:
1	Top And Left
2	Bottom and Left
3	Top and Left
4	Bottom and Left

Fig. 12.5 PTTC/EBCD partial card-code assignments

During the design of EBCDIC, this same typewriter arithmetic had been reviewed in the context of the console typewriter for the System/360. It had been decided that, in EBCDIC, the 62 numerics, specials, and capital letters would be assigned to the hole patterns of the 62 interpretable graphics on the 029 Keypunch.

It was now decided that these 62 029 Keypunch hole patterns would be assigned to the 62 printable graphic positions of PTTC/EBCD referred to above. Since the set of 62 hole patterns included 12-8-7, 0-8-5, and 8-7, these three hole patterns should not be assigned in PTTC/EBCD to nonprintable positions (as they had been in PTTC/BCD); and since they did not include 8-1, 0-8-1, and 12-8-1 (assigned to printable positions in PTTC/BCD), these should not be assigned to printable positions in PTTC/EBCD. Once again, Criterion 3, EBCDIC compatibility, outweighed Criterion 2, PTTC/BCD compatibility. Hole patterns 0-8-1, 12-8-1, and 8-1 were assigned to the nonprintable positions in upper case designated by [24], [27], and [30] in Fig. 12.5; 12-8-7, 0-8-5, and 8-7 would be assigned somewhere to printable positions.

The situation on assignment for PTTC/EBCD, Duocase Alphabet Set, and Extended Card-Code Set is shown in Fig. 12.5, where X indicates code positions with assigned hole patterns. Eighteen printable code positions remained for assignment of hole patterns and graphics.

12.9 ALPHABETIC EXTENDERS

At this point another factor was taken into consideration. In EBCDIC, graphics # $ @ were designated as upper-case alphabetic extenders for European and South American countries. That is to say, on printing, display, and interpreting devices, these graphics would be replaced by alphabetics as required. For example, the German language requires 29 alphabetics—the 26 alphabetics of English-speaking countries and three more alphabetics, Ä, Ü, and Ö. On equipment designed for Germany, therefore, Ä, Ü, and Ö would replace # $ @ respectively. Also in EBCDIC, three graphics " ! ¢ were designated as lower-case alphabetic extenders, to be replaced, in Germany for example, by ä, ü, and ö.

Certainly, provision must be made in PTTC/EBCD for alphabetic extenders, both lower and upper case. The dilemma was that upper-case alphabetic extenders were to replace graphics # $ @ which had been assigned in PTTC/EBCD to *lower-case* shift. The actual assignment in PTTC/EBCD was not significant, but the *reason* for the assignment was. In accordance with the long-established U.S.A. electric typewriter practice, # $ @ were in upper-case shift. Which should take precedence, the U.S.A. electric typewriter practice or the alphabetic extender requirement for Europe? A very interesting decision was made. For the new 1050s implementing PTTC/EBCD, # $ @ and their respective hole patterns 8-3, 11-8-3, 8-4 would indeed be in lower-case shift for the U.S.A. But, for new 1050's for Europe, modifications would be made so that the hole patterns assigned to # $ @ in upper case for the U.S.A. would be in lower case for Europe.

		Lower Case				Upper Case			
Bit Pattern →			A	B	BA		A	B	BA
↓	Hole Pattern → ↓								
			@				¢		
1									
2									
2 1									
4									
4 1									
4 2									
4 2 1									
8									
8 1									
8 2									
8 2 1		#		$		"		!	
8 4									
8 4 1									
8 4 2									
8 4 2 1									
	↑ Hole Pattern →								

1	3
2	4

Block	Hole Patterns at:
1	Top And Left
2	Bottom and Left
3	Top and Left
4	Bottom and Left

Fig. 12.6 PTTC/EBCD alphabetic extender positions

Clearly, for Europe, one would require that alphabetic extender keys on the 1050 that provide capital alphabetic extenders in upper case would provide the equivalent small alphabetic extenders in lower case. Relating this to the U.S.A., the 1050 keys with # $ @ in lower case must provide " ! ¢ in lower case. This dictated, for PTTC/EBCD, that " ! ¢ be in upper-case code positions corresponding to the lower-case code positions of # $ @ . That is to say, the PTTC/EBCD code positions for # $ @ already having been assigned, " ! ¢ must be assigned as shown in Fig. 12.6.

12.10 DIFFERENCES WITH PTTC/BCD

As described in Chapter 9, hole patterns 8-7, 11-8-2, and 12-8-2 had been assigned to " ! ¢. The assignment of ! (as shown in Fig. 12.6) with its hole pattern of 11-8-2 coincidentally matched PTTC/BCD (Fig. 6.6), but the assignment of hole patterns 8-7 and 12-8-2 to graphics " ¢ (as shown in Fig. 12.6) would displace the PTTC/BCD hole patterns 0-8-7 and 11-8-7 assigned to these positions. Also, as previously noted, the assignment of 0-8-1, 12-8-1, and 8-1 to positions ^{24}, ^{27}, and ^{30} (Fig. 12.5) would displace hole patterns 12-8-7, 0-8-5, and 8-7 from PTTC/BCD positions ^{16}, ^{19}, and ^{22} as shown in Fig. 6.6.

In short, five hole patterns had been assigned to PTTC/EBCD differently than to PTTC/BCD as shown below:

Move 1. 8-7 assigned to ^{21}, displacing 0-8-7

Move 2. 12-8-2 assigned to ^{22}, displacing 11-8-7

Move 3. 8-1 assigned to ^{30}, displacing 8-7

Move 4. 0-8-1 assigned to ^{24}, displacing 12-8-7

Move 5. 12-8-1 assigned to ^{27}, displacing 0-8-5

(Code table position references above are in respect to Fig. 12.5)

12.11 "MUSICAL-CHAIRS" PHENOMENON

These moves, by the "musical chairs" phenomenon, necessarily led to further moves as shown below:

Move 6. 12-8-7, displaced by Move 4, replaced 0-8-1 in ^{25}, moved from ^{25} by Move 4

Move 7. 11-8-7, displaced by Move 2, replaced 12-8-1 in ^{31}, moved from ^{31} by Move 5

Move 8. 0-8-7, displaced by Move 1, replaced 12-8-2 in [23], moved from [23] by Move 2

Move 9. 0-8-5, displaced by Move 5, replaced 0-8-6 in [26]

Move 10. 0-8-6, displaced by Move 9,* replaced 8-1 in [17], moved from [17] by Move 3

Of course, these code positions, having hole patterns assigned to them by these moves, also took their EBCDIC graphics with them under Criterion 3, EBCDIC compatibility.

No further moves were made. Under Criterion 2, PTTC/BCD compatibility, the remainder of the code positions in upper-case shift took the hole patterns from PTTC/BCD (Fig. 6.6 in Chapter 6), but under

	PTTC/BCD		PTTC/EBCD		
Hole pattern	Graphic	Code position (Figure 6.6)	Graphic	Code position (Figure 12.8)	EBCDIC graphic
8-6	>	3	=	11	=
12-8-4	¤ or)	4	<	12	<
11-8-6	;	5	;	13	;
8-2	ƀ	6	:	14	:
0-8-4	% or (	7	%	15	%
8-5	:	8	'	16	'
11-8-4	*	10	*	18	*
12-8-5	[	11	(	19	(
11-8-5	]	12	)	20	)
11-8-2	!	20	!	28	!
12-8-6	<	21	+	29	+

Fig. 12.7 PTTC/BCD compatibility

* The author cannot recall why, in Move 9, 0-8-5 displaced 0-8-6 in code position [26]. It would seem to have been reasonable for 0-8-5 to have replaced 8-1, moved from code position [17] in Move 3. Such a move would then have completed the "musical chairs" moves. However, Move 9 *was* made, for whatever reason, and led to Move 10, which *did* complete the moves.

Criterion 3, EBCDIC compatibility, they took for those hole patterns the EBCDIC graphics as shown in Fig. 12.7.

This, then, completed the assignment for the Duocase Alphabet Set and Extended Card-Code Set for PTTC/EBCD, as shown in Fig. 12.8.

Bit Pattern ↓	Hole Pattern ↓	Lower Case				Upper Case			
Bit Pattern →			A	B	BA		A	B	BA
	Hole Pattern →		11-0	12-11	11-0		0	11	12
		SP	@ [1]	– [5]	& [8]	SP	¢ [22]	_ [26]	+ [29]
1	1	1	/ [2]	j	a	= [11]	? [23]	J	A
2	2	2	s	k	b	< [12]	S	K	B
2 1	3	3	t	l	c	; [13]	T	L	C
4	4	4	u	m	d	: [14]	U	M	D
4 1	5	5	v	n	e	% [15]	V	N	E
4 2	6	6	w	o	f	' [16]	W	O	F
4 2 1	7	7	x	p	g	> [17]	X	P	G
8	8	8	y	q	h	* [18]	Y	Q	H
8 1	9	9	z	r	i	([19]	Z	R	I
8 2	0	0	[3]	[6]	[9]	) [20]	[24]	[27]	[30]
8 2	8-3	# [32]	, [4]	$ [7]	. [10]	" [21]	| [25]	! [28]	¬ [31]
8 4	4	PN	BYP	RES	PF	PN	BYP	RES	PF
8 4 1	5	RS	LF	NL	HT	RS	LF	NL	HT
8 4 2	6	UC	EOB	BS	LC	UC	EOB	BS	LC
8 4 2 1	7	EOT	PRE	IL	DEL	EOT	PRE	IL	DEL
	↑ Hole Pattern →	9	9-0	9-11	9-12	9	9-0	9-11	9-12

Hole Patterns

[1] 8-4	[8] 12	[15] 0-8-4	[22] 12-8-2	[29] 12-8-6
[2] 0-1	[9] 12-0	[16] 8-5	[23] 0-8-7	[30] 8-1
[3] 0-8-2	[10] 12-8-3	[17] 0-8-6	[24] 0-8-1	[31] 11-8-7
[4] 0-8-3	[11] 8-6	[18] 11-8-4	[25] 12-8-7	[32] 8-3
[5] 11	[12] 12-8-4	[19] 12-8-5	[26] 0-8-5	
[6] 11-0	[13] 11-8-6	[20] 11-8-5	[27] 12-8-1	
[7] 11-8-3	[14] 8-2	[21] 8-7	[28] 11-8-2	

1	3
2	4

Block	Hole Patterns at:
1	Top And Left
2	Bottom and Left
3	Top and Left
4	Bottom and Left

Fig. 12.8 PTTC/EBCD, Duocase Alphabet Set, Final Version, Extended Card-Code Set

The Basic Card-Code Set had been completed (see Fig. 12.3), but the Monocase Alphabet Set had not. The Monocase Alphabet Set could now be completed since it, as previously described, would be different from the Duocase Alphabet Set only in that capital letters would appear in both lower- and upper-case shift. That is to say, the specials in upper-case shift for the Duocase Alphabet Set (Fig. 12.8) would also

		Lower Case				Upper Case			
Bit Pattern →			A	B	BA		A	B	BA
↓	Hole Pattern →		0	11	12		0	11	12
		SP	@ [1]	-	&	SP	¢ [1]	_	+
1	1	1	/ [2]	J	A	=	? [2]	J	A
2	2	2	S	K	B	<	S	K	B
2 1	3	3	T	L	C	;	T	L	C
4	4	4	U	M	D	:	U	M	D
4 1	5	5	V	N	E	%	V	N	E
4 2	6	6	W	O	F	'	W	O	F
4 2 1	7	7	X	P	G	>	X	P	G
8	8	8	Y	Q	H	*	Y	Q	H
8 1	9	9	Z	R	I	(	Z	R	I
8 2	0	0	N.P.	N.P.	N.P.	)	N.P.	N.P.	N.P.
8 2 1	8-3	#	,	$	.	"	|	!	¬
8 4	4	PN	BYP	RES	PF	PN	BYP	RES	PF
8 4 1	5	RS	LF	NL	HT	RS	LF	NL	HT
8 4 2	6	UC	EOB	BS	LC	UC	EOB	BS	LC
8 4 2 1	7	EOT	PRE	IL	DEL	EOT	PRE	IL	DEL
	↑ Hole Pattern →	9	9-0	9-11	9-12	9	9-0	9-11	9-12

N. P. - Non-Printing

Hole Patterns:
[1] 8-4
[2] 0-1

1	3
2	4

Block	Hole Patterns at:
1	Top And Left
2	Bottom and Left
3	Top and Left
4	Bottom and Left

Fig. 12.9 PTTC/EBCD, Final Version, Monocase Alphabet Set, Basic Card-Code Set

appear in upper-case shift for the Monocase Alphabet Set. The final version of PTTC/EBCD Monocase Alphabet Set and Basic Card-Code Set, is shown in Fig. 12.9.

At this point it should be emphasized that the 64-character Basic Card-Code Set and the 63-character Monocase Alphabet Set do not consist of the same set of characters. The Basic Card-Code Set consists of the hole patterns for the Space character, 47 graphic characters (3 nonprinting), and 16 control characters. The Monocase Alphabet Set consists of the Space character (in upper- and lower-case shift), 10 numerics in lower-case shift, 8 specials in lower-case shift, 18 specials in upper-case shift, and 26 capital letters in upper- and lower-case shift.

12.12 INTERACTIONS, BASIC AND EXTENDED SETS

Some examples are now given to illustrate interactions between the Basic and Extended Card-Code Sets and the Monocase and Duocase Alphabet Sets. Applications are straightforward when the 1050 is configured either with the Basic Card-Code Set feature and the Monocase Alphabet Set or with the Extended Card-Code Set and the Duocase Alphabet Set. Other combinations were possible, such as the Basic Card-Code Set feature with the Duocase Alphabet Set, but ingenuity and a knowledge of the codes was necessary in such cases in order to make the 1050 produce the desired result.

In the examples that follow, the "desired result" is a line of printed characters, whether produced by reading a punched card on the 1050 card reader or produced by the operator keying from the 1050 keyboard.

Example 1

This example illustrates the use of the 1050 Basic Card-Code Set feature to drive the Monocase Alphabet Set, with capital letters, but with lower-case shift only; that is, no shift characters are required.

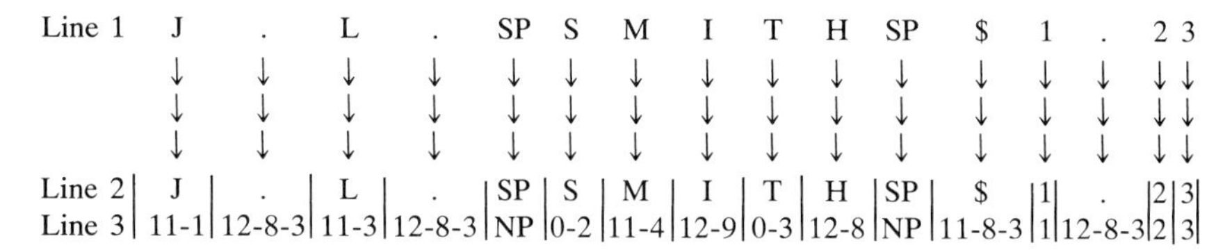

Note: NP means No Punches. SP means Space

Line 1 is the print line—J. L. Smith $1.23.
Line 2 are the Basic Card-Code Set characters.
Line 3 are the Basic Card-Code Set hole patterns.

Example 2

This example illustrates the use of the 1050 keyboard to drive the Monocase Alphabet Set, with capital letters but with lower-case shift only; that is, no shifting is required during the keyboard operation.

Line 1	J	.	L	.	SP	S	M	I	T	H	SP	$	1	.	2	3
Line 2	J	.	L	.	SP	S	M	I	T	H	SP	$	1	.	2	3
Line 3	B1	A821	B21	A821	SP	A2	B4	BA81	A21	BA8	SP	B821	1	A821	2	3

Line 1 is the print line—J. L. Smith $1.23.
Line 2 are the keys.
Line 3 are the PTTC bit patterns.

Example 3

This example illustrates the use of the 1050 Basic Card-Code Set feature to drive the Monocase Alphabet Set, with capital letters only, but with upper- and lower-case shift.

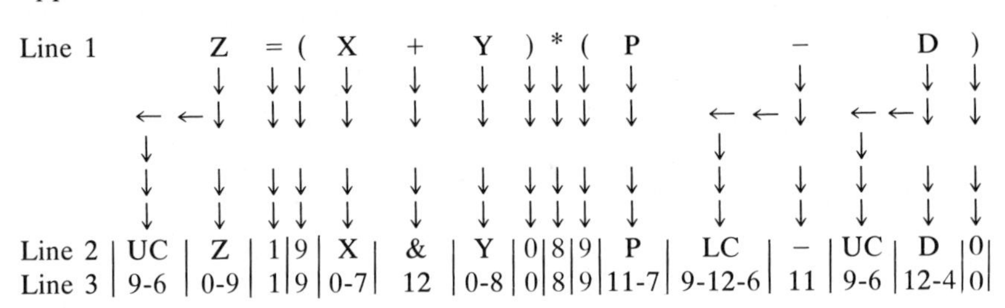

Line 1 is the print line—Z=(X+Y)*(P−D).
Line 2 are the Basic Card-Code Set characters.
Line 3 are the Basic Card-Code Set hole patterns.

Example 4

This example illustrates the use of the 1050 keyboard to drive the Monocase Alphabet Set, with capital letters only, but with upper- and lower-case shift.

Line 1		Z	=	(	X	+	Y	)	*	(	P		−		D	)
Line 2	UC	Z	=	(	X	+	Y	)	*	(	P	LC	−	UC	D	)
Line 3	842	A81	1	1	A421	BA	A8	82	8	81	B421	BA842	B	842	BA4	82

Line 1 is the print line—Z=(X+Y)*(P−D).
Line 2 are the keys.
Line 3 are the PTTC bit patterns.

Example 5

This example illustrates the use of the Basic Card-Code Set feature to drive the Duocase Alphabet Set, with small and capital letters.

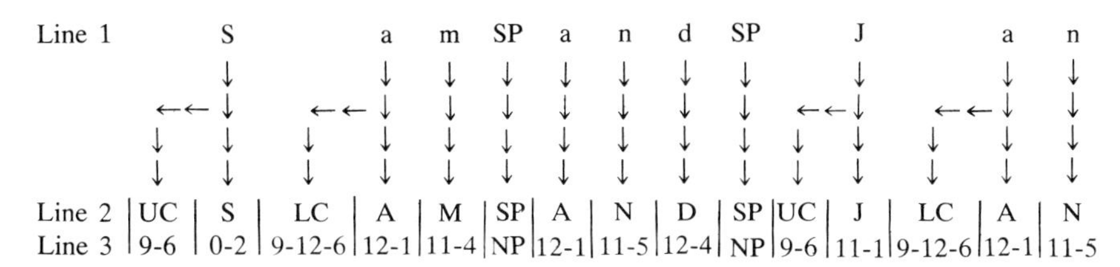

Line 1 is the print line—Sam and Jan.
Line 2 are the Basic Card-Code Set characters.
Line 3 are the Basic Card-Code Set hole patterns.

Example 6

This example illustrates the use of the 1050 keyboard to drive the Duocase Alphabet Set, with small and capital letters.

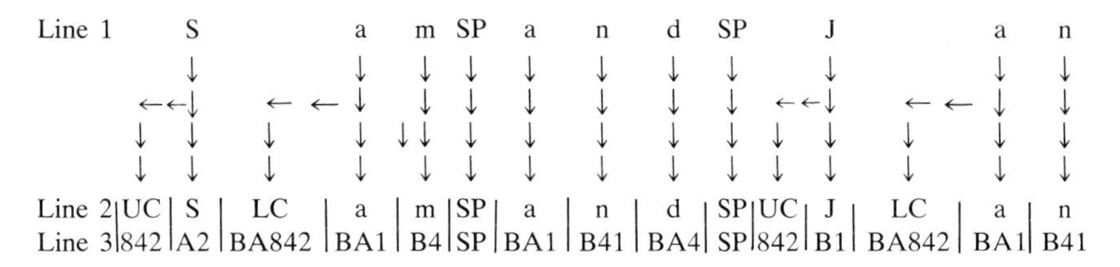

Line 1 is the print line—Sam and Jan.
Line 2 are the keys.
Line 3 are the PTTC bit patterns.

Example 7

This example illustrates the use of the Extended Card-Code Set feature to drive the Duocase Alphabet Set, with small and capital letters. Shift characters are not required.

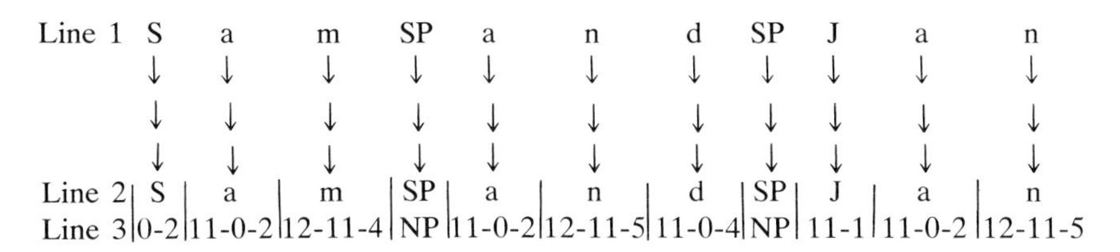

Line 1 is the print line—Sam and Jan.
Line 2 are the Extended Card-Code Set characters.
Line 3 are the Extended Card-Code Set hole patterns.

12.13 PTTC AND EBCDIC

One final aspect of PTTC/EBCD needs to be covered—the translation of the shifted 6-bit PTTC into the unshifted 8-bit EBCDIC.

The translation of a shifted code into an unshifted code, and vice versa, is an interesting problem. For example, a and A both have the same bit pattern, BA1, in PTTC/EBCD, but have different bit patterns, 10000001 and 11000001, respectively, in EBCDIC. Of course, whether BA1 means a or A in PTTC/EBCD depends on whether it was preceded in the data stream by BA842, lower case, or by 842, upper case. Another complication is that in the EBCDIC data stream equivalent to a PTTC/EBCD data stream, the UC and LC bit patterns should not be present.

The solution, for the 1050 and System/360, was to transform the PTTC/EBCD data stream first into an intermediate 8-bit data stream, with shift characters replaced by "shift bits" in each 8-bit byte, and then to translate this string of 8-bit bytes into a string of EBCDIC bytes. The transformation process from shifted to intermediate form was effected by hardware (by the IBM 2701 Data Adapter Unit, which stood between the data transmission lines and the System/360). The translation from the intermediate form, which was called the "System/360 Oriented Form" in IBM literature, was effected, if necessary, by software in the System/360.

PTTC/EBCD as actually transmitted was a 9-bit byte as follows:

Start B A 8 4 2 1 C Stop

The start–stop bits are deleted by the 2701 on receive operations and inserted on transmit operations. The C-bit is a "check bit," actually an odd-parity check.

The transformation process of the resultant 7-bit byte (start and stop bits deleted) into the 8-bit "System/360 Oriented Form" proceeded as follows:

S B A 8 4 2 1 C

As the data stream goes through the transformation process, UC bit patterns are removed from the data stream, and a one-bit is set into the S bit position of each succeeding 8-bit byte until an LC bit pattern is detected (and removed from the data stream). Then a zero-bit is set into the S bit position of each succeeding 8-bit byte until a UC bit pattern is detected, and so on.

The bit positions of a System/360 8-bit byte are numbered as follows:

0 1 2 3 4 5 6 7

Then the S-bit as described above, the 6 bits BA9421, and the check-bit C of the PTTC/EBCD byte are set into the System/360 byte as follows:

S	B	A	8	4	2	1	C
↓	↓	↓	↓	↓	↓	↓	↓
↓	↓	↓	↓	↓	↓	↓	↓
0	1	2	3	4	5	6	7

The resulting code table, PTTC/EBCD, System/360 Oriented, is shown in Fig. 12.10. The six shaded code positions come from the six nonprinting code positions of PTTC/EBCD.

The System/360 Oriented Form of PTTC/EBCD could then be translated (if necessary) by software. The "if necessary" aspect should be noted. There are applications, store and forward, for example, where translation into EBCDIC would be unnecessary.

	Column	0	1	2	3	4	5	6	7	8	9	A	B	C	D	E	F
	Bit Pat.	00				01				10				11			
Row		00	01	10	11	00	01	10	11	00	01	10	11	00	01	10	11
0	0000		8	@		-			h		*	¢		_			H
1	0001	SP			y		q	&		SP			Y		Q	+	
2	0010	1			z		r	a		=			Z		R	A	
3	0011		9	/		j			i		(	?		J			I
4	0100	2						b		<						B	
5	0101		0	s		k					)	S		K			
6	0110		#	t		l			.		"	T		L			¬
7	0111	3			,		$	c		;			|		!	C	
8	1000	4			BYP		RES	d		:			BYP		RES	D	
9	1001		PN	u		m			PF		PN	U		M			PF
A	1010		RS	v		n			HT		RS	V		N			HT
B	1011	5			LF		NL	e		%			LF		NL	E	
C	1100		UC	w		o			LC		UC	W		O			LC
D	1101	6			EOB		BS	f		-			EOB		BS	F	
E	1110	7			PRE		IL	g		>			PRE		IL	G	
F	1111		EOT	x		p			DEL		EOT	X		P			DEL

Fig. 12.10 PTTC/EBCD, System/360 Oriented Form

A similar, but opposite, process took place on transmitting from the System/360 to a 1050. A string of bytes of the System/360 Oriented Form was processed through the 2701. The first byte was inspected for its S-bit. If S-bit is one, a UC bit pattern is injected. If S-bit is zero, an LC bit pattern is injected into the data stream. S-bits of succeeding bytes were inspected for a *change*: if a change was from zero to one, a UC bit pattern is injected; if a change was from one to zero, an LC bit pattern is injected into the data stream. For all bytes, the S-bit was deleted, yielding PTTC/EBCD bytes. Of course, start–stop bits were appended to each PTTC/EBCD byte before transmission from the 2701.

12.14 DIFFERENCES, PTTC/BCD AND PTTC/EBCD

As has been pointed out, there were a number of differences between PTTC/BCD and PTTC/EBCD. These stemmed mainly from changes going from BCDIC to EBCDIC and from different principles for assignment of card hole patterns to the small and capital alphabetics. Before the design of either PTTC, the well-established hole patterns for alphabetics had been assigned to capital letters, the only kind of letters then available on monocase data processing equipment. The use of a duocase electric typewriter for PTTC/BCD and for PTTC/EBCD introduced the capability for both small and capital letters. The principles established for PTTC/BCD and for PTTC/EBCD were as follows:

PTTC/BCD. The card hole patterns previously associated with monocase alphabetics will be assigned to the lower-case shift (of the typewriter), regardless of whether small or capital letters appear in that case shift.

PTTC/EBCD. The card hole patterns previously associated with monocase alphabetics will be assigned to capital letters, regardless of whether capital letters appear in upper- or lower-case shift.

The principle for PTTC/BCD had as its objective maximum simplicity of the logic circuitry between the keys of the keyboard and the hole patterns of the punched card. The principle for PTTC/EBCD had as its objective unvarying (for the future) hole patterns for small and capital letters, even though this would increase the complexity of the logic circuitry between the keys of the keyboard and the hole patterns of the punched card.

13 The Size and Structure of ASCII

During the late 1950s, the need was recognized for a standard code for the communications industry not only in the U.S.A. but also in Europe and in Japan. Internationally, the development work was carried out in ISO/TC97/SC2. In the U.S.A., the work started under the auspices of E.I.A. (Electronic Industries Association). With the formation of the X3 Committee under the auspices of the A.S.A. (American Standards Association) to develop standards for the data processing industry, the X3.2 Subcommittee was established to develop a standard code, standard media (magnetic tape, punched cards, paper tape), and the representation of the standard code on those media.

13.1 NAME OF THE CODE

Since a code was to be developed as an American Standard, it would be called the American Standard Code. It was thought well to qualify the name of the code, according to its purpose, and it came to be titled the American Standard Code for Information Interchange. From the initials of this title emerged the acronym ASCII. Later the American Standards Association changed its name to the United States of American Standards Institute (U.S.A.S.I.). The code then came to be titled the United States of America Standard Code for Information Interchange, from which emerged the acronym USASCII. Both acronyms, ASCII and USASCII, enjoyed currency and were eventually written into the standard itself as co-equal "standard" acronyms. Then U.S.A.S.I. once again changed its name, to the American National Standards Institute (A.N.S.I.). Needless to say, the code again changed its name, to the American National

Standard Code for Information Interchange. However, the suggestion that the third acronym, ANSCII, be adopted, met with opposition, and was rejected. The code is now commonly referred to as ASCII, the acronym USASCII having fallen into disuse.

During the initial development of ASCII, the developers went through the same process that the developers of PTTC/BCD went through (see Chapter 6). The first thing to determine was the functional requirements of the code, how many graphic characters and how many control characters. It was at this time that the American Telephone and Telegraph Company stated its official requirements on the code. There should be an all-zeros character, Null, and an all-ones character, Delete.

13.2 GRAPHIC REQUIREMENTS

The standards committee first tackled the question of graphic characters. Existing codes were studied.

CCITT #2 had 26 alphabetics, 10 numerics, 3 code positions for national use, and 11 specials

. , : = ( ) ′ + ? / −

for a total of 50 graphics.

The Western Union Telegraph Company, using equipment from the Teletype Corporation, had substituted

; ″ for = +

The punched card code of the day commonly (there were some variations) provided 52 graphics; 26 alphabetics, 10 numerics, 6 unique specials . , * $ / − and 10 specials as duals:

% ⌑ & # @
( ) + = ′

Fieldata, a code developed by the United States Army (later to become a military standard) for telecommunications, had 10 numerics, 26 alphabetics, and 19 specials:

. , * $ / − ( ) + =
_ < > ; : ? ! ′ ″

BCDIC had 10 numerics, 26 alphabetics, and 32 specials (5 dual pairs)

. , * / $ − \ % (⌑) & + # = @
: ; ! ? [ ] ǂ ⧧ ⧻ ƀ γ Δ √ ′ < >

for a total of 68. Of these 68 specials, 7 were for the representation of control functions, leaving 25 as true graphics.

It seemed, therefore, that widely used codes of the day had a requirement for 10 numerics, 26 alphabetics, and from 11 to 25 specials. A total across these codes comes out as follows:

Punctuation and correspondence	, , ; : ! ? ′ ″ _	9
Bracketing	( ) [ ]	4
Commerical	& % # @ % ¤	6
Mathematical	+ − = * / \ < >	8
		27

From this preliminary survey then, there appeared to be a requirement for at least 46 graphics and maybe for as many as 64 graphics. Other graphics in wide use were fractions $\frac{1}{4}$ and $\frac{1}{2}$, commonly provided on electric typewriters, and small letters (as well as capital letters).

13.3 CONTROL FUNCTION REQUIREMENTS

It began to appear that upward of 64 graphics should be provided in the standard code for information interchange. The standards committees were also studying the requirements for control characters.

CCITT #2 had provided 7:

Space	Letter Shift
Carriage Return	Answer Back
Line Feed	Audible Signal
Figure Shift	

Fieldata had provided 9 specific control characters and code positions for 64 unspecified control characters:

Master Space	Space
Upper Case	Stop
Lower Case	Special
Line Feed	Idle
Carriage Return	

BCDIC provided 7:

Record Mark	Mode Change
Group Mark	Word Separator
Tape Mark	Substitute Blank
Segment Mark	

The standards committee working on data transmission was studying the question of characters purely for data transmission control. A requirement for about 10 data transmission control characters seemed to be emerging.

13.4 MORE THAN 64 CHARACTERS!

Putting the two tentative requirements for graphic and control characters together, one fact seemed to be very clear. More than 64 characters would be required for a code to span the needs of computing and of communications. The figure 64 was a key figure because it pointed at a code of more than 6 bits—at least 7 bits. This fact was very significant because nearly all the computers of the day had essentially a 6-bit architecture. In order to implement the standard code for information interchange, therefore, it was very desirable that it be 6 bits (or less). But try as it could, the standards committee could not reduce the character requirement to 64 or less.

13.5 SHIFTED CODES

At this point, the possibility of a shifted or precedence code was raised. The concept is explained in Chapter 2. The world-wide telegraphic code, CCITT #2 (see Chapter 3), was a shifted 5-bit code. IBM had made a decision to provide a shifted 6-bit code (see Chapter 6, the Size and Structure of PTTC).

The great virtue of a shifted code is the capability of providing more characters than the byte size of a code would normally permit. The formula (given in Chapter 2) for the number of different characters is

$$2^{x+1} - y$$

where x is the number of bits in the code byte and y is the number of shift-dependent characters.

CCITT #2, a shifted 5-bit code, provided 58 different characters. PTTC, a shifted 6-bit code, provided 111 different characters. It appeared, therefore, that the character requirements for the standard code for information interchange would be accommodated by a shifted 6-bit code.

A strong argument arose against adopting the concept of a shifted or precedence code for the code. In those days, telecommunication lines were not wonderfully reliable. A phenomenon known as a "hit" occurred not infrequently. When a one-bit was hit, it turned into a zero-bit. When a zero-bit was hit, it turned into a one-bit. If an individual graphic bit pattern was hit, the individual graphic would be garbled, but the word in which it appeared would generally be intelligible to a human. For example, suppose the bit pattern 11101 for Q was hit and turned into 11001, the bit pattern for W. Then a word REQUIRE would be received and printed as REWUIRE. But, from context of the sentence and message in which REWUIRE appeared, it would usually be possible for the recipient to reason out that REQUIRE had been intended.

If the graphic bit pattern that was hit was a numeric, and it was in consequence turned into another numeric or into a special, it was virtually impossible to reason out from context what the numeric had been. Indeed, if a numeric was changed into another numeric, it was not even evident to a human reader that a hit had taken place. To compensate for this, telegraphists would commonly take all numerics that had occurred in a telegram and rekey them in sequence at the end of the message. Provided there were no hits on this sequence of numerics, a comparison by the recipient showed what numerics, if any, had been hit in the message.

Consider an example using CCITT #2 with its two shift characters, FS for Figure Shift and LS for Letter Shift. Consider a data stream

| LS | X | X | X | FS | X | X | LS | X | X | X | X | FS | X | X |, etc.

where X stands for a 5-bit graphic bit pattern.

If a hit occurred in the bit pattern of either a Figure Shift character itself or a Letter Shift character itself, the message would generally be so garbled as to be incomprehensible, and retransmission would have to be requested.

There were two situations then. If a graphic bit pattern were hit, the individual graphic would be garbled, but could sometimes be reasoned out. If a Figure Shift or Letter Shift bit pattern were hit, the message or a portion of the message was generally incomprehensible.

The first controversy on the standards committee was the economy of a shifted code versus the potential occasional garble of a message on the telecommunication lines. Giving more weight to reliability than to cost, the standards committee decided against the concept of a shifted or precedence code. (Interestingly, much later, the committees nevertheless did decide to place two shift characters in the code, Shift In and Shift Out.)

13.6 7 BITS OR 8 BITS?

However, at that time, the consequence of the decision against a shifted code was that the code would apparently require at least 7 bits. At this point, it was proposed that the code should be 8 bits. The central basis of this proposal was the efficiency of representation of numerics. Each of the 10 numerics can be represented by 4 bits (the BCD representation). The 26 alphabetics can be represented by 5 bits. A 36-character code set consisting of the numerics and alphabetics requires 6 bits. But consider. If 4 bits *can* represent numerics, but numerics *are* represented by 6 bits, then there are 2 bits of "overhead." Two bits more than strictly necessary are used to represent numerics, and this is clearly inefficient. The alphabetics also, then, are inefficiently represented, with 1 bit of overhead.

At the time the standard code was being developed, it was estimated that 75 percent of the data in data processing operations was numeric data. In short, 75 percent of the data was inefficiently represented. And now it was being suggested on the standards committees to use a 7-bit code. This would bring about 3 bits of overhead for numerics and 2 bits of overhead for alphabetics—even more inefficiency than in 6-bit representation.

Into an 8-bit byte, two 4-bit bit patterns can be packed; that is, two numerics can be represented in an 8-bit byte. And there is zero overhead. An 8-bit byte provides optimum efficiency of representation of numeric data. Of course, the consequent 3 bits of overhead for alphabetics is more inefficient than the 2 bits of a 7-bit representation.

The argument for an 8-bit byte, therefore, was that numeric data, 75 percent of all data, could be represented with optimum efficiency.

There were arguments against an 8-bit byte. One argument was a cost argument. In those days of relay logic and vacuum-tube logic, a "bit" cost an appreciable amount. Seven-bit registers were appreciably more costly than 6-bit registers, and 8-bit registers were appreciably more costly than 7-bit registers. Also, given a data communications line speed of a fixed number of bits per second, it would take more time to transmit 1000 8-bit characters than 1000 7-bit characters. And time of use of data communication lines bears directly on cost of use of the lines.

Another argument bore on the reliability of perforated tape. A common perforated tape of the day was 1-inch, 8-track. Representing a 7-bit byte on such perforated tape meant 7 tracks for data, 1 track for parity. Representing an 8-bit byte on such tape meant 8 tracks for data, and no parity track. In short, a 7-bit byte could be represented more reliably on 8-track perforated tape than could an 8-bit byte.

13.7 A 7-BIT CODE!

The arguments for a 7-bit byte—cost of communications products, cost of data communication, and reliability of perforated tape—were weighed by the committee against the argument for an 8-bit byte—efficiency of representation of numeric data. This technical controversy was decided, as all technical controversies on standards committees are decided, by the democratic process of taking a vote. The majority voted for the 7-bit byte. The decision was thus made that the standard code for information interchange would be a 7-bit code. The words set down by Subcommittee X3.2 are interesting (set down after the character set had been developed, but essentially justifying the 7-bit decision):

> Consideration led the Subcommittee to a seven bit code set providing 128 combinations. This character set contains a graphic subset adequate for both data processing and communication purposes. The character set also provides control characters for use in controlling transmission terminal equipment and input/output devices; data delimiting characters for segregating and formatting data; and selected characters for special purposes.
>
> The Subcommittee recognizes that computer manufacturers are unlikely to design computers that use 7-bit codes internally. They are more likely to use 4-bit, 6-bit, and 8-bit codes. There is no widespread need at present for interchange of more than 128 separate and distinct characters between computers, and between computers and associated input/output equipment. However, an eight bit code structure does have distinct advantages in that two 4-bit numeric characters can be packed into an 8-bit frame. And larger code sets reduce the number of multicharacter symbols required for problem definition and programming.
>
> The Subcommittee concluded that a set larger than seven bits should not be recommended as a standard. Some of the primary factors which led to this conclusion were as follows:
>
> a) The 128 combinations available in a 7-bit set satisfy the information and control interchange requirements for the large majority of users.
>
> b) Utilizing an 8-bit set which provides 256 combinations would require recording and transmission of 8-bits by all input–output and transmission systems even though the great majority of requirements are satisfied by a code of fewer bits.
>
> c) A redundancy (parity) bit is employed in most read/write operations and may be used in transmission of data for error control purposes. 8-bits (7 coded bits plus one redundancy bit) are the maximum that can be recorded in a single frame or character position on one inch perforated tape under present recording practices.

13.8 STRUCTURE OF THE CODE

With the decision on size out of the way, the standards committee now went on to consider the structure of the code. The first decision to be made was more of an administrative than of a technical nature. How should the code be exhibited in documents? The committee opted for a matrix or tableau of eight columns and sixteen rows. The three high-order bits of the seven bits 000, 001, . . . , 111 would be used to distinguish the eight columns. And the four low-order bits 0000, 0001, . . . , 1111 would be used to distinguish the sixteen rows.

The next administrative decision was how to name or number the seven bit positions. It was decided to name them

b7, b6, b5, b4, b3, b2, b1

from high-order bit position to low-order bit position.

These administrative decisions are shown in Fig. 13.1.

Some facts were now reviewed.

1. AT&T had stated a functional requirement for an all-zeros character, Null, and an all-ones character, Delete.
2. The Subcommittee's surveys had shown a requirement for 10 numerics, 26 alphabetics, and up to 27 specials; that is, up to 63 graphic characters.
3. There might or might not be a requirement for small letters, as well as for capital letters.
4. From the data transmission standards committee was emerging a requirement for 10 or more data transmission control characters.
5. There was a requirement for a number of format-effector characters, such as Space, Carriage Return, Line Feed, New Line, Horizontal Tab, Vertical Tab, Form Feed.
6. There was a need for data-delimiter or information-separator characters. How many would be required was far from clear.
7. Looking to the future, it would be wise to include characters, such as Escape, Shift In, Shift Out, that could be used to extend the repertoire of control and graphic characters without increasing the byte size of the code.
8. There would be a requirement for a number of specific or general control characters to control either devices or functions of devices.

Two conclusions were drawn from these facts.

Conclusion 1. About 64 graphic characters might be adequate.

Conclusion 2. More than 16 control characters would be needed.

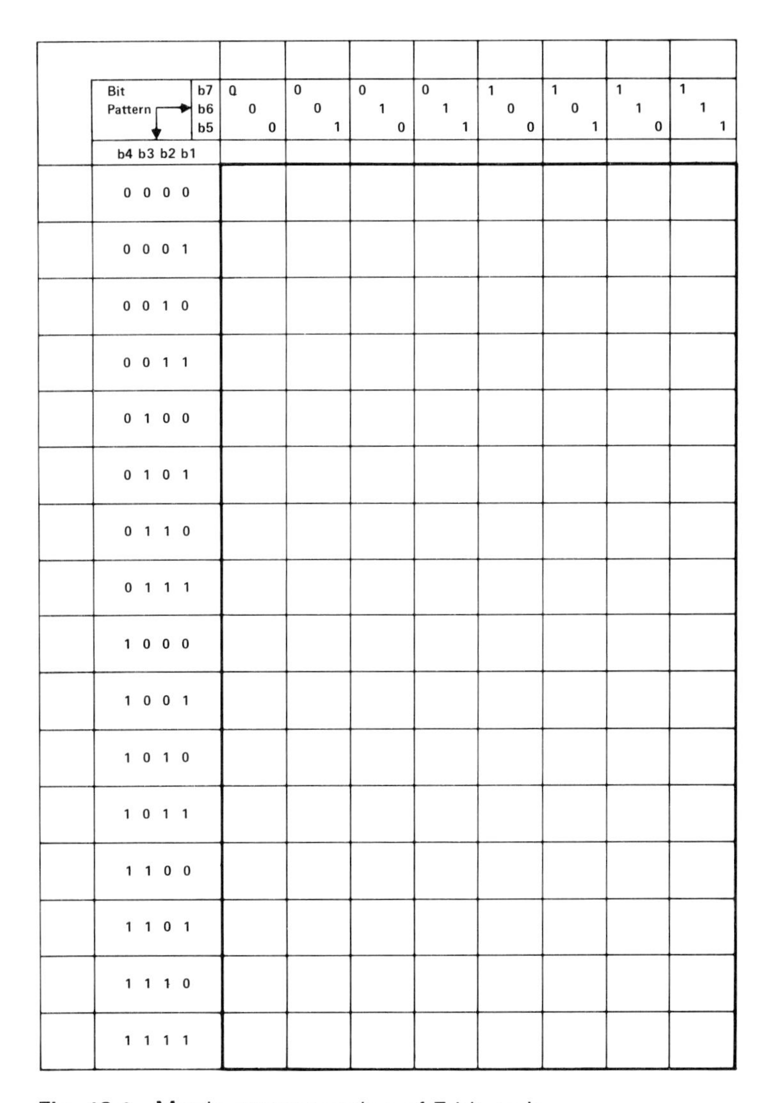

Fig. 13.1 Matrix representation of 7-bit code

The numbers 64 and 16 (above) were used because the standards committee was beginning to think of the code in terms of the code table (see Fig. 13.1) with its 8 columns of 16 characters each—16 and 64 are multiples of 16.

At this point, the first criterion relative to the structuring of the code emerged.

	Column	0	1	2	3	4	5	6	7
	Bit Pattern b7	0	0	0	0	1	1	1	1
	b6	0	0	1	1	0	0	1	1
	b5	0	1	0	1	0	1	0	1
Row	b4 b3 b2 b1								
0	0 0 0 0								
1	0 0 0 1								
2	0 0 1 0								
3	0 0 1 1								
4	0 1 0 0								
5	0 1 0 1								
6	0 1 1 0								
7	0 1 1 1								
8	1 0 0 0								
9	1 0 0 1								
10	1 0 1 0								
11	1 0 1 1								
12	1 1 0 0								
13	1 1 0 1								
14	1 1 1 0								
15	1 1 1 1								

Fig. 13.2 7-bit code table

13.8.1 Criterion 1

Control characters and graphic characters should not be intermingled. Control characters should be grouped contiguously, and graphic characters should be grouped contiguously.

Conclusion 3. A further review of facts 4, 5, 6, 7, 8 above led to the conclusion that more than 16 control character positions were needed, but 32 positions (that is, two columns) might be sufficient.

At this point, for purposes of easy reference, another administrative decision was made—to number the code columns 0, 1, 2, . . . , 7, and to number the code rows 0, 1, 2, . . . , 15, as shown in Fig. 13.2.

Conclusions 1 and 3 said that four columns of graphics and two columns of controls should be assumed as an initial basis for structuring the code. There were then twelve possibilities, as shown in Fig. 13.3, where

c stands for a column of control characters,
g stands for a column of graphic characters, and
x stands for a column of as yet undefined function.

Table Columns → Possibilities ↓	0	1	2	3	4	5	6	7
1	c	c	g	g	g	g	x	x
2	c	c	x	g	g	g	g	x
3	c	c	x	x	g	g	g	g
4	x	c	c	g	g	g	g	x
5	x	c	c	x	g	g	g	g
6	x	x	c	c	g	g	g	g
7	g	g	g	g	c	c	x	x
8	g	g	g	g	x	c	c	x
9	g	g	g	g	x	x	c	c
10	x	g	g	g	g	c	c	x
11	x	g	g	g	g	x	c	c
12	x	x	g	g	g	g	c	c

Figure 13.3

Some of these possibilities were eliminated because of the requirement for a control character of all-zeros, Null, and because of Criterion 1 (not intermingling controls and graphics). The committee put two interpretations on Criterion 1.

Interpretation 1. Within a column, there should not be both controls and graphics.

Interpretation 2. A column of controls should not be positioned between columns of graphics, and a column of graphics should not be positioned between columns of controls.

Given Null, a control character, in column 0, Interpretation 1 ruled out possibilities 7, 8, and 9.

All possibilities satisfied Interpretation 2. But if the x columns ultimately were defined to be graphic columns, possibilities 4, 5, and 8 were ruled out, and probably 10. And if the x columns ultimately were defined as control columns, possibilities 2, 4, 10, and 11 were ruled out.

Essentially, then, all possibilities with separated single x columns were ruled out, leaving possibilities 1, 3, 6, 12 with the two x columns always appearing as contiguous column pairs.

The four remaining possibilities each gave rise to two possibilities, depending on whether both x columns were defined as controls or as graphics, as shown in Fig. 13.4.

Table Columns → Possibilities ↓	0	1	2	3	4	5	6	7	
1a	c	c	g	g	g	g	c	c	
1b	c	c	g	g	g	g	g	g	
3a	c	c	c	c	g	g	g	g	
3b	c	c	g	g	g	g	g	g	←ruled out same as 1b
6a	c	c	c	c	g	g	g	g	←ruled out same as 3a
6b	g	g	c	c	g	g	g	g	
12a	c	c	g	g	g	g	c	c	←ruled out same as 1a
12b	g	g	g	g	g	g	c	c	

Figure 13.4

We see that 6b and 12b were ruled out by Interpretation 1 and Null, a control character, in column 0; 3a was ruled out by Interpretation 1 and Delete, a control character, in column 7.

Strictly speaking, the ruling that ruled out 3a also should rule out 1b, leaving only 1a as a possibility. However, the standards committee was reluctant, at this time, to rule out possibility 1b. The committee wanted to retain the possible configuration ccggggxx, with xx not yet decided as to controls or graphics. To rule out 1b would rule out the graphic possibility for xx for configuration ccggggxx, and the committee was not yet ready to decide to rule that possibility out. However, possibility 3a was ruled out, for another, somewhat more torturous reason.

The Space Character

One character that was definitely going to be included in the final set was the Space character. But was the Space character a control character or a graphic character? Is it the nonvisible or nonprinting graphic in a set of graphics or is it the control character that moves the carriage of a serial printer one character position forward? It is, of course, both. However, from the point of view of a parallel printer, it is only one of these things, the invisible graphic. By this rather hair-splitting reasoning, the standards committee persuaded itself that the Space character must be regarded as a graphic character; that is, it must be positioned in a column of graphics, not in a column of controls.

Now an interesting conclusion could be drawn. It was a well-established data processing practice that in sorting and collating operations, the Space character should collate low to all other graphic characters, specials. numerics, and alphabetics. Consider then the two Possibilities 1 and 3:

	0	1	2	3	4	5	6	7
1	c	c	g	g	g	g	x	x
3	c	c	x	x	g	g	g	g

13.8.2 Criterion 2

The Space character should collate low to all graphic characters.

For Possibility 1, the Space character would clearly be positioned in column 2, row 0, thus preceding all graphics, and this precedence would hold regardless of whether the two x columns were subsequently decided to be graphic or control columns. But for Possibility 3, the situation was different. The Space character would be positioned in column 4, row 0, thus preceding all graphics. If the x columns, columns 2 and 3, were subsequently decided to be control columns, the precedence would still hold. But if the x columns, columns 2 and 3, were subsequently decided to be graphic columns, then the graphics in these two columns would collate low to the Space character, thus violating Criterion 2; that is to say, this possibility would be ruled out. In short, positioning the two x columns as columns 2 and 3 preempted the choice that the x columns might in the future be decided to be graphic columns.

The standards committee did not at this time want the future choice of the two x columns as graphic columns or control columns to be preempted. Possibility 3 would really preempt this decision in advance (as outlined in the preceding paragraph). Therefore, the committee ruled out Possibility 3. This left Possibility 1 as the committee's decision for structure of the code:

0	1	2	3	4	5	6	7
c	c	g	g	g	g	x	x

The basic structure of ASCII had now been decided. See Fig. 13.5.

Row	Column →	0	1	2	3	4	5	6	7
	Bit Pattern b7	0	0	0	0	1	1	1	1
	b6	0	0	1	1	0	0	1	1
	b5	0	1	0	1	0	1	0	1
Row	b4 b3 b2 b1								
0	0 0 0 0								
1	0 0 0 1								
2	0 0 1 0								
3	0 0 1 1								
4	0 1 0 0								
5	0 1 0 1								
6	0 1 1 0								
7	0 1 1 1								
		CONTROLS			GRAPHICS			UNDEFINED	
8	1 0 0 0								
9	1 0 0 1								
10	1 0 1 0								
11	1 0 1 1								
12	1 1 0 0								
13	1 1 0 1								
14	1 1 1 0								
15	1 1 1 1								

Fig. 13.5 ASCII structure

14 The Sequence of ASCII

In the previous chapter, the basic structure of ASCII was defined: Controls in Columns 0, 1; Graphics in Columns 2, 3, 4, 5; Undefined for Columns 6, 7. The Null character would be in code position 0/0, the Space character in code position 2/0, and the Delete character in code position 7/15. The standards committee now turned to the definition of ASCII in finer detail.

During the discussion of ASCII structure, four kinds of control characters had been discussed; Transmission Controls, Formal Effectors, Device Controls, and Information Separators. As a preliminary step, the committee decided to apportion the 32 control code positions equally among these four categories. It was recognized that it was very unlikely that, in the final analysis, there would be exactly eight of each kind of control character. It was also recognized that there were control characters, such as Escape, Shift In, Shift Out, that would not fit into any of the four categories. Nevertheless, it was decided to make this preliminary categorization of control characters, as shown in Fig. 14.1, and see what would befall.

Attention now focused on the question of collating sequence of graphics. As described in the previous chapter, Space, by being positioned in code position 2/0, would collate low to all graphics.

Column 2 was chosen for specials for two reasons:

1. Numerics could not be located in this column, because if so, "0" would require the row 0 position, and this was already preempted by the Space character.

	Column	0	1	2	3	4	5	6	7
	Bit Pattern b7	0	0	0	0	1	1	1	1
	b6	0	0	1	1	0	0	1	1
	b5	0	1	0	1	0	1	0	1
Row	b4 b3 b2 b1								
0	0 0 0 0	NUL		SP					
1	0 0 0 1								
2	0 0 1 0								
3	0 0 1 1								
4	0 1 0 0	TC	DC						
5	0 1 0 1								
6	0 1 1 0								
7	0 1 1 1			GRAPHICS				UNDEFINED	
8	1 0 0 0								
9	1 0 0 1								
10	1 0 1 0								
11	1 0 1 1	FE	IS						
12	1 1 0 0								
13	1 1 0 1								
14	1 1 1 0								
15	1 1 1 1								DEL

TC - Transmission Controls
FE - Format Effectors
DC - Device Controls
IS - Information Separators

Fig. 14.1 ASCII, basic structure

2. Alphabetics should not be in Column 2, because if so, specials in columns 3, 4, or 5 would then necessarily collate high to alphabetics. But there were some specials, such as period and hyphen, which should collate low to all alphabetics in sorting operations on names of people.

14.1 SEPARATE OR INTERLEAVED ALPHABETS?

This apparently left columns 3, 4, and 5 for numerics and alphabetics. But other questions had to be settled first. In the event that it was eventually decided to include both small and capital letters, should the two alphabets be *separate*, or *interleaved*? And if separate, should small letters collate low to capital letters or vice versa?

The question of separate or interleaved alphabets was approached first. Two possibilities were apparent for interleaving:

Possibility 1	Possibility 2
a	A
A	a
b	B
B	b
.	.
.	.
.	.
z	Z
Z	z

The choice between these two possibilities was clear, and stemmed from the very reason for having interleaved alphabets. In sorting names, it is conventional for capital letters to precede small letters. Thus, the AA Company precedes the Aardvark Company. But in sorting names of peoples, the rules become more subtle and complex. Does MacKenzie precede Mackenzie? In some telephone directories, yes, but in other telephone directories, the capitalization or noncapitalization of the K will be ignored in MacKenzie and Mackenzie—such names being blocked together, and ordered on the basis of the first names or initials. Indeed, the proponents of alphabet separation cited the fact that different telephone directories had different rules as evidence that alphabet interleaving would really not accomplish anything tangible.

In any event, there was a more compelling argument against interleaving. In columns 2, 3, 4, and 5 there are 64 code positions, sufficient to accommodate the Space character, specials, numerics, and alphabetics; that is to say, a graphic set sufficient for most data processing applications. And this set of 63 graphics and Space is derivable from the 7-bit code by dropping b6. The four columns 2, 3, 4, and 5 then form a 6-bit subset.

If, however, the alphabets were interleaved, then it would clearly take columns 2 through 7 to contain Space, specials, numerics, small letters, and capital letters. With the alphabets interleaved, the derivation of a 64-character, 6-bit subset containing Space, specials, numerics, and capital letters, would require more complex logic. Suppose, for example, that Fig. 14.2 exhibits a 7-bit code with interleaved alphabets, and Fig. 14.3 exhibits the 6-bit, 64-character subset to be derived. Let the bit positions of the 6-bit subset be named a6, a5, a4, a3, a2, a1 from high-order bit position to low-order bit position. Then the transformation equations, from 7 bits to 6 bits, are as follows:

$$a6 = b7 \wedge \overline{b1}$$

$$a5 = (\overline{b7} \wedge b6 \wedge b5) \dot{\vee} (b7 \wedge b6 \wedge \overline{b1})$$

$$a4 = (\overline{b7} \wedge b6 \wedge b4) \dot{\vee} (b7 \wedge b5 \wedge \overline{b1})$$

$$a3 = (\overline{b7} \wedge b6 \wedge b3) \dot{\vee} (b7 \wedge b4 \wedge \overline{b1})$$

$$a2 = (\overline{b7} \wedge b6 \wedge b2) \dot{\vee} (b7 \wedge b3 \wedge \overline{b1})$$

$$a1 = (\overline{b7} \wedge b6 \wedge b1) \dot{\vee} (b7 \wedge b2 \wedge \overline{b1})$$

The consideration of a 6-bit, 64-character graphic subset was important to the standards committee. If the ultimate decision was that columns 6 and 7 would be for graphics, then columns 2 through 7 would contain Space, 94 graphics, and Delete. But, even with the code providing 94 graphics, a major assumption of the standards committee was that data processing applications would, for the foreseeable future, be satisfied with a monocase alphabet (that is, a 64- or less graphic subset) as they had in the past—that 64-character printers would predominate. So it was important to be able to derive a 64-character, monocase alphabet, graphic subset from the code by simple, not complex, logic.

It was this consideration that weighted the decision against interleaved alphabets. Interestingly, consideration of this example led to another, and unexpected, conclusion. In the example, the capital alphabet was contained in two columns. Clearly, two alphabets, small and capital letters, could be contained in four columns; that is, the two undefined columns, 6 and 7, could contain an alphabet of small letters, if it was eventually decided to include that alphabet.

		Column	0	1	2	3	4	5	6	7
	Bit Pattern	b7	0	0	0	0	1	1	1	1
		b6	0	0	1	1	0	0	1	1
		b5	0	1	0	1	0	1	0	1
Row	b4 b3 b2 b1									
0	0 0 0 0		NUL		SP	0	A	I	Q	Y
1	0 0 0 1					1	a	i	q	y
2	0 0 1 0					2	B	J	R	Z
3	0 0 1 1					3	b	j	r	z
4	0 1 0 0					4	C	K	S	
5	0 1 0 1					5	c	k	s	
6	0 1 1 0					6	D	L	T	
7	0 1 1 1					7	d	l	t	
8	1 0 0 0					8	E	M	U	
9	1 0 0 1					9	e	m	u	
10	1 0 1 0						F	N	V	
11	1 0 1 1						f	n	v	
12	1 1 0 0						G	O	W	
13	1 1 0 1						g	o	w	
14	1 1 1 0						H	P	X	
15	1 1 1 1						h	p	x	DEL

Fig. 14.2 Interleaved alphabets

Row	Bit Pattern b7 b6 b5 → / b4 b3 b2 b1 ↓	Column 0 (b7=0, b6=0, b5=0)	Column 1 (b7=0, b6=0, b5=1)	Column 2 (b7=0, b6=1, b5=0)	Column 3 (b7=0, b6=1, b5=1)
0	0 0 0 0	SP	0	A	Q
1	0 0 0 1		2	B	R
2	0 0 1 0		3	C	S
3	0 0 1 1		4	D	T
4	0 1 0 0		5	E	U
5	0 1 0 1		6	F	V
6	0 1 1 0		7	G	W
7	0 1 1 1		8	H	X
8	1 0 0 0		9	I	Y
9	1 0 0 1			J	Z
10	1 0 1 0			K	
11	1 0 1 1			L	
12	1 1 0 0			M	
13	1 1 0 1			N	
14	1 1 1 0			O	
15	1 1 1 1			P	

Fig. 14.3 6-bit subset

14.2 THREE COLUMNS FOR ALPHABETICS?

But consider the kind of code structure where the alphabet is contained in three columns (see Fig. 2.29). In order to provide two alphabets, small and capital letters, as in EBCDIC (see Fig. 2.28), six columns are required. And to provide two alphabets and also a column for numerics, seven columns are required.

Row \ Column			2	3	4	5		
0			SP	†	†	0		
1			A	J	†	1		
2			B	K	S	2		
3			C	L	T	3		
4			D	M	U	4		
5			E	N	V	5		
6			F	O	W	6		
7			G	P	X	7		
8			H	Q	Y	8		
9			I	R	Z	9		
10			†	†	†	†		
11			†	†	†	†		
12			†	†	†	†		
13			†	†	†	†		
14			†	†	†	†		
15			†	†	†	†		

† - Special

Fig. 14.4 BCD arrangement

Even if it was eventually decided to assign graphics to columns 6 and 7, there would be only six columns available for graphics. Given the basic structure of ASCII, as defined in Fig. 14.1, two alphabets structured noncontiguously, as in BCDIC and in EBCDIC, and a column of numerics could not be accommodated in the 7-bit code table. At least an 8-bit code table is necessary to accommodate a column of numerics, three columns of small letters, and three columns of capital letters. And the standards committee had already decided for a 7-bit code and against an 8-bit code.

The conclusion of the preceding paragraph is based on the assumption that two alphabets, small letters and capital letters, would be included in the 7-bit code and that decision had not yet been made. If the decision was ultimately made that columns 6 and 7 would contain controls, then small letters would not be included in the 7-bit code.* If only capital letters were to be included in the code, it would be quite feasible to have a BCD arrangement, such as shown in Fig. 14.4, for the graphic subset. Such an arrangement was, in fact, proposed to the standards committee, but it was rejected because it intermingled specials with alphabetics, and the subcommittee deemed this to be unwise, for collating reasons.

The standards committee at this time made a fundamental decision. The 26 letters of the alphabet should be grouped contiguously in the code and should occupy two contiguous columns.

The standards committee had, as described above, decided that column 2 would contain specials. With respect to the assignment of numerics and alphabetics, there were two possibilities:

Possibility 1. Numerics in column 3

Alphabetics in columns 4 and 5

Possibility 2. Alphabetics in columns 3 and 4

Numerics in column 5

14.3 EXISTING COLLATING SEQUENCE

The committee recognized that an existing collating practice was that alphabetics collate low to numerics. So Possibility 2 seemed the clear choice. But there was an argument against this choice.

If the ultimate decision for columns 6 and 7 was for graphics, then there would be two choices for the graphics: specials, or small letters. Suppose the choice was for small letters. Then the two possibilities above became as shown in Fig. 14.5. Assume, for purposes of discussion, that the alphabets are positioned as shown. Then, for Possibility 1, the bit patterns of the capital letters and the bit patterns of the small letters have a single bit difference, b6, for all corresponding small and capital letters. For Possibility 2, three bits, b7, b6, b5, are different and the bit differences between A and a, for example, are not the same as the bit differences between Q and q.

---

* If the committee did decide for controls in columns 6 and 7, it is still likely that they would have wanted an alphabet of small letters to be provided. Presumably, the small letter alphabet would then have been provided by a caseshift approach.

Columns	2	3	4	5	6	7		2	3	4	5	6	7
b7	0	0	1	1	1	1		0	0	1	1	1	1
b6	1	1	0	0	1	1		1	1	0	0	1	1
b5	0	1	0	1	0	1		0	1	0	1	0	1
	SP	0	A	Q	a	q		SP	A	Q	0	a	q
	.	1	B	R	b	r		.	B	R	1	b	r
	.	2	C	S	c	s		.	C	S	2	c	s
	.	3	D	T	d	t		.	D	T	3	d	t
	.	.	.	.	.	.		.	.	.	.	.	.
	.	.	.	.	.	.		.	.	.	.	.	.
	Possibility 1							Possibility 2					

Fig. 14.5 Positioning of alphabetics

Clearly then, in order to keep open the choice for columns 6 and 7 between graphics and controls, and between small letters and specials, Possibility 1 was preferable.

Possibility 1, of course, would provide a collating sequence, specials, numerics, alphabetics, from low to high, contrary to the existing practice, specials, alphabetics, numerics. But to the standards committee the argument above, keeping choices open at this time, was more compelling.

The committee rationalized the decision against accepting the existing collating sequence somewhat along the following lines:

> If it is necessary to achieve the de facto collating sequence (specials, alphabetics, numerics), it may be achieved, during comparison operations, by inverting b7 if b6 = b5 = 1. That is, the three high-order bits of the column of numerics would then become 111, which would make them collate high to the alphabetics, with high-order bits of 100 and 101.

Out of this discussion, the committee established a major criterion.

Criterion

There should be a single bit difference between capital and small letters.

The standards committee had now made its final decision with respect to the sequence of ASCII. The result was as follows:

0	1	2	3	4	5	6	7
Controls		Specials	Numerics	Alphabetics		Undefined	

We have, then, a code structure as shown in Fig. 14.6.

	Column	0	1	2	3	4	5	6	7
	Bit Pattern b7	0	0	0	0	1	1	1	1
	b6	0	0	1	1	0	0	1	1
	b5	0	1	0	1	0	1	0	1
Row	b4 b3 b2 b1								
0	0 0 0 0	NUL		SP					
1	0 0 0 1								
2	0 0 1 0								
3	0 0 1 1								
4	0 1 0 0	TC'S	DC'S	SPEC-IALS	NUM-ERICS	ALPHABETICS			
5	0 1 0 1								
6	0 1 1 0								
7	0 1 1 1								
8	1 0 0 0								
9	1 0 0 1								
10	1 0 1 0								
11	1 0 1 1	EF'S	IS'S						
12	1 1 0 0								
13	1 1 0 1								
14	1 1 1 0								
15	1 1 1 1								DEL

TC - Transmission Control
FE - Format Effector
DC - Device Control
IS - Information Separator

Fig. 14.6 ASCII structure

14.4 CRITERIA

Up to this point, the standards committee had made a number of decisions, based on criteria. Three of those criteria have been stated so far in this chapter. In fact, the committee formulated 20 criteria. It should be noted that some of these criteria are conflicting, so not all can be met.

Criterion 1. All bit patterns in the code should consist of the same number of bit positions.

Criterion 2. The structure of the code should be such that logically related subsets or supersets are derivable simply; that is, by simple bit dropping, bit adding, or bit inversion.

Criterion 3. All possible bit patterns of the code should be considered valid. For illustration, on 7-track magnetic tape with even parity, the all-zeroes 6-bit bit pattern was considered invalid, as being indistinguishable from unrecorded tape, with the recording practice used at that time.

Criterion 4. The code size, that is, the number of different possible character positions, should be sufficient to accommodate alphabetics, numerics, specials, and control characters needed for information interchange.

Criterion 5. The numerics 0 through 9 should be contained in a 4-bit subset.

Criterion 6. The numerics should have bit patterns such that the four low-order bits shall be the binary coded decimal representation of numerics.

Criterion 7. The intermingling of control and graphic characters should be avoided. The bit patterns of control characters should be distinguishable from those of graphics by some simple test of the high-order bits.

Criterion 8. The meaning associated with a bit pattern should depend on only the bit pattern itself, and not on any preceding bit patterns.

Criterion 9. The alphabetics A through Z, and some code positions contiguous to the code position of Z, should be contained in a 5-bit subset.

Criterion 10. The alphabetics should have contiguous bit patterns.

Criterion 11. Such control characters are as required for communication and data processing should be included.

Criterion 12. An Escape character, to allow for code extension, should be included.

Criterion 13. The class of specials, the class of numerics, and the class of alphabetics should be distinguishable one from the other by simple binary comparison tests.

Criterion 14. The Space character should be positioned so as to collate low to all other graphics.

Criterion 15. Specials that are involved in sorting and collating operations should be positioned so as to collate low to both numerics and alphabetics.

Criterion 16. Specials should be grouped according to their functions; for example, punctuation and mathematical symbols.

Criterion 17. Graphics that are normally paired on typewriter keytops should differ only in a common single bit position.

Criterion 18. The graphics of the principal programming languages should be included.

Criterion 19. The bit patterns of all control characters should have a common, distinguishing subpattern of bits.

Criterion 20. The all-zeroes characters, Null, and the all-ones character, Delete, should be included.

14.5 DECISIONS FROM CRITERIA

Up to this point in the deliberations of the standards, the standards committee had made some 17 decisions. These are now presented, and in parentheses, the criteria which affected the decisions.

Decision 1. There would be at least 64 graphic characters in the code (Criterion 4).

Decision 2. There would be a total of more than 64 characters in the code (Criterion 4).

Decision 3. There would be upwards of 16 control characters in the code (Criterion 11).

Decision 4. The code would be an unshifted code—therefore, at least 7 bits (Criterion 8).

Decision 5. The code would not be 8 bits—therefore, 7 bits.

Decision 6. (a) The code table would be exhibited in a tableau of 8 columns and 16 rows. (b) Columns would be numbered 0, 1, 2, 3, . . . , 7. (c) Rows would be numbered 0, 1, 2, 3, . . . , 15. (d) Bit positions would be named b7, b6, b5, b4, b3, b2, b1, from high to low.

Decision 7. There would be an all-zeroes character, Null, and an all-ones character, Delete, in the code (Criterion 20).

Decision 8. Tentatively, more than 16, but less than or equal to 32 control characters would be sufficient (Criterion 11).

Decision 9. Columns 0 and 1 would be for control characters, columns 2, 3, 4, and 5 for graphic characters, and columns 6 and 7 undefined at this time (Criteria 7, 13, 19).

Decision 10. The Space character would be in code position 2/0 (Criterion 14).

Decision 11. Tentatively, code positions would be reserved for 8 Transmission-Control characters, 8 Format-Effector characters, 8 Device-Control characters, and 8 Information-Separator characters (Criterion 11).

Decision 12. Column 2 would be reserved for Specials (Criterion 13).

Decision 13. Small and capital letters, if provided, would be provided as separate alphabets, not as interleaved alphabets.

Decision 14. A 6-bit, 64-character graphic subset should be collapsible out by dropping one of the seven bits (Criterion 2).

Decision 15. The 3-column BCD arrangement for alphabetics is rejected (Criteria 9, 10).

Decision 16. Alphabetics would be contiguous (Criterion 10).

Decision 17. The structure of the code would be

Columns 0 and 1, controls
Column 2, specials
Column 3, numerics
Columns 4 and 5, alphabetics
Columns 6 and 7, undefined at this time.

As stated above, the standards committee had decided that the alphabet(s) would be contiguous and positioned in two contiguous columns of the code. For English-speaking countries, there are 26 alphabetics. There are 32 contiguous code positions in two columns. The first letter, A, could therefore be positioned in any of seven positions of column 4, as shown in Fig. 14.7.

The standards committee noted that in Fieldata (see Fig. 3.3) the contiguous alphabet had been positioned with A in 4/7 down to Z in 5/15. One factor precluded the position of the alphabet of the standard code into the Fieldata positions. The alphabets of some European countries (Germany, Sweden, Norway, Denmark, Finland) require 29 letters—the 26 letters of the English-speaking countries and 3 diacritic letters. The Portuguese and Spanish languages require one or more diacritic letters. The French and Italian languages require accented letters.

14.6 NATIONAL USE POSITIONS

It was generally recognized by American manufacturers marketing equipment in Europe that these additional diacritic or accented letters must be provided. It was a natural decision, therefore, to provide code positions in the standard code to meet such requirements. In some of the continental European countries, from a collating sequence point of view, the diacritic letters are interspersed among the other letters. But in Sweden they follow the letter Z. It was a natural decision, therefore, to assign the three code positions following the code position of Z to accommodate the alphabetic extender requirement. These three code positions came to be called National Use positions.

It should be noted that this consideration rules out the last three possibilities shown in Fig. 14.7. In any event, the Fieldata positioning was ruled out by this consideration. But this still left four possibilities—positioning the letter A in code positions 4/0, 4/1, 4/2, or 4/3. Which of these should be chosen?

The American standards committee decided on position 4/1 for the letter A because that code position had been decided for a draft British Standard and also for a draft ECMA Standard being developed at that time. So the American decision was based on the sensible desire for international accord on this point. (But the author does not know on what factors the British and ECMA decision was based.) This decision, *Decision 18*, then, was the first on the specific positioning of graphics.

Row \ Column	4	4	4	4	4	4	4
0	A						
1	B	A					
2	C	B	A				
3	.	C	B	A			
4	.	.	C	B	A		
5	.	.	.	C	B	A	
6		.	.	.	C	B	A
7			.	.	.	C	B
8				.	.	.	C
9					.	.	.
.						.	.
.							.
.							

Fig. 14.7 Positioning of A

14.7 POSITIONING OF NUMERICS

Decision 19. The next decision of the standards committee had to do with the positioning of the numerics. It had already been decided to position the numerics in column 3. Criterion 6 clearly required that the numerics 0 through 9 should be in code positions 3/0 through 3/9, respectively. The specifics of the code were now beginning to shape up, as shown in Fig. 14.8.

	Column	0	1	2	3	4	5	6	7
	Bit Pattern b7	0	0	0	0	1	1	1	1
	b6	0	0	1	1	0	0	1	1
	b5	0	1	0	1	0	1	0	1
Row	b4 b3 b2 b1								
0	0 0 0 0	NUL		SP	0		P		
1	0 0 0 1				1	A	Q		
2	0 0 1 0				2	B	R		
3	0 0 1 1				3	C	S		
4	0 1 0 0				4	D	T		
5	0 1 0 1				5	E	U		
6	0 1 1 0				6	F	V		
7	0 1 1 1				7	G	W		
8	1 0 0 0				8	H	X		
9	1 0 0 1				9	I	Y		
10	1 0 1 0					J	Z		
11	1 0 1 1					K			
12	1 1 0 0					L			
13	1 1 0 1					M			
14	1 1 1 0					N			
15	1 1 1 1					O			DEL

Fig. 14.8 ASCII, initial specifics

14.8 ASSIGNMENT OF SPECIAL CHARACTERS

Decision 20. The standards committee now turned its attention to assignment of specials. After much discussion, the standards committee decided on the 27 graphics to go in the available code positions in columns 2, 3, 4, and 5. The specials are classified by function.*

Punctuation and Correspondence	. , : ; ! ? ' "	8
Commercial Usage	# $ % & @	5
Bracketing (Programming)	( ) [ ]	4
Mathematical (Programming)	+ − * / \ = < >	8
Flow Charting (Programming)	↑ ←	2

Clearly, Criteria 4 and 18 bore on this decision.

Decision 21. The standards committee now considered specific code positions for these specials. A number of criteria bore on this decision, Criteria 13, 15, 16, and 17. Actually, Criterion 13 was of little significance here, because the sets of available bit patterns had already been established by previous decisions on the positioning of Space, numerics, and alphabetics.

It was soon evident that Criterion 16, which spoke to grouping of specials by function, would conflict with Criteria 15 and 17, which spoke to collating considerations and to typewriter-keytop-pairing considerations. Criterion 16 was considered to be of less importance than the other two.

Criterion 17 spoke to positioning graphics in the code table to correspond to their positioning on typewriter keys. From this criterion, some decisions stemmed easily.

The specials # $ % were positioned 2/3, 2/4, 2/5, respectively, in correspondence with 4, 5, 6 in 3/3, 3/4, 3/5, respectively, thus providing the typewriter-keytop pairing. The specials / and ? were positioned in 2/15 and 3/15, respectively, thus providing correspondence with typewriter-keytop pairing.

14.8 Period and Comma

On electric typewriters both the period and the comma appear in both lower and upper-case shift. It was decided to correspond these two

* Note that these classifications are not mutually exclusive. Bracketing symbols, the hyphen, and the asterisk are used in business correspondence. Period, comma, semicolon, apostrophe are used in programming languages. And so on.

graphics with two others that typewriter manufacturers would reckon were unneeded in normal business correspondence; that is, the period and the comma in one case shift would be replaced. The specials < and > seemed to fill the bill. Accordingly, it was decided to pair , and < and to pair . and > in the code, but it was not yet clear where these four should specifically go.

It was noted that the specials , . – frequently appear in sorting or collating situations. Under Criterion 15, then, these should be positioned so as to collate low to numerics and alphabetics. Clearly, this meant they would have to be positioned in Column 2. Of these three specials, it had been decided, as related in the previous paragraph, to pair , . with < >. To satisfy these two conditions, specials , . < > were positioned in code positions 2/12, 2/14, 3/12, 3/14, respectively.

14.8.2 Left and Right Parentheses

The graphics (and) are paired with 9 and 0 on electric typewriters. But no graphic could be paired with 0 in the code, since the Space character had already been assigned to the pair position of 0. It was decided to pair them in the code with 8 and 9 because then, if the code were implemented on a keyboard, they would be located as close as possible to their usual electric typewriter positions; that is, paired with 9 and 0. Also, on many European typewriter keyboards, (and) appear paired with 8 and 9. Therefore, (and) were positioned in 2/8 and 2/9, respectively.

It was now pointed out that in the United Kingdom the monetary system required not only the numerics 0 through 9 but also 10 and 11. Clearly, if these numerics were provided in the code for implementations for the United Kingdom, they would occupy the two code positions in the column of numerics under the 9; that is, code positions 3/10 and 3/11. It was deemed wise to assign to these two code positions graphics that could be replaced in the United Kingdom with minimum anguish. Eventually, the standards committee decided to assign : and ; to code positions 3/10 and 3/11, respectively.

14.8.3 Alphabetic Extenders

Attention was now focussed on the three code positions 5/11, 5/12, 5/13 that, as was explained above, would receive alphabetic extenders in European implementations. As in the preceding paragraph, the search was for graphics whose replacement would cause minimum anguish. The standards committee decided for [\ ] for code positions 5/11, 5/12, 5/13, respectively.

14.8.4 Further Special Characters

There now remained 10 specials to be assigned:

! ' " & @ + * = ↑ ←

It seemed apt to position = in the code position between those occupied

Row	Column	0	1	2	3	4	5	6	7
	Bit Pattern b7	0	0	0	0	1	1	1	1
	b6	0	0	1	1	0	0	1	1
	b5	0	1	0	1	0	1	0	1
Row	b4 b3 b2 b1								
0	0 0 0 0	NUL	DC0	SP	0	@	P		
1	0 0 0 1	TC1	DC1	!	1	A	Q		
2	0 0 1 0	TC2	DC2	'	2	B	R		
3	0 0 1 1	TC3	DC3	#	3	C	S		
4	0 1 0 0	TC4	DC4	$	4	D	T		
5	0 1 0 1	TC5	DC5	%	5	E	U		
6	0 1 1 0	TC6	DC6	&	6	F	V		
7	0 1 1 1	TC7	DC7	'	7	G	W	UNASSIGNED	
8	1 0 0 0	FE0	IS0	(	8	H	X		
9	1 0 0 1	FE1	IS1	)	9	I	Y		
10	1 0 1 0	FE2	IS2	*	:	J	Z		
11	1 0 1 1	FE3	IS3	+	;	K	[		
12	1 1 0 0	FE4	IS4	,	<	L	\		
13	1 1 0 1	FE5	IS5	-	=	M	]		
14	1 1 1 0	FE6	IS6	.	>	N	↑		
15	1 1 1 1	FE7	IS7	/	?	O	←		DEL

TC - Transmission Control
FE - Format Effector
DC - Device Control
IS - Information Separator

Fig. 14.9 ASCII, sequence of 63 graphics

by $<$ and $>$. Thus these three mathematical symbols would be in code sequence $< \; = \; >$, which might aid human beings to remember their code positions. Therefore, = was assigned to 3/13.

Because the special @ is not used in continental Europe, it seemed likely to be replaced with an accented letter à in France and Italy. This letter should be in proximity to other letters in the code table. Code position 4/0 filled the bill, and @ was assigned thereto.

For the eight specials remaining, no reasons could be found for any particular code position. They were therefore positioned in the remaining eight code positions, more or less arbitrarily. The code table now looked like that shown in Fig. 14.9.

14.9 CONTROL CHARACTERS

The standards committee responsible for coded character sets discussed with the standards committee responsible for data communications the control characters necessary for data transmission control.

Nine functions were identified as being required for data transmission control:

SOM	Start of Message
EOA	End of Address
EOM	End of Message
EOT	End of Transmission
WRU	Who Are You?
RU	Are You. . .?
DC0	Device control reserved for Data Link Escape
SYNC	Synchronous Idle
ACK	Acknowledge

When it came to decisions to position these characters in the code table, the concept of "Hamming distance" came into play. On transmission lines transmitting binary digital data, what was called a "hit" could occur. If a 0-bit was hit, it changed into a 1-bit. If a 1-bit was hit, it changed into a 0-bit.

As a result of hits, with resultant changes to bit patterns, changes in meaning could occur. Consider the following:

Graphic meaning	Bit pattern
B	1000010
C	1000011

If the bit pattern 1000010 meaning B received a hit in its last bit, changing it to 1000011, the meaning would be C. Hits on graphic bit patterns would result in garbled messages. But if hits occurred to data transmission control characters changing them into other data transmission control characters, the transmission system could go out of control. This was clearly to be guarded against to the maximum extent possible.

Consider two bit patterns:

b7	b6	b5	b4	b3	b2	b1
1	0	0	0	0	1	0
0	1	0	1	1	0	0

	Column	0	1	2	3	4	5	6	7
	Bit Pattern b7	0	0	0	0	1	1	1	1
	b6	0	0	1	1	0	0	1	1
	b5	0	1	0	1	0	1	0	1
Row	b4 b3 b2 b1								
0	0 0 0 0		DCO						
1	0 0 0 1	SOM							
2	0 0 1 0	EOA							
3	0 0 1 1	EOM							
4	0 1 0 0	EOT							
5	0 1 0 1	WRU							
6	0 1 1 0	RU	SYNC						
7	0 1 1 1								
8	1 0 0 0								
9	1 0 0 1								
10	1 0 1 0								
11	1 0 1 1								
12	1 1 0 0								ACK
13	1 1 0 1								
14	1 1 1 0								
15	1 1 1 1								

Fig. 14.10 Data transmission control characters

For these two bit patterns, b7, b6, b4, b3, and b2 are different. That is to say, five hits would have to occur to change one bit pattern into the other. The number of bits different between two bit patterns is known as their "Hamming distance."

Clearly, to minimize the possibility of one data transmission control character being hit and turning into another data transmission control character, the Hamming distance between the two characters must be maximized. The set of data transmission control characters, therefore, should be positioned in the code table to maximize the hamming distances between and among them.

Many combinations were studied and, ultimately, agreement was reached to position them as shown in Fig. 14.10.

The standards committee eventually came into agreement to include the following control characters:

Format Effectors	
HT/SK	Horizontal Tabulation, Skip (punched card)
LF	Line feed
VT	Vertical Tabulation
FF	Form Feed
CR	Carriage Return
FE0	Format Effector

Device Control	
DC1	Device Control 1
DC2	Device Control 2
DC3	Device Control 3
DC4	Device Control 4

Code Extension	
ESC	Escape
SO	Shift Out
SI	Shift In

Information Separators	
S0	Separator 0
S1	Separator 1
S2	Separator 2
S3	Separator 3
S4	Separator 4
S5	Separator 5
S6	Separator 6
S7	Separator 7

Miscellaneous	
BELL	Audible Signal
ERR	Error
NULL	Null
DEL	Delete
①	Unassigned Control

The final ASCII code table, as of 1963, is shown in Fig. 14.11.

	Column	0	1	2	3	4	5	6	7
	Bit Pattern b7	0	0	0	0	1	1	1	1
	b6	0	0	1	1	0	0	1	1
	b5	0	1	0	1	0	1	0	1
Row	b4 b3 b2 b1								
0	0 0 0 0	NULL	DC0	SP	0	@	P		
1	0 0 0 1	SOM	DC1	!	1	A	Q		
2	0 0 1 0	EOA	DC2	"	2	B	R		
3	0 0 1 1	EOM	DC3	#	3	C	S		
4	0 1 0 0	EOT	DC4	$	4	D	T		
5	0 1 0 1	WRU	ERP	%	5	E	U		
6	0 1 1 0	RU	SYNC	&	6	F	V	UNASSIGNED	UNASSIGNED
7	0 1 1 1	BELL	LEM	'	7	G	W		
8	1 0 0 0	FE0	S0	(	8	H	X		
9	1 0 0 1	HT/SK	S1	)	9	I	Y		
10	1 0 1 0	LF	S2	*	:	J	Z		
11	1 0 1 1	VT	S3	+	;	K	[		
12	1 1 0 0	FF	S4	,	<	L	\		ACK
13	1 1 0 1	CR	S5	-	=	M	]		①
14	1 1 1 0	SO	S6	.	>	N	↑		ESC
15	1 1 1 1	SI	S7	/	?	O	←		DEL

Fig. 14.11 ASCII, 1963

14.10 ASCII, 1967

At the first meeting of ISO/TC97/SC2 in 1963 October 29–31, a resolution was passed that the lower-case alphabet should be assigned to columns 6 and 7. Of course, with the assignment of three code positions 7/11, 7/12, 7/13 for National Use, this meant that ACK must be removed from code position 7/13.

	Column	0	1	2	3	4	5	6	7
	Bit Pattern b7	0	0	0	0	1	1	1	1
	b6	0	0	1	1	0	0	1	1
	b5	0	1	0	1	0	1	0	1
Row	b4 b3 b2 b1								
0	0 0 0 0	NUL	DLE [1]	SP	0	@	P	` [2]	p
1	0 0 0 1	SOH [1]	DC1	!	1	A	Q	a	q
2	0 0 1 0	STX [1]	DC2	"	2	B	R	b	r
3	0 0 1 1	ETX [1]	DC3	#	3	C	S	c	s
4	0 1 0 0	EOT [2]	DC4	$	4	D	T	d	t
5	0 1 0 1	ENQ [3]	NAK [1]	%	5	E	U	e	u
6	0 1 1 0	ACK [3]	SYN	&	6	F	V	f	v
7	0 1 1 1	BEL	ETB [2]	'	7	G	W	g	w
8	1 0 0 0	BS [2]	CAN [2]	(	8	H	X	h	x
9	1 0 0 1	HT	EM [1,2]	)	9	I	Y	i	y
10	1 0 1 0	LF	SUB [2]	*	:	J	Z	j	z
11	1 0 1 1	VT	ESC [3]	+	;	K	[	k	{ [2]
12	1 1 0 0	FF	FS [2]	,	<	L	\	l	¦ [2]
13	1 1 0 1	CR	GS [2]	-	=	M	]	m	} [2]
14	1 1 1 0	SO	RS [2]	.	>	N	^ [2]	n	~ [2]
15	1 1 1 1	SI	US [2]	/	?	O	_ [2]	o	DEL

[1] Change of name
[2] New character
[3] Moved character

Fig. 14.12 ASCII, 1967 and 1968

This decision was in due course accepted by X3.2 for ASCII. Interaction between members of X3.2 and delegations at ISO/TC97/SC2 ultimately led to further changes in ASCII. The final code table, as embodied in USAS X3.4–1967, is shown in Fig. 14.12.

Changes were of four kinds:

- Changes of name. For example, Start of Message (1963) became Start of Header (1967).
- Characters moved. For example, Escape, in position 7/14 (1963), was moved to position 1/11 (1967).
- Introduction of new characters. For example, grave accent and the opening brace {. For example, SUB (Substitute), CAN (Cancel).
- Deletion completely of some characters. For example, RU (Are You . . . ?) and ERR (Error) in the 1963 version are not in the 1967 version at all.

15 Which Bit First?

Following the approval of the American Standard Code for Information Interchange (ASCII) in 1963, the data transmission standards committee turned its attention to determining the bit sequence in which the bit patterns of ASCII should be transmitted for serial-by-bit–serial-by-character data transmission. The committee soon decided that the ASCII bit patterns should be transmitted consecutively. As well as considering problems of character framing and parity on data transmission lines (which problems are not discussed in this book), the committee considered the problem of whether the ASCII bit patterns should be transmitted high-order bit first or low-order bit first.

15.1 SPECIFIC CRITERIA

The committee developed a set of ten specific criteria* pertinent to the decision of bit sequencing. Not all of the criteria were satisfied by the committee's final decision. Some of the criteria are conflicting. The final decision on bit sequencing was based on a detailed analysis and weighting

---

*The ten Specific Criteria are reproduced with permission from American National Standard for Bit Sequencing of the American National Standard Code for Information Interchange in Serial-by-Bit Data Transmission, X3.15-1966, copyright 1966, by the American National Standards Institute at 1430 Broadway, New York, New York 10018. The Criteria are reproduced from Appendix A2 of the Standard X3.15-1966. The Standard is available from the American National Standards Institute at 1430 Broadway, New York, New York 10018.

of the criteria. The Specific Criteria follow:

1. The transmission bit sequence should be in consecutive numerical order (ascending or descending) in terms of ASCII nomenclature.
2. The transmission bit sequence should minimize the amount and complexity of existing and future hardware.
3. The transmission bit sequence should be selected to minimize adverse consequences of equipment or system malfunction.
4. The transmission of a binary bit stream should not be precluded.
5. The transmission of encrypted material should not be precluded.
6. There should be a correspondence among media track (channel or row) designation, ASCII bit number, and transmission bit sequence, in order to minimize training and reduce confusion of operating, maintenance, and engineering personnel.
7. The transmission bit sequence should allow a logical extension of supersets of ASCII.
8. The transmission bit sequence of any subset or superset of ASCII should provide that any designated bit be immutable in its position in the transmission sequence as well as in its logical order and media representation.
9. The character parity bit should be positioned to allow it to be generated "on the fly," following the data bits.
10. The transmission bit sequence should allow maximum design flexibility in future systems utilizing ASCII.

The two bit-sequencing choices, high-order bit first, or low-order bit first, were then investigated to determine their influence on data interchange from the following points of view:

a) flexibility of hardware design,
b) hardware efficiency,
c) ease of maintenance,
d) contraction of ASCII to subsets,
e) expansion of ASCII to supersets, and
f) system reliability.

The arguments that were advanced to the committee are now reproduced. It is to be emphasized that the author does not testify to the validity or significance of the arguments. He merely reports the arguments. The arguments are grouped under the last five of the above points of view. In parentheses after each argument is indicated whether the argument was in favor of transmission high- or low-order bit first.

15.2 HARDWARE EFFICIENCY

1. Although the bit sequence is immaterial in a great majority of today's applications, nevertheless specific cases were considered in which either one or the other bit order was advantageous.
2. When ASCII is transmitted high-order bit first, it is possible to determine, by the first two or three bits received, the general use of the character and, in certain classes of equipments, thereby know the routing and final disposition of the remaining bits. In particular, this can reduce the necessary bit storage in I/O typewriters where reduction in bit-storage requirements can be a reasonably significant portion of the total cost. (high)
3. The problem of mapping the 7-bit ASCII code into a 6-bit data processor character code can be simplified if the high-order bit is placed first.

   In particular, the first two bits received may be sufficent to generate an "escape" character prior to reception of the complete ASCII character, thus allowing a considerably longer effective time upon completion of reception of the ASCII character with consequent increase in traffic handling capacity for a given equipment. (high)
4. Time (clock) codes are transmitted low-order bit first and low-order character first so that the fine detail will appear earlier, and the redundant, infrequently changing coarse portions will appear later in each time code group. If it is desired to intersperse time codes in general interchange data, less confusion should arise, and less hardware should be required, if the interchange data is also transmitted low-order bit first. (low)

15.3 EASE OF MAINTENANCE

1. With low-order bit transmitted first, the first data pulse can correspond to ASCII bit b1, the second to bit b2, etc. Thus "third" will mean third pulse as well as bit b3. It can also mean third track (or channel or row) in media. This extremely simple relationship among media track number, pulse number, and bit designation number is highly desirable in the maintenance of communication equipment, especially in discussions between remote technicians or between technicians and engineers. (low)

2. This correspondence argument was at least partially offset in asynchronous systems where serially received bits are accompanied by synchronization bits. Thus the received ASCII bit b1 could actually be the second received bit, bit b2 could be the third received bit, etc. (high)

15.4 CONTRACTION TO ASCII SUBSETS

1. Logic for serial recognition of characters limited to specific coding groups of the 7-bit ASCII is expected to be implemented with less total hardware where transmission is sequential with high-order bit first. (high)
2. If subsets of ASCII, such as a 4-bit numeric subset or a 6-bit graphic subset, are used, then the low-order-bit-first arrangement allows high-order bits to be appended "on the fly," according to logical rules, for transmission of the full 7-bit ASCII. (low)
3. Equipment receiving the full 7-bit ASCII, but operating on only a subset, may, with the low-order bit first, obtain the subset by simply ignoring bits received after the prescribed number for each ASCII character received. (low)

15.5 EXPANSION TO ASCII SUPERSETS

It has not been decided just how the 7 bits (b7 through b1) of ASCII will be represented in an 8-bit environment. If a superset takes the form of an 8th bit which is higher in order than bit 7, then

1. In the expansion and contraction between both 7- and 8-bit sets and 6- and 8-bit sets, only the data contained in the high-order bits will be needed to determine the transformation. The transmission of high-order bit first provides the maximum time to convert between the sets. (high)
2. With low-order bit transmitted first, compatibility between terminal equipments using ASCII and terminal equipments using an 8-bit superset of ASCII may be simplified, and transmission switching equipment may more readily handle either mode of transmission. (low)

15.6 RELIABILITY

1. Asynchronous transmission of characters results in a greater probability of error in the later bits transmitted.

2. If low-order bits are transmitted first, an error in the later bits would tend to convert some graphics to control characters. (high)
3. If high-order bits are transmitted first, numerics may be converted to other numerics, and control characters to other control characters. (low)

After many committee meetings, long discussion, and the consideration of over seventy technical papers on the subject, the standards committee decided in favor of low-order bit first for serial-by-bit–serial-by-character data transmission.

Author's Note

All data is transmitted high-order *character* first, and it may be observed that similar arguments for order-of-*character* transmission could be made as for order-of-*bit* transmission. That is to say, it might have been argued that since similar reasons could have been advanced for order-of-*character* transmission but that nevertheless high-order-*character* transmission is universally practiced, it would seem to be logical to conclude that high-order-*bit* transmission first should become the practice. This argument, however, was not introduced into the discussion.

16
Decimal ASCII

After the bit code ASCII became an approved American National Standard in 1963 (actually termed "American Standard" then), the attention of the standards committee turned to developing standards for the representation of the code on the principal media, perforated tape, magnetic tape, and punched card.

16.1 PERFORATED TAPE

The representation for perforated tape presented no technical problems. A common form of perforated tape of the day was one-inch, eight-track paper tape. It was soon agreed

a) To number the tracks of the tape 1, 2, 3, 4, 5, 6, 7, 8.

b) To record the seven bits of the code:
   b1 in track 1
   b2 in track 2
   .
   .
   .
   b7 in track 7.

c) To use track 8 as a parity track.

16.2 MAGNETIC TAPE

The problem for magnetic tape was not quite so simple. First, the committee decided to reject as a candidate the existing magnetic tape of the day—half-inch, seven tracks. One of the seven tracks was dedicated to parity—odd parity for some computing systems, even parity for other computing systems.

If a track were to be dedicated to parity for the standard on magnetic tape, and the standards committee agreed that it should be, then only six tracks would remain to record the seven bits of the code. While it is feasible to devise a theoretical scheme for recording the 128 characters of a 7-bit code on 6 tracks (and, indeed an ISO Recommendation for just such a scheme was eventually approved), the American standards committee deemed such a scheme unacceptable for an American standard. As described in Chapter 20, the standards committee proposed a recording format of nine tracks, and eventually, the representation of ASCII on magnetic tape became an approved American standard.

16.3 PUNCHED CARDS

The problem of deciding how to record ASCII on punched cards turned out to be extremely troublesome.

The most common form of punched cards in use in the U.S.A. at the time used a 12-row, rectangular-holes representation (which came to be called the Hollerith Card Code in the U.S.A.). A less common representation, provided by the UNIVAC Division of the Sperry-Rand Corporation, used a punched card of virtually the same size, and twelve rows of punching, but the holes were circular. The initial draft American standard specified both the rectangular-hole and circular-hole representations.

Eventually, the standards committee voted to exclude the circular-hole representation from further consideration, for an interesting reason. The circular-hole card had 45 columns of punching. The encoding format divided the twelve rows into two tiers of 6 holes per tier. The card was visualized as having 90 columns, and was frequently called the 90-column card (the rectangular hole card had 80 columns of punching, and was frequently called the 80-column card). But these 90 columns had only 6 punchable rows per column and therefore could record a maximum of 64 different characters. The problem with the 90-column card was the same as the problem for magnetic tape (6 data tracks plus 1 parity track). How could the 128 characters of ASCII be recorded on the 6 rows of the card? Of course, physically, the card had 12 rows. The alternatives for the circular-hole card were

a) Using the 12 rows as necessary, record all 128 characters of ASCII, but then have a capacity of only 45 columns per card.
b) Record only 64 characters of ASCII, and have a capacity of 90 columns per card.

c) Use 90 columns, record all 128 characters of ASCII by some complicated recording scheme involving the concept of a shifted or precedence code.

None of these alternatives was attractive, and the committee dropped the circular-hole card from further consideration.

Attention then focused solely on the rectangular-hole card. Incidentally, the "name" of this kind of card enjoyed some changes. During the discussion of the circular-hole card, it was necessary to differentiate between the two kinds of cards. "Circular-hole card" and "rectangular-hole card" were two differentiating names; "90-column card" and "80-column card" were more commonly used differentiating names (both kinds of card had 12 rows, so this characteristic could not be used to differentiate). After the circular-hole card was dropped from further consideration, the remaining card was referred to as the 80-column card for a while. But it was pointed out that this name was a misnomer, because different lengths of the card (that is, different numbers of columns) were available in the market. At this point, therefore, the standards committee began to refer to the rectangular-hole card as the 12-row card (12 rows being a characteristic of such cards regardless of length). The ISO Recommendations on punched cards refer to the card as the 12-Row Card.

16.4 BINARY REPRESENTATION

The standards committee now focussed on the 12-row card. At first, the problem seemed simply solvable. The card has twelve rows, commonly named the 12-row, the 11-row, the 0-row, the 1-row, the 2-row, . . . , the 9-row, as shown in Fig. 16.1. Some members of the standards committee

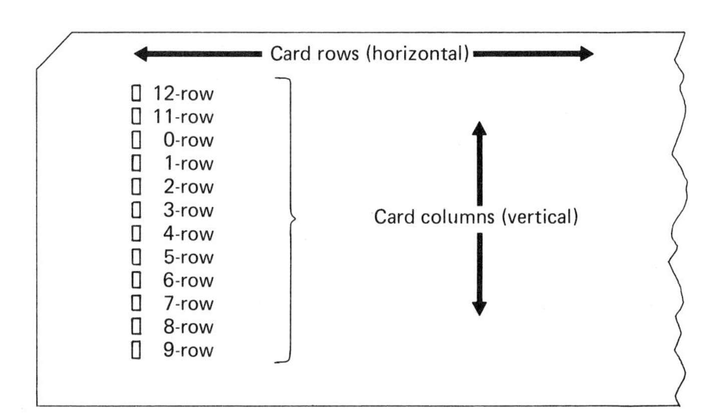

Fig. 16.1 Punched card

suggested that the code be recorded according to a very simple algorithm:

a) b1 would be recorded in the 1-row,
   b2 would be recorded in the 2-row,
   b3 would be recorded in the 3-row,
   b4 would be recorded in the 4-row,
   b5 would be recorded in the 5-row,
   b6 would be recorded in the 6-row,
   b7 would be recorded in the 7-row.

b) When a bit of the bit pattern is 1, punch a hole.
   When a bit of the bit pattern is 0, leave the hole position unpunched.

This proposed representation on punched cards came to be called the Binary Representation (and later came to be called the Direct Binary Representation). The advocates of the Binary Representation pointed out its advantages:

1. It was a simple, direct representation (no translation required).
2. If it became necessary some day in the future to expand the 7-bit code to an 8-bit code, the eighth bit of such a code could be recorded in the 8-row.
3. The 12-row, 11-row, and 9-row would be available so that error-checking, and even error-correcting, schemes could be implemented, a facility not previously available with the punched card medium.

The initial argument *against* the Binary Representation was that it was completely different from the existing Hollerith Card Code. This argument was discounted by the Binary Representation proponents. After all, they argued, ASCII was different from any existing code; the representation of ASCII on magnetic tape would be different from any existing magnetic tape code; the representation of ASCII on paper tape would be different from any existing paper tape code. So what was alarming about the suggestion that the representation of ASCII be different from the existing punched card code? While the opponents of the Binary Representation grappled with this argument, a much more telling objection emerged.

16.5 NUMBERS OF HOLES

Observe, said the Direct Binary opponents, what happens when the numerics of ASCII are recorded on the punched card under such a scheme (see Fig. 16.2).

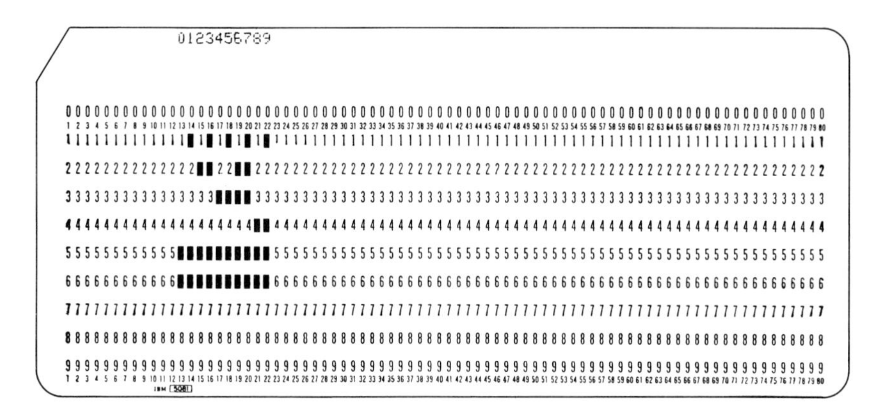

Fig. 16.2 Numerics, Binary Representation

Two facts emerge with respect to numerics:

1. 0 requires 2 holes; 1, 2, 4, and 8 require 3 holes; 3, 5, 6, and 9 require 4 holes; 7 requires 5 holes; this gives an average of 3.5 holes per numeric.
2. For all numerics, both the 5-row and the 6-row have punched holes.

These two facts contrast with the equivalent facts for the Hollerith Representation of numerics (Fig. 16.3).

1. Each numeric requires exactly one hole.
2. There is a different row punched for each different numeric.

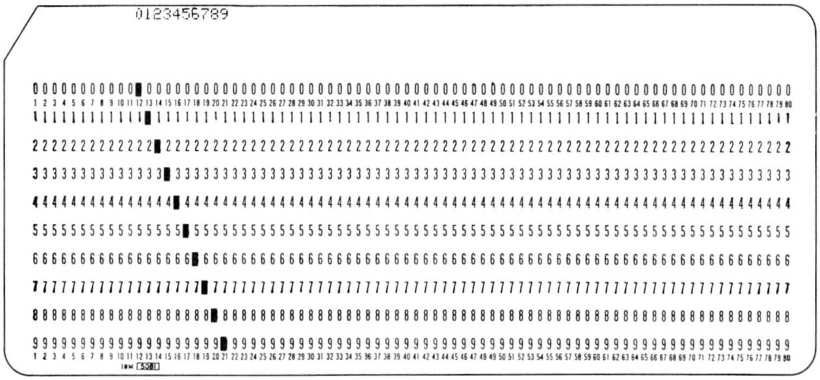

Fig. 16.3 Numerics, Hollerith Representation

The significance of fact (1) is that if more holes are required per character, the dies that punch the holes will wear out sooner, and thus maintenance costs will be higher. The statistics above for fact (1) relate to numerics only. Consider the statistics for all 64 characters in columns 2, 3, 4, 5 of the code:

2	characters require 1 hole =	2
10	characters require 2 holes =	20
20	characters require 3 holes =	60
20	characters require 4 holes =	80
10	characters require 5 holes =	50
2	characters require 6 holes =	12
64		224

The average is 224/64 = 3.4 holes per character.

By contrast, consider the Hollerith Card Code associated with BCDIC (see Fig. 16.4):

1	character requires 0 holes =	0
12	characters require 2 holes =	24
35	characters require 2 holes =	70
16	characters require 3 holes =	48
64		142

The average is 142/64 = 2.2 holes per character.

We have, then, average number of holes per character, as shown below:

Kind of characters	Binary Representation	Hollerith Representation
Numerics	3.5	1
All 64 characters	3.4	2.2

For numeric data, which was estimated at that time to constitute 75 percent of all data punched, we have 3.5 holes per character compared to 1 hole per character. For the 64 graphic characters, we have 3.4 holes per character compared to 2.2 holes per character.

Bit Pattern →			A	B	BA
↓	Hole Pattern →		0	11	12
	↓	SP	b̸ [1]	-	& or +
1	1	1	/	J	A
2	2	2	S	K	B
2 1	3	3	T	L	C
4	4	4	U	M	D
4 1	5	5	V	N	E
4 2	6	6	W	O	F
4 2 1	7	7	X	P	G
8	8	8	Y	Q	H
8 1	9	9	Z	R	I
8 2	0	0	‡ [2]	!	?
8 2 1	8-3	# or =	,	$	.
8 4	8-4	@ or '	% or (	*	⌑ or)
8 4 1	8-5	:	γ	]	[
8 4 2	8-6	>	\	;	<
8 4 2 1	8-7	√	⧻	Δ	‡

SP - Space

Hole Patterns:
[1] 8-2
[2] 0-8-2

Fig. 16.4 BCDIC, Hollerith Card Code

Fact (1) led to the conclusion that the Binary Representation would result in higher maintenance costs for punched card equipment than the existing Hollerith Representation. And, it is important to note, this would not be a one-time conversion cost (because of converting from one code to another); it would be a continuing cost.

16.6 LACING

Fact (2), however, led to an even more compelling argument against the Binary Representation. Observe Fig. 16.5. For numeric data, for *all* numerics, rows 5 and 6 are punched. If a card was punched with numeric

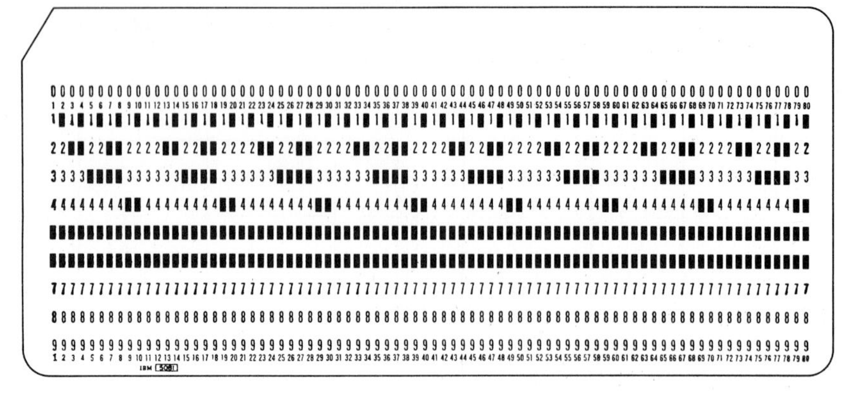

Fig. 16.5 Laced card

data only, the rows 5 and 6 would be punched completely across the length of the card. The technical term used for this phenomenon is "lacing." For the Binary Representation, the card would be "laced" completely across two rows for numeric data.

The punched card is unique among the physical recording media in one very significant aspect, the way in which it is handled by human beings. Of course, reels of magnetic tape are also handled by human beings. And rolls of paper tape or lengths of paper tape are also handled by human beings. But these human beings are operators in a computing-room environment who handle the magnetic tape reels, or the paper tape, with some care. Punched cards, by contrast, go out of the computing-room environment into the hands of people who, not infrequently, treat the card with considerable roughness. The punched card is used for pay checks, for insurance premium billing, for utility billing, etc. In many cases, the punched card goes to people outside the computing-room environment, and is then subsequently returned for further computer processing. The cards may be folded, crumpled, wetted, scraped, torn, spindled, etc. (The famous phrase, "Do not spindle, fold, or mutilate" was devised by Mr. Charles A. Phillips in the hope that people, so advised, would treat cards more carefully.)

The punched card is made of a fairly stiff paper stock. To some extent, it resists folding, wrinkling, tearing, etc. The presence in a punched card of two rows laced across the length of the card clearly make it much more susceptible to damage when casually or roughly treated by human beings. The thrust of this argument was that the Binary Representation would make the card unreliable. On the standards committee, manufacturers of punched card equipment were unanimous in their opposition to the Binary Representation; partly because of the potential increase in continuing maintenance costs but mainly because of the potential unreliability of the punched card which would result.

The proponents of the Binary Representation offset the cost argument with a counter argument on cost. A hardware translator to translate ASCII to/from the Hollerith Representation would be very much more costly than a hardware translator to translate ASCII to/from the Binary Representation. But the reliability argument could not be offset. At first, it was suggested that using the 12- and/or 11-rows for error checking or error correcting would partly compensate for the unreliability aspect. But, punching 12 or 11 rows would add even more holes per character, which would worsen the maintenance cost situation.

16.7 MODIFIED BINARY REPRESENTATIONS

The reliability defect of the Binary Representation stemmed from the lacing phenomenon, which stemmed from the three high-order zone bits of ASCII. This defect could clearly be removed if the numerics had no zone bits. The solution now advanced by the Binary Representation proponents was to modify the binary representation as punched on the card by modifying the zone holes. Two representations were proposed for consideration—the Modified Binary Representation and the Optimum Modified Binary Representation. In both these representations, the numerics had no zone punches in the 5-row or 6-row, so the lacing phenomenon disappeared for numerics.

The zone bits for the three binary representations are shown in Fig. 16.6. The three binary representations and the Hollerith Representation are compared in Fig. 16.7, which shows the average number of holes per numeric and the average number of holes for the 64 characters of table-columns 2, 3, 4, and 5 of ASCII.

While the Optimum Modified Binary Representation came the closest to Hollerith in average number of holes per numeric or character, it suffered from some other defects:

1. A 64-character, 6-bit subset from columns 2, 3, 4, and 5 of the 7-bit code cannot be generated by simply dropping one bit.
2. The translation algorithm, ASCII to/from Representation, is somewhat complex (although not as complex as the one to/from Hollerith).

If the three high-order bits of the Optimum Modified Binary Representation are b_7', b_6', b_5' and the three high-order bits of ASCII are b7, b6, b5,

Representation \ Table column	0	1	2	3	4	5	6	7	
Direct Binary	0	0	0	0	1	1	1	1	b7
	0	0	1	1	0	0	1	1	b6
	0	1	0	1	0	1	0	1	b5
Modified Binary	1	1	0	0	0	0	1	1	b7
	1	0	1	0	0	1	0	1	b6
	1	0	1	0	1	0	1	0	b5
Optimum Modified Binary	1	0	1	0	0	0	1	1	b7
	1	1	0	0	0	1	0	1	b6
	1	1	0	0	1	0	1	0	b5

Fig. 16.6 Binary representation

Representation	Average number of holes per character	
	Numerics	All 64 characters
Direct Binary	3.5	3.4
Modified Binary	1.4	3
Optimum Modified Binary	1.4	2.7
Hollerith	1	2.2

Fig. 16.7 Average number of holes per character

then the translation equations are

$$b_7' = (\overline{b7} \wedge \overline{b5}) \veebar (b7 \wedge b6)$$
$$b_6' = (\overline{b7} \wedge \overline{b6}) \veebar (b7 \wedge b5)$$
$$b_5' = (\overline{b7} \wedge \overline{b6}) \veebar (b7 \wedge \overline{b5})$$

With respect to all three Binary Representations, two more problems arose, which came to be called the Null/Space/Blank Problem, and the Plus and Minus Zero Problem.

16.8 NULL/SPACE/BLANK PROBLEM

In punched card applications, a blank card column, with no holes punched, represented one of three things, depending on the application:

1. A card column not used in the application.
2. A card column not punched in the initial keypunching operation but punched in a subsequent card-punching operation.
3. A space; that is, if the card is listed on either serial or parallel printers, blank card columns would be represented by unprinted printing positions on the paper.

In practice, the blank card column was equated to the Space character. In keypunching, blank card columns are created by depressing the Space bar, or by skipping the card to a subsequent card column, or by ejecting the card. These operations are precisely analagous to the typing operations of Space, Horizontal Tabulation, and Carriage Return. The format of data on the punched card is precisely analogous to the format of data printed from the card.

Observe, however, the hole patterns for Null, Space, and Zero in the Binary Representations (Fig. 16.8).

Character \ Representation	Direct Binary	Modified Binary	Optimum Modified Binary
Null	Blank column	7-6-5 punches	7-6-5 punches
Space	6 punch	6-5 punches	7 punch
Zero	6-5 punches	Blank column	Blank column

Fig. 16.8 Null/Space/Zero hole patterns

The blank card column is associated with the Null character in the Direct Binary Representation, and with the Zero character in the Modified and Optimum Binary Representation. In no case is blank card column associated with the Space character.

At first, the Binary proponents took the following lines:

For the Direct Binary Representation. In the future, associate the blank card column on punched cards with the Null character. On keypunches the "old" Space bar would now have to be called the Null bar, but a change in nomenclature should not be too distressing to users.

The Binary opponents held that this proposal would be unacceptable.

The proposal might be acceptable purely in the context of the punched card environment. But punched cards do not exist in a vacuum. A common punched card application is to read a deck of punched cards into a card reader, translate the data to a transmission code, and then transmit the data to some other location for further processing. But it was known that some communications products, when receiving the Null character, would not transmit it further. Also, it was known that, for various reasons in some data transmission systems, Null characters are injected into the data stream.

In short, Null characters might be injected into, or removed from, the data stream. In the context of the punched card used in a data transmission application, if the Null character was equated to a blank card column, this would mean that, under data transmission, blank card columns would be added to, or removed from, the punched cards. Even the Binary proponents had to concede that such a consequence would be intolerable.

For the Modified and Optimum Modified Binary Representations. In the future, associate the blank card column on punched cards with the Zero character.

The Binary opponents held that this proposal would be unacceptable.

The proposal to equate the blank card column with the zero character would lead to a dilemma. Consider a card punched as shown in Fig. 16.9. Card-columns 1, 2, 4, 7, 12, 13, 16, and 19 are punched with

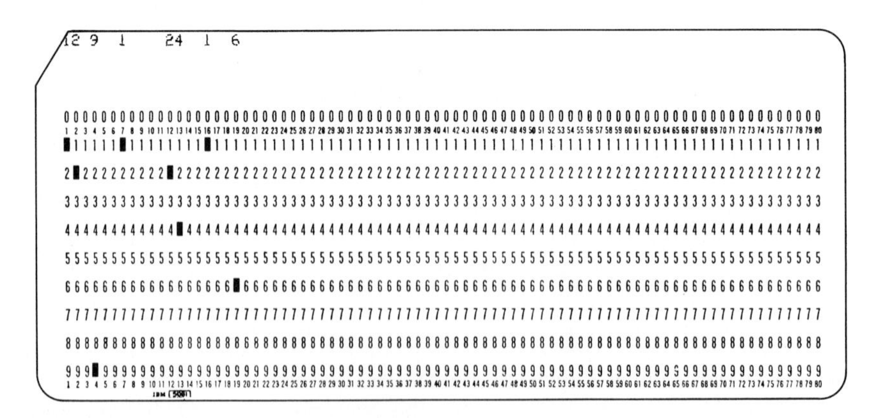

Fig. 16.9 Card with blank columns

numerics. Card-columns 3, 5, 6, 8, 9, 10, 11, 14, 15, 17, 18 and 20 through 80 are blank. This punching represents numeric fields 12809001 in card-columns 1 through 8, and 24001006 in card-columns 12 through 19.

The dilemma is how to list such a card. Card-columns 3, 5, 6, 14, 15, 17, and 18 were Zeros in the data, and should be listed as Zeros. But card-columns 8 through 11 and 20 through 80, although blank card columns, should *not* be listed as Zeros, but as Spaces. And there is no way for a printer to tell when a blank card column means Zero, and when it means Space.

The Binary proponents responded that the problem is that card columns 8 through 11 and 20 through 80 should not have been blank card columns, which is equated to the Zero character, but should have been punched whatever hole pattern would be associated with the Space character.

The Binary opponents labeled this unacceptable for two reasons:

1. Card-columns 8 through 11 would normally be created as blank card columns in keypunching by skipping, and card-columns 20 through 80 by ejecting. Now, while it might be feasible to modify keypunches so that they would create the specific hole pattern for the Space character on skipping or on ejecting, the modification would reduce the relatively fast card motion of skipping or ejecting to the relatively slow card motion of punching. That is to say, the consequence of such a keypunch would be a substantial reduction in keypunch productivity.
2. How would one provide the traditional capability of leaving certain card columns unpunched (blank card columns) during keypunching to be filled with punched data on subsequent card processing operations? Such card columns would, in fact, have to be created by punching the Zero character that is equated to blank card column. In normal keypunching operations, such card columns are created by spacing, skipping, or ejecting. Under this proposal, then, the relatively fast card motion of skipping or ejecting would be replaced by the relatively slow card motion of manual keying by an operator. As in the previous argument, key punching productivity would be substantially reduced.

After much discussion, it was accepted that none of the Binary Representations, as shown in Fig. 16.6, would be viable, because of the Null/Space/Blank Problem. The Binary proponents then made some new proposals. Under these proposals, the zone hole patterns shown in Fig.

Representation \ Table column	0	1	2	3	4	5	6	7	
Direct Binary	0	0	0	0	1	1	1	1	b7
	1	0	0	1	0	0	1	1	b6
	0	1	0	1	0	1	0	1	b5
Modified Binary	1	1	0	0	0	0	1	1	b7
	0	1	0	0	1	1	1	0	b6
	0	1	0	1	0	1	1	1	b5
Optimum Modified Binary	1	0	0	1	0	0	1	1	b7
	1	1	0	0	1	0	0	1	b6
	1	1	0	0	0	1	1	0	b5

Fig. 16.10 Row 0

16.6 would hold for rows 1 through 15 of the code table, but for row 0 of the code table, some changes should be made, as shown in Fig. 16.10.

Note, in Fig. 16.10, that for all three Binary Representations, the Space character, which is Column 2, Row 0 of the ASCII code table, is equated to blank card column. The three high-order bits in Fig. 16.10 have been chosen to preserve the desirable characteristics of each of the Binary Representations and, at the same time, to minimize the translation complexity—ASCII to/from Binary card-code representation.

This proposal would, of course, introduce translation complexity into the translation of ASCII to/from Binary card code. And translation simplicity, or requirement for no translation at all, was the primary and in fact the only argument in favor of a binary card-code representation over the de facto Hollerith card-code representation. The Binary opponents pointed out this undesirable consequence.

The Null/Space/Blank Problem in the context of Binary Representation was *not* resolved by the standards committee, for a reason that will emerge later in this chapter.

16.9 PLUS AND MINUS ZERO PROBLEM

The capability to store greater and greater quantities of data has been a requirement since the very beginning of data processing. Insufficient memory capacity, data records overrunning magnetic tape reels or paper tape reels, etc., have plagued, and will probably always plague, the data processing industry. Punched cards as a medium for storing data are not exempt from, and in fact are particularly prone to, this aggravation. How many readers of this book have experienced the aggravation of trying to squeeze 81 characters into an 80-column card? Indeed, most modern schemes of packing or compacting data had their forerunners in punched card applications.

One very common "trick" was to make a single card column do double, triple, or multiple duty. This was particularly evident in statistical applications. For example, the 12-punch could be used to signify male or female; the 11-punch, married or single; and the numerics 0 through 9 could be used to specify some other statistical characteristic.

A widespread convention was the use of a 12-punch, an 11-punch, or neither of these, to signify positive, negative, or absolute numerics, respectively. Usually the units position of a numeric field on a card was the sign position. Either the 12- or 11-punch was punched over the appropriate units position of a numeric field (as well as punching the actual digit for the units position). Since 12- and 11-punches, in conjunction with numeric punches, also had the meanings of alphabetics, the result was dual meanings for these hole patterns, as shown in Fig. 16.11.

A crucial aspect of this convention for signed numerics was that they must be keypunchable by the technique of overpunching. A skilled keypunch operator, being required to keypunch −3, for example, would

Hole pattern	Meaning		Hole pattern	Meaning		Hole pattern	Meaning	
	Alphabetic	Numeric		Alphabetic	Numeric		Alphabetic	Numeric
0		0	12-0	*	+0	11-0	*	−0
1		1	12-1	A	+1	11-1	J	−1
2		2	12-2	B	+2	11-2	K	−2
3		3	12-3	C	+3	11-3	L	−3
4		4	12-4	D	+4	11-4	M	−4
5		5	12-5	E	+5	11-5	N	−5
6		6	12-6	F	+6	11-6	O	−6
7		7	12-7	G	+7	11-7	P	−7
8		8	12-8	H	+8	11-8	Q	−8
9		9	12-9	I	+9	11-9	R	−9

* In BCDIC, 12-0 and 11-0 have the meanings of ? and !, respectively. In EBCDIC, they have the meanings of { and }, respectively.

Fig. 16.11 Overpunched numerics

know that this was equivalent to the alphabetic L, and would depress the L key. However, a less skilled operator would use the multipunch key that had the function that, when depressed, would allow further key depressions of alphabetic, numeric, or special keys that would generate the appropriate hole patterns in the card, but the card would not advance longitudinally until the multipunch key was released; that is to say, multiple punches could be created in a single card column. The operator, then, being required to generate the 11-3 hole pattern, for −3, would depress the multipunch key, would then depress the 3 key, then depress the − key (which generates an 11-punch), then release the multipunch key. Similarly to generate the hole pattern 12-3 for +3, the sequence would be depress multipunch key, depress 3 key, depress + key (which generates a 12-punch), release multipunch key.*

The requirement that signed numerics be keypunchable in this fashion places an interesting constraint on hole patterns for signed numerics. The hole pattern for positive, or for negative, must not conflict with the hole patterns for numerics. In the case of the Hollerith Card Code, where numerics had hole patterns 0 through 9, this constraint was met by the hole patterns for numerics. In the case of the Hollerith Card Code, where respectively.

What would this constraint say with respect to a Binary Representation? Given that the numerics are represented by BCD equivalents, that is, punches in rows 1, 2, 3, 4 of the card, the hole patterns for positive and negative must be restricted to rows 5, 6, 7, that is, to the zone rows. Further, if the same convention would be used—minus sign for negative zone and plus sign for positive zone—then the hole patterns for plus sign and minus sign must *not* have any holes in card rows 1, 2, 3, 4, for they would then conflict with hole patterns for numerics. But this constraint cannot be met, since ASCII plus sign and minus sign are in table-rows 11 and 13; that is, they would have hole patterns in card-rows 1, 2, 3, 4.

There is, then, no way in which the sign-overpunch-numeric convention can be incorporated into a Binary card code, unless the minus sign and plus sign had zone bits only, no digit bits; that is, plus sign and minus sign to be in row 0 of the ASCII code table. Such a change to ASCII itself was not acceptable.

The Binary proponents proposed that this problem be solved by making the problem go away. They proposed that, with a Binary Representation, algebraic sign be represented not by overpunching but by carrying the algebraic sign in a separate card column. The Binary propo-

* Whether the sequence was first 3 key and then − key or first − key and then 3 key was immaterial.

nents then were proposing not only that the user change his card code, from Hollerith to Binary, but also that he give up the widespread practice of overpunching numerics for algebraic sign.

This problem, as with the Null/Space/Blank Problem, was *not* resolved, for a reason that will emerge later in this chapter.

In the remainder of this chapter, various card codes are illustrated and described. Some of these card codes have 128 hole patterns and are taken in conjunction with a 7-bit code. Other card codes have 256 hole patterns and are taken in conjuction with an 8-bit code. Both the Null/Space/Blank Problem and the Plus and Minus Zero Problem emerged with respect to some of these codes, and they became major points of technical controversy on the standards committees.

16.10 TRANSLATION SIMPLICITY

An aspect of these card codes that became crucial in discussions was the translation relationship, card code to/from bit code. The relative simplicity or complexity of translation became a factor for decision between candidate card codes. Boolean equations for the various card codes are set down in this chapter, using the notation described in Chapter 2. When comparing equations, the three simplifying assumptions of Chapter 2 (repeated here for emphasis) are made.

Assumption 1. The circuit complexity is equal to implement each of the four Boolean operators:

AND $\wedge$
Inclusive OR $\vee$
Exclusive OR $\veebar$
IDENTITY $\equiv$

Assumption 2. The circuitry which generates a bit generates the inverse of a bit with no additional complexity.

Assumption 3. Given two sets of Boolean equations representing two sets of translation relationships, the relative circuit complexity of implementing the relationships is proportional to the number of Boolean operators in the equations.

It should be understood that, to implement a hardware translator, bit code to/from card code, *two* sets of equations are necessary; the equations for deriving bit patterns from hole patterns, and the equations for deriving hole patterns from bit patterns. However, in order to compare two card codes for relative complexity, one set of equations is sufficient.

Accordingly, in this chapter, we set down only the equations for deriving bit patterns from hole patterns.

At this point it must be stated that the equations for deriving EBCDIC bit patterns from EBCDIC hole patterns (to be given later) are not necessarily the actual set of equations used in implementing hardware translators. The equations for EBCDIC were derived by the author purely for purposes of illustration and comparison in this chapter. The

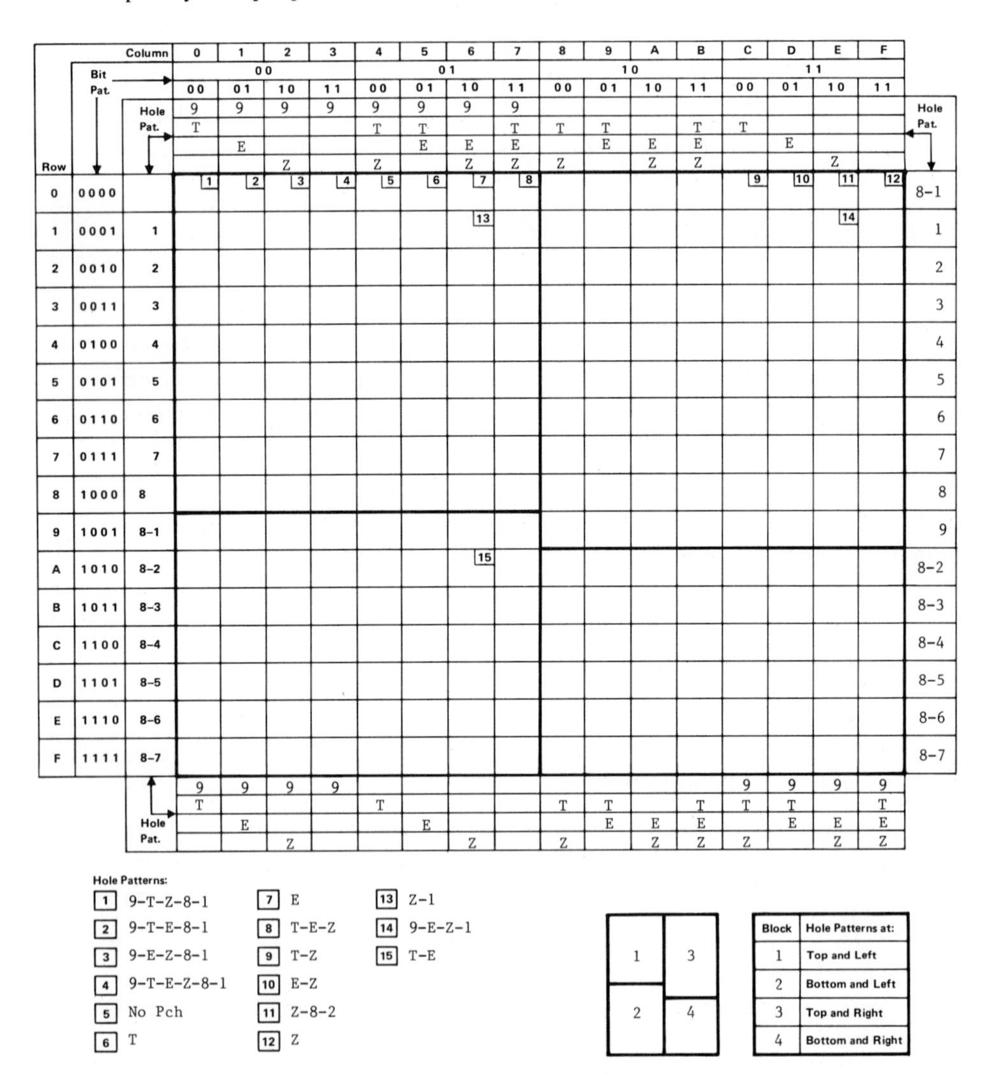

Fig. 16.12 EBCDIC, 1963

optimization of Boolean equations is an art. It is quite possible that the EBCDIC equations given here could be optimized further. However, they are adequate for the purposes of this chapter.

During the early part of 1963, the author had been evolving the bit code and card code that came to be called EBCDIC. As described in Chapter 8, two criteria were of major importance; the embedment of BCDIC collating sequence in the EBCDIC collating sequence, and upward compatibility of the BCDIC card code to the EBCDIC card code. These two requirements together resulted in less than optimal simplicity in the translation relationships, EBCDIC card code to/from bit code. In consequence, at that time, the EBCDIC card code had not been adopted in IBM. The EBCDIC bit code and card code then under consideration are shown in Fig. 16.12.

16.11 BENDIX PRIME

The author had been requested to review a card code provided on some card equipment by the Bendix Corporation, to see if it might lead to a card code with simpler translation relationships to EBCDIC. Also, the "Bendix card code" did not suffer from the defects described above for binary card codes.

The "Bendix card code," per se, will not be described in this book. However, the principle of the Bendix card code is interesting and will be described. It will be called "Bendix Prime" for purposes of reference.

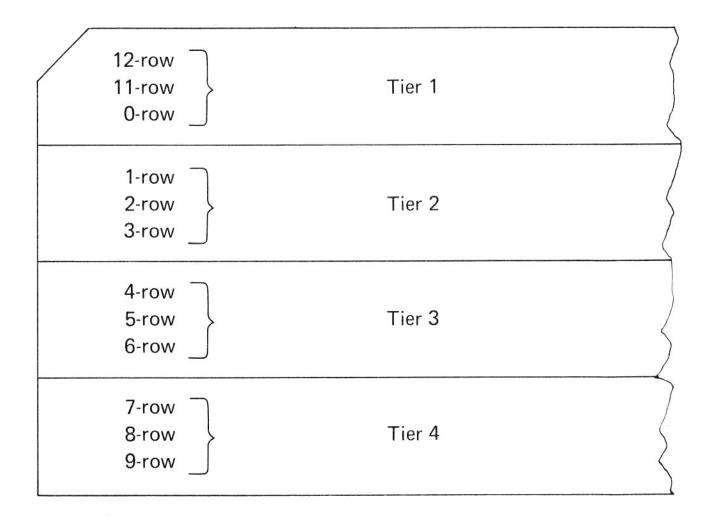

Fig. 16.13 Bendix card

There are twelve punching rows in the punched card. In Bendix Prime, these are grouped in four tiers of three rows each, as shown in Fig. 16.13.

Within a tier within a card column, only one of the three rows may be punched, or none may be punched. For example, within a card column, in the third tier, there are four possible hole patterns; 4-hole, 5-hole, 6-hole, or no holes. There are therefore four possible hole patterns for each tier, and there are four tiers. Hence, within a card column, there are $4 \times 4 \times 4 \times 4 = 256$ different possible hole patterns. That is to say, the Bendix Prime card code could be used to represent 256 characters.

One possible Bendix Prime representation is shown in Fig. 16.14. For convenience, the twelve-, eleven-, and zero-rows are represented by

		Column	0	1	2	3	4	5	6	7	8	9	A	B	C	D	E	F	
	Bit Pat.		00				01				10				11				
			00	01	10	11	00	01	10	11	00	01	10	11	00	01	10	11	
		Hole Pat.					T	T	T	T									Hole Pat.
											E	E	E	E					
															Z	Z	Z	Z	
Row				1	2	3		1	2	3		1	2	3		1	2	3	
0	0000																		
1	0001	7																	
2	0010	8																	
3	0011	9																	
4	0100	4																	
5	0101	4–7																	
6	0110	4–8																	
7	0111	4–9																	
8	1000	5																	
9	1001	5–7																	
A	1010	5–8																	
B	1011	5–9																	
C	1100	6																	
D	1101	6–7																	
E	1110	6–8																	
F	1111	6–9																	
		Hole Pat.																	

Fig. 16.14 Bendix Prime

$$
\begin{aligned}
\mathrm{E0} &= E \veebar Z \\
\mathrm{E1} &= T \veebar Z \\
\mathrm{E2} &= 2 \veebar 3 \\
\mathrm{E3} &= 1 \veebar 3 \\
\mathrm{E4} &= 5 \veebar 6 \\
\mathrm{E5} &= 4 \veebar 6 \\
\mathrm{E6} &= 8 \veebar 9 \\
\mathrm{E7} &= 7 \veebar 9
\end{aligned}
$$

Fig. 16.15 Bendix Prime equations

Common expressions

$$A = 2 \veebar 3$$

$$R = 4 \veebar 5$$

$$C = 6 \veebar 7$$

$$D = \bar{9} \wedge \bar{8} \wedge \bar{1}$$

$$F = A \veebar R \veebar C$$

$$G = (T \wedge \bar{E} \wedge \bar{Z}) \veebar (\bar{T} \wedge E \wedge \bar{Z}) \veebar (\bar{T} \wedge \bar{E} \wedge Z) \veebar (\bar{T} \wedge \bar{E} \wedge \bar{Z})$$

$$\bar{G} = (T \wedge \bar{E} \wedge Z) \veebar (T \wedge E \wedge \bar{Z}) \veebar (\bar{T} \wedge E \wedge Z) \veebar (T \wedge E \wedge Z)$$

$$J = \overline{(T \wedge E)} \wedge (1 \veebar Z)$$

$$L = \bar{T} \wedge \bar{E} \wedge Z \wedge 8 \wedge 2 \wedge \bar{9}$$

$$K = (1 \wedge \bar{F} \wedge \bar{8} \wedge \bar{T} \wedge Z) \wedge (9 \equiv E)$$

$$U = \bar{1}F$$

Equations

$$
\begin{aligned}
\mathrm{E0} = {} & \{(\bar{F} \wedge \bar{9}) \wedge [(1 \wedge \bar{G}) \veebar (\bar{8} \wedge J)]\} \veebar \{(\bar{1} \wedge \bar{F}) \wedge (9 \veebar 8)\} \\
& \veebar \{U \wedge (\bar{9} \wedge \bar{8}) \veebar (8 \wedge \bar{G})\} \veebar \{(9 \wedge \bar{T} \wedge E \wedge Z \wedge 1 \wedge \bar{8})\} \veebar L
\end{aligned}
$$

$$
\begin{aligned}
\mathrm{E1} = {} & \{\bar{9} \wedge \bar{8} \wedge \bar{1} \wedge \bar{F}\} \veebar \{(9 \veebar G) \wedge \{[\bar{F} \wedge (1 \veebar 8)] \veebar U\}\} \veebar \{(G \wedge \bar{F}) \\
& \wedge [(\bar{9} \wedge 8 \wedge 1) \veebar (9 \wedge \bar{8} \wedge \bar{1})]\}
\end{aligned}
$$

$$
\begin{aligned}
\mathrm{E2} = {} & \{[(\bar{T} \wedge \bar{E}) \veebar (E \wedge Z)] \wedge [(\bar{F} \wedge \bar{D}) \veebar U]\} \veebar \{(\bar{F} \wedge D) \\
& \wedge \{(E \wedge \bar{Z}) \vee -[(Z \wedge 1) \wedge (T \equiv E)]\}\}
\end{aligned}
$$

$$
\begin{aligned}
\mathrm{E3} = {} & \{\{T \wedge E) \veebar (\bar{T} \wedge \bar{Z})\} \wedge \{[\bar{F} \wedge (8 \vee 1)] \veebar [(\bar{1} \wedge \bar{8} \wedge 9) \veebar U]\}\} \veebar \{[\bar{F} \wedge D] \\
& \wedge \{(E \wedge Z) \veebar [\bar{E} \wedge (T \veebar Z)]\}\}
\end{aligned}
$$

$$
\begin{aligned}
\mathrm{E4} = {} & \{\bar{F} \wedge [8 \wedge \overline{(1 \wedge \bar{G})}]\} \veebar \{[\bar{8} \wedge \bar{1}] \wedge [9 \vee (T \wedge E \wedge \bar{Z}]\} \\
& \veebar \{8 \wedge \{[3 \veebar R \veebar C] \veebar [2 \wedge \overline{(\bar{9} \wedge \bar{T} \wedge \bar{E} \wedge Z)}]\}\}
\end{aligned}
$$

$$\mathrm{E5} = R \veebar C$$

$$\mathrm{E6} = \{[3 \veebar 6 \veebar 7] \veebar [2 \wedge (\bar{9} \wedge 8 \wedge \bar{T} \wedge \bar{E} \wedge Z]\} \veebar \{D \wedge \bar{F} \wedge T \wedge E \wedge \bar{Z}\}$$

$$\mathrm{E7} = \{3 \veebar 5 \veebar 7\} \veebar \{\bar{F} \wedge \{1 \wedge [\bar{8} \veebar (8 \wedge G)] \veebar [9 \wedge \bar{8} \wedge \bar{1}]\}\}$$

Fig. 16.16 EBCDIC equations

T, E, Z, respectively. The bits of an 8-bit byte are named E0, E1, E2, . . . , E7, from high to low order.

Using Boolean notation, the translation equations may be derived as shown in Fig. 16.15. These translation relations for Bendix Prime card code to/from EBCDIC bit code are considerably less complex than those for EBCDIC card code to/from EBCDIC bit code, which are shown in Fig. 16.16.

16.12 EBCDIC PRIME

While the author was reviewing Bendix Prime, it occurred to him that it would be useful to have some basic card-code–to–bit-code relationship against which other relationships could be compared for simplicity or complexity. Such a basic relationship is shown in Fig. 16.17. It is called

Row	Bit Pat.	Hole Pat.	0	1	2	3	4	5	6	7	8	9	A	B	C	D	E	F	Hole Pat.
	Bit Pat.		00				01				10				11				
			00	01	10	11	00	01	10	11	00	01	10	11	00	01	10	11	
		Hole Pat.									9	9	9	9	9	9	9	9	Hole Pat.
							T	T	T	T					T	T	T	T	
					E	E			E	E			E	E			E	E	
				Z		Z		Z		Z		Z		Z		Z		Z	
0	0000																		
1	0001	1																	
2	0010	2																	
3	0011	3																	
4	0100	4																	
5	0101	5																	
6	0110	6																	
7	0111	7																	
8	1000	8																	
9	1001	8–1																	
A	1010	8–2																	
B	1011	8–3																	
C	1100	8–4																	
D	1101	8–5																	
E	1110	8–6																	
F	1111	8–7																	
		Hole Pat.																	

Fig. 16.17 EBCDIC Prime

EBCDIC Prime for purposes of reference. The letters T, E, and Z are used to represent the 12-row, 11-row, and 0-row. The Boolean relations for EBCDIC Prime, card code to bit code, are shown in Fig. 16.18.

It may be noted, then, that Bendix Prime equations and EBCDIC Prime equations both require 8 Boolean operators.

Common expressions

$A = 2 \veebar 3$

$B = 4 \veebar 5$

$C = 6 \veebar 7$

Equations

$\text{E0} = 9$

$\text{E1} = T$

$\text{E2} = E$

$\text{E3} = Z$

$\text{E4} = 8$

$\text{E5} = B \veebar C$

$\text{E6} = A \veebar C$

$\text{E7} = 1 \veebar 3 \veebar 5 \veebar 7$

Fig. 16.18 EBCDIC Prime equations

16.13 COMPARISON OF BENDIX PRIME AND EBCDIC PRIME

The possibility of using either Bendix Prime or EBCDIC Prime, or some version of them, as the card code for ASCII was then considered. Neither card code manifests the undesirable trait of lacing. In order to arrive at figures of comparison for the average number of holes per character, we observe that the figures in Fig. 16.7 were in terms of 64 characters; that is, we would have to decide *which* 64 characters of Bendix Prime, or of EBCDIC Prime, were to be considered. If we want to optimize on the minimum number of holes per character, for Bendix Prime (Fig. 16.14), we would select table-columns 0, 1, 2, and 3; and for EBCDIC Prime (Fig. 16.17), we would select table-columns 0, 1, 2, and 4. For these selections, the figures for 64 characters are as follows:

	Average number of holes per character
Bendix Prime	1.12
EBCDIC Prime	0.98

Both Bendix Prime and EBCDIC Prime, for 64 characters, have an average of far fewer holes per character than do the Binary Representations (Fig. 16.7).

It is to be noted that Figs. 16.14 and 16.17 do not represent codes per se; that is, a set of meanings assigned to a set of bit patterns or hole patterns. Figures 16.14 and 16.17 show a relationship between a set of hole patterns and a set of bit patterns.

These sets of hole patterns for Bendix Prime and EBCDIC Prime have interesting characteristics in contrast to the Binary Representations described above:

1. No card lacing.
2. On the average, fewer holes per character than Binary Representations.
3. Simple translation relationships, bit patterns to/from hole patterns, although *slightly* more complex than the Binary Representations.

16.14 THE PLOMONDON PROPOSAL

Such a card code would seem to be the obvious candidate for the card code for ASCII. In November 1963, a card code based on the principle of EBCDIC Prime was proposed for study to the standards committee by E. E. Plomondon. This card code (although not this actual version) came to be called Decimal ASCII.

The Plomondon proposals were for a 128-character version and a 256-character version, shown in Figs. 16.19 and 16.20, respectively. It should be noted that the 256-character proposal is, strictly speaking, not the one that was actually made. As described in Chapter 20, the algorithm for embedding the 7-bits of ASCII in an 8-bit byte had not actually been decided at that time by the standards committees. The algorithm E6 = b7 had been implemented on the System/360.

Ultimately, the standards committees decided for the algorithm E8 = 0. The actual embedment algorithm does not affect any of the discussion that follows in this chapter. In consequence, since the E8 = 0 algorithm *was* the one chosen, the author has used that algorithm in this chapter, even though the actual proposal at that time assumed the E6 = b7 algorithm. What is meant by the E8 = 0 algorithm is that the 8 columns of the 7-bit ASCII code table were embedded in the first 8 columns, the high-order bit, E0, is zero; hence the algorithm was characterized as E8 = 0.

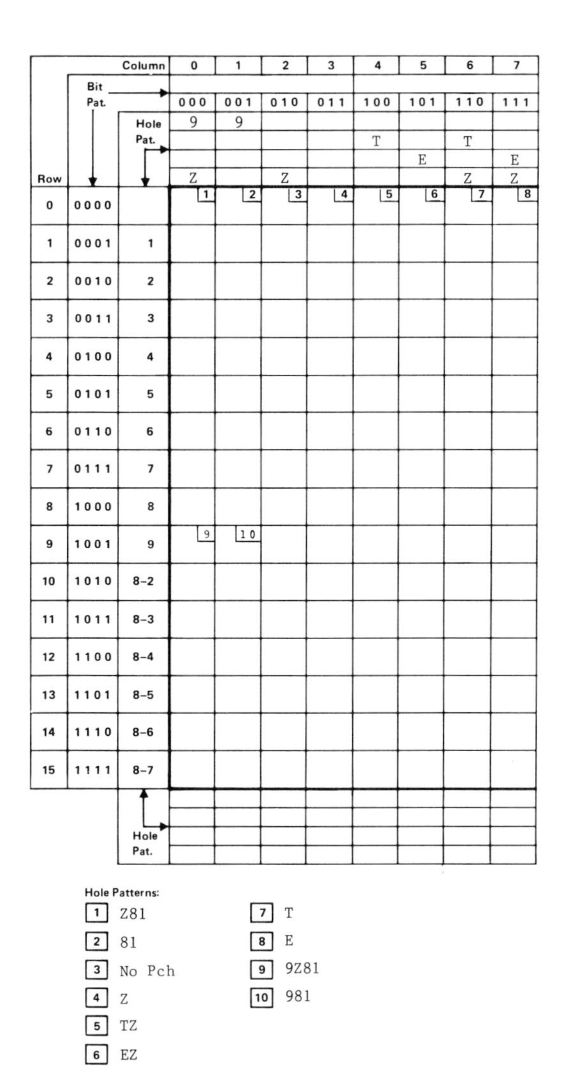

Fig. 16.19 Decimal ASCII-128, Plomondon proposal

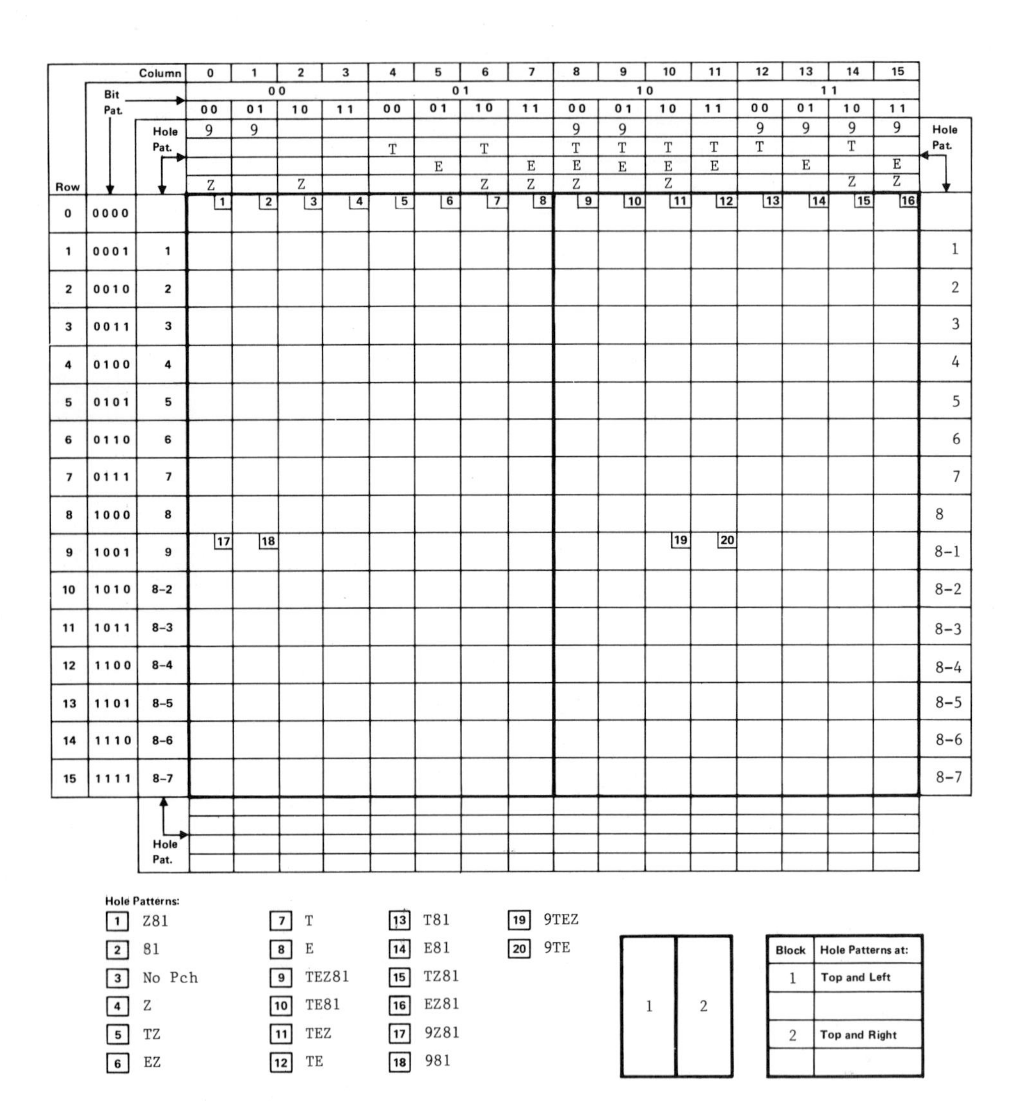

Fig. 16.20 Decimal ASCII-256, Plomondon proposal

16.15 DECIMAL ASCII, VERSIONS 1 AND 2

It was pointed out, in connection with these proposals, that the 9-punch was functioning as a zone punch. In the 128-character proposals, the 9 as a zone punch was assigned to columns 0 and 1 of the code table; that is, to control characters. And this was cited as desirable with respect to circuitry in terminals where a clear differentiability between control characters and graphic characters would be desirable.

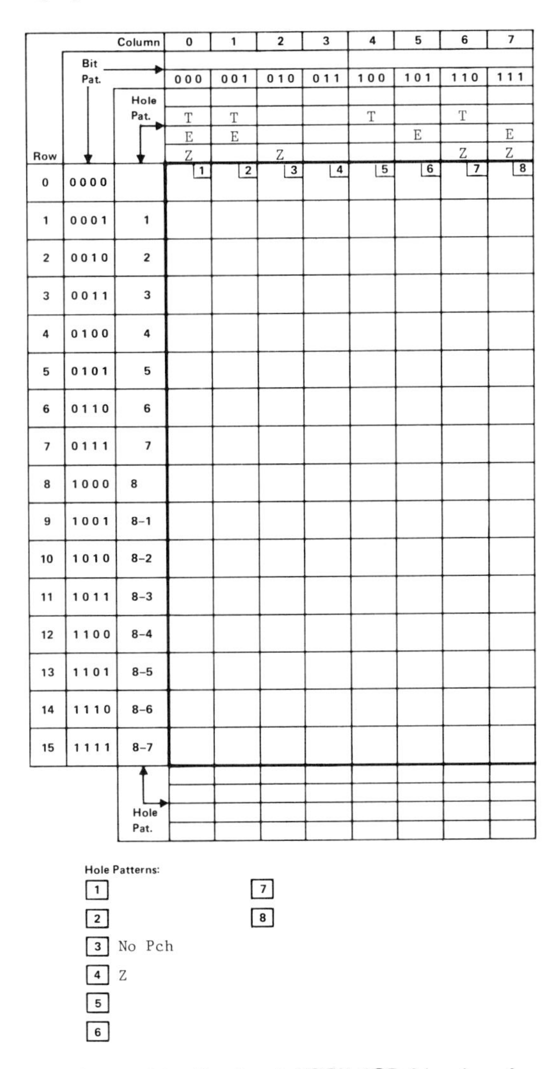

Fig. 16.21 Decimal ASCII-128, Version 1

Another member of the standards committee (Task Group X3.2.3) suggested that it seemed preferable to use the 9-punch, when used as a zone punch, to differentiate between the E8 = 0 and E8 = 1 halves of the 8-bit code table, as shown in Fig. 16.22. If a distinguishing punch (or punches) was desirable for control characters, then the 12-11 combination could serve as well as the 9 proposed by Plomondon. This suggestion

Row	Bit Pat.	Hole Pat.	0	1	2	3	4	5	6	7	8	9	10	11	12	13	14	15	Hole Pat.
	Bit Pat.		00				01				10				11				
			00	01	10	11	00	01	10	11	00	01	10	11	00	01	10	11	
		Hole Pat.									9	9	9	9	9	9	9	9	Hole Pat.
			T	T			T		T		T	T			T		T		
			E	E				E		E	E	E				E		E	
			Z		Z				Z	Z	Z		Z				Z	Z	
0	0000		1	2	3	4	5	6	7	8	9	10	11	12	13	14	15	16	81
1	0001	1																	1
2	0010	2																	2
3	0011	3																	3
4	0100	4																	4
5	0101	5																	5
6	0110	6																	6
7	0111	7																	7
8	1000	8																	8
9	1001	9																	8–1
10	1010	8–2																	8–2
11	1011	8–3																	8–3
12	1100	8–4																	8–4
13	1101	8–5																	8–5
14	1110	8–6																	8–6
15	1111	8–7																	8–7
		Hole Pat.																	

Hole Patterns:

1		7		13	T81
2		8		14	E81
3	No Pch	9	TEZ81	15	TZ81
4	Z	10	TE81	16	EZ81
5		11	Z81		
6		12	81		

1	2

Block	Hole Patterns at:
1	Top and Left
2	Top and Right

Fig. 16.22 Decimal ASCII-256, Version 2

seemed good, and was accepted by the standards committee. The result was Version 1, 128 characters (Fig. 16.21) and Version 2, 256 characters (Fig. 16.22).

Versions 1 and 2 *were* superior to the initial Plomondon proposal in one respect. The translation equations (which are shown later in this chapter), card code to/from bit code, are less complex.

16.16 THE NULL/SPACE/BLANK PROBLEM (AGAIN)

For Version 1, as for the IBM 128-character proposal, the No punches hole pattern was assigned to code position 2/0 (the Space character), and the Zero hole pattern was assigned to code position 3/0 (the zero character). (This is a reflection of the Null/Blank/Zero Problem referred to previously in this chapter.) In the Plomondon proposal and in Versions 1 and 2 the assignment of No punches to Space was made. This later became a matter of contention in the standards committee because, if No punches had been assigned instead to code position 3/0, and if Zero punch had been assigned instead to code position 2/0, the translation relationships, card code to/from bit code, would have been simpler. And simplicity of translation relationships was desirable. The assignment, however, was ultimately accepted by the committee.

It should be borne in mind that, at the time Decimal ASCII was proposed, there were two contenders for standardization—a Binary Representation of one kind or another and Hollerith Representation.

The following comparison of the merits of Decimal ASCII and of Binary Representation shows clearly that Decimal ASCII suffered from none of the defects previously described for the Binary Representations, and enjoyed a reasonably simple translation relationship, to/from ASCII.

If

A = complexity of translation, Binary card code to/from ASCII bit code,

and if

B = complexity of translation, Decimal ASCII card code to/from ASCII bit code.

and if

C = complexity of translation. Hollerith card code to/from ASCII bit code,

then

$$A < B < C.$$

And in fact, A and B are *very much* less than C.

Technically, then, Decimal ASCII appeared to the standards committee as superior to Binary Representations. Indeed, the standards committee soon dropped Binary Representations from further consideration. (Recall that the Null/Blank/Zero Problem and the Plus and Minus Zero Problem were previously stated not to have been resolved for Binary Representations. The reason, of course, is because the Binary Representation card codes were themselves dropped from further consideration.)

16.17 EUROPEAN CARD CODES

There was another important point in favor of Decimal ASCII. In Europe, three manufacturers of punched card equipment, IBM, ICT (now ICL), and Bull, employed card codes in their equipment radically different one from another (see Fig. 16.23). In the European standards committee responsible for codes, ECMA/TC1, card code standardization was at an impasse.

Each of the three manufacturers advocated his own code as a candidate for standardization. More significantly, if the punched card code of one manufacturer was accepted for standardization, then that manufacturer could enjoy an advantage in the market place. The other members of ECMA/TC1 felt that, until the three punched card manufacturers came into agreement on some proposal, it was useless to try to arrive at a consensus on a standard card code.

These European card codes deserve comment. Their common area of agreement is the original Hollerith numerics. The card codes used by IBM and by ICT also agreed on alphabetics. But the method of extending the repertoire of hole patterns beyond this point was different. For the IBM card code, the extension was achieved by using the 8-punch as a zone punch. As has been described elsewhere in this book, this had the merit of preserving the BCD characteristic of the code. By contrast, the ICT card code was extended by using the 1-punch as a zone punch.

And for the Bull code, to extend the repertoire of hole patterns beyond the numerics, the 7-, 8-, and 9-punches were used as zone punches. This undoubtedly had to do with the method of feeding a card through a card reader. If a card is fed 12-edge first (IBM), then punches toward that edge of the card (12, 11, 0) serve best as zone punches. But if the card is fed 9-edge first (ICT), then punches toward that edge of the card (7, 8, 9) serve best as zone punches.

Not long after E. Plomondon proposed Decimal ASCII to X3.2.3, W. F. Bohn proposed it to ECMA/TC1. It was perceived that Decimal ASCII was not implemented on any equipment. In consequence, all three

Hole Pat.		12					12									
			11					11					9	9	9	
				0					0			8			8	
					1						7			7		
12	10					& +				11				*	−	
11	11					−				10	O	I		/	+	
	SP					SP				SP	A	J	S	,	=	
0	0					0	?			0	B	K	T	'	(	
1	1	A	J			1	A	J		1	C	L	U	%	)	
2	2	B	K	S	%	2	B	K	S	2	D	M	V	$	x	
3	3	C	L	T	¼	3	C	L	T	3	E	N	W	£	:	
4	4	D	M	U	−	4	D	M	U	4	F	P	X	◇	<	
5	5	E	N	V	/	5	E	N	V	5	G	Q	Y	□	>	
6	6	F	O	W	½	6	F	O	W	6	H	R	Z	△	‡	
7	7	G	P	X	.	7	G	P	X	7						
8	8	H	Q	Y	@	8	H	Q	Y	8						
9	9	I	R	Z	¾	9	I	R	Z	9						
8-2						ƀ			‡							
8-3						# =	.	$	,							
8-4						@ '	⌑)	*	% (							
8-5						:	[	]	γ							
8-6						>	<	;	\							
8-7						√	ǂ	△	⧻							
	ICT NEW HOLLERITH					IBM BCDIC				BULL 300 SERIES						

Fig. 16.23 European card codes

manufacturers could begin to design and develop Decimal ASCII card equipment from an equal start. Decimal ASCII was seen by ECMA/TC1 as a proposal which would remove the impasse, and Decimal ASCII was accepted. Decimal ASCII was now accepted in principle both by ECMA/TC1 and by ASA X3.2.

16.18 THE PLUS AND MINUS ZERO PROBLEM (AGAIN)

The Plus and Minus Zero Problem now arose to plague the committees. It will be observed in the original Plomondon proposals (Figs. 16.19 and

16.20) that, although the general translation relationship for the code table would have prescribed T, E, TZ, and EZ for code positions 4/0, 5/0 6/0, and 7/0, respectively, hole patterns TZ, EZ, T, and E, respectively, were assigned instead.

It will further be observed that in the Decimal ASCII Version 1 proposals (Figs. 16.21 and 16.22) these translation exceptions were removed. What was behind this?

The intent behind the Plomondon proposals was to provide the overpunched numeric capability in Decimal ASCII. Hole patterns T1 through T9 are assigned to code table positions 4/1 through 4/9, and E1 through E9 to 5/1 through 5/9. Assuming that the overpunched numeric convention prevalent with Hollerith punched card applications would be continued by users in Decimal ASCII punched card applications, it would be necessary also to provide for plus zero and for minus zero. And the TZ hole pattern must correspond to the same ASCII bit code table column as T1 through T9, and the EZ hole pattern must correspond to the same table column as E1 through E9. This would displace, in the 0-row of the code table, T and E, which would be moved to code positions 6/0 and 7/0.

These four translation exceptions were the solution to the Plus and Minus Problem in the Plomondon proposals. But they were not provided in the Decimal ASCII Version 1 Proposals. Why not?

They were not provided precisely because they *were* translation exceptions. For those members of the standards committees who felt that translation simplicity was the primary criterion, it had been hard to accept the two previously mentioned translation exceptions to solve the Null/Blank/Zero Problem. And these members would *not* accept four more translation exceptions, as proposed by IBM to resolve the Plus and Minus Zero Problem.

16.19 DECIMAL ASCII, VERSIONS 3 AND 4

Representatives to the standards committee did urge the solution of the Plus and Minus Zero Problem, and submitted proposals incorporating the solutions, Decimal ASCII Version 3 (128 characters) and Version 4 (256 characters), as shown in Figs. 16.24 and 16.25, respectively.

The arguments for Versions 1 and 2 versus Versions 3 and 4 then centered on the relative importance of translation simplicity versus provision for Plus and Minus Zero.

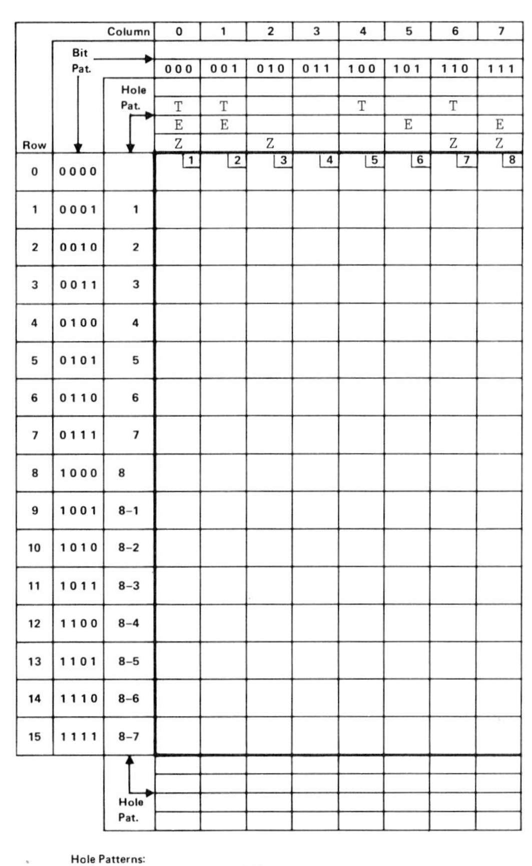

Hole Patterns:

[1]

[2]

[3] No Pch

[4] Z

[5] TZ

[6] EZ

[7] T

[8] E

Fig. 16.24 Decimal ASCII-128, Version 3

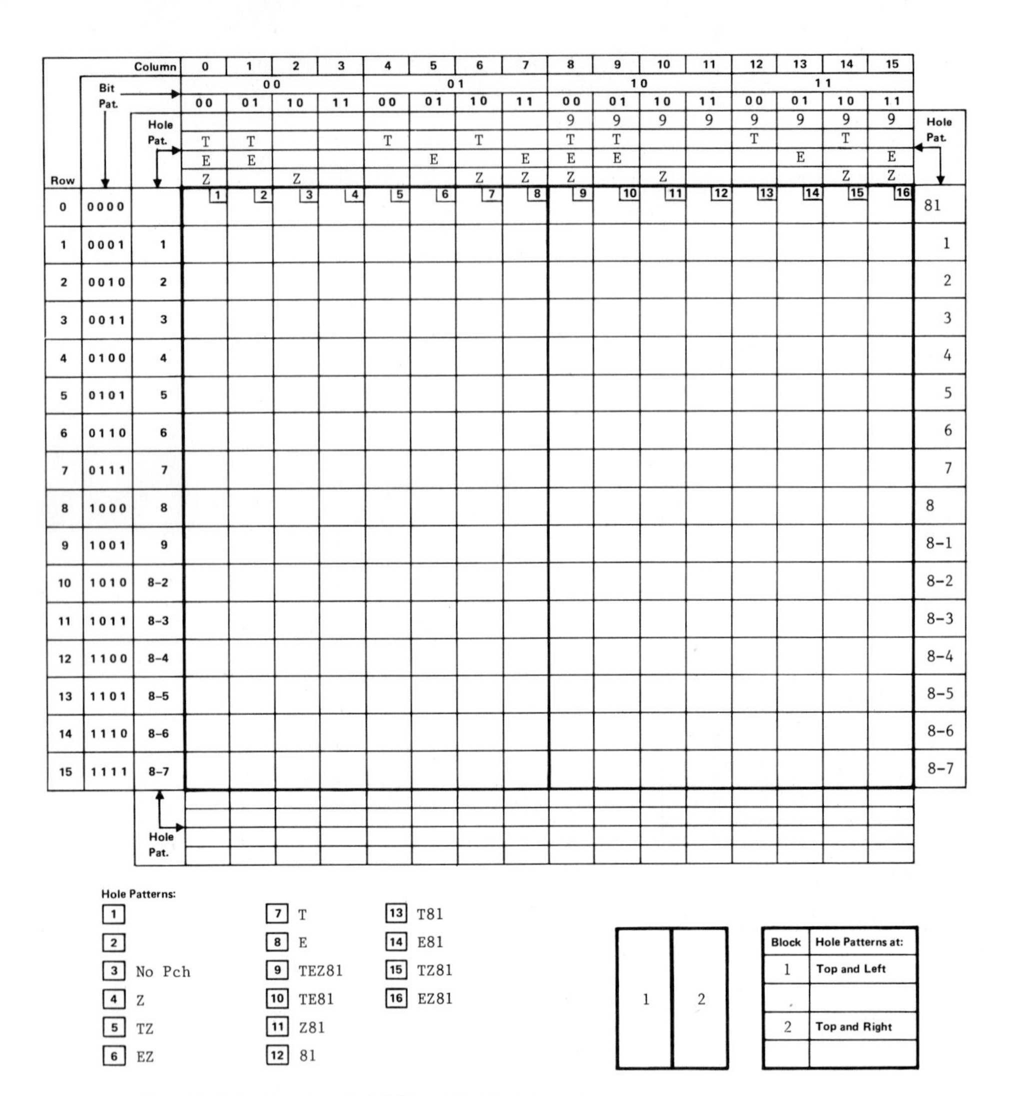

Fig. 16.25 Decimal ASCII-256, Version 4

16.20 DECIMAL ASCII PRIME

In order to assess the relative translation complexity/simplicity of Versions 1 and 2 versus Versions 3 and 4, Boolean equations are derived. To have a base against which comparisons can be made, Decimal ASCII,

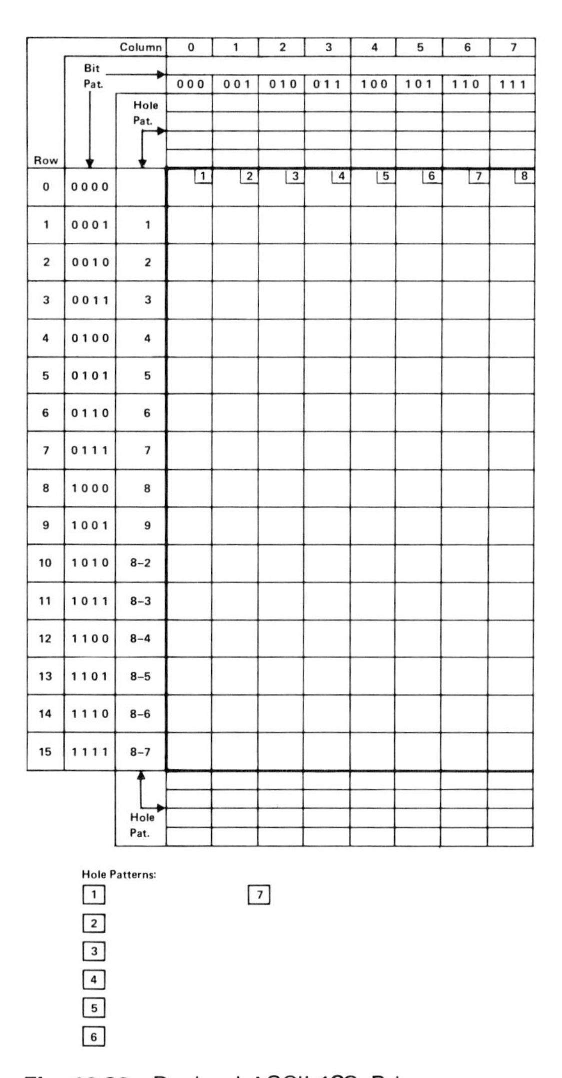

Fig. 16.26 Decimal ASCII-128, Prime

Prime (128 characters) and Decimal ASCII, Prime (256 characters) are shown in Figs. 16.26 and 16.27, respectively. There are *no* translation exceptions in these latter two card codes, neither the exceptions to solve the Null/Blank/Zero Problem nor the exceptions to solve the Plus and Minus Zero Problem.

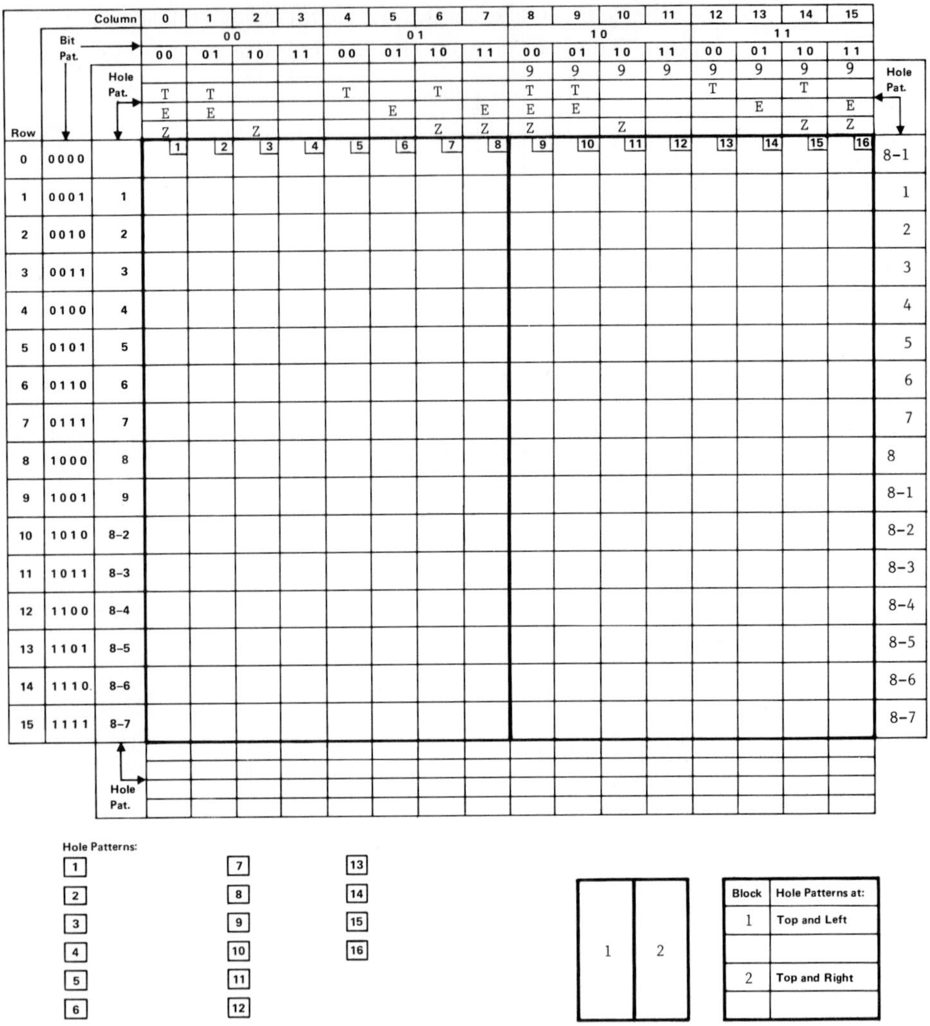

Row	Bit Pat.	Hole Pat.	0	1	2	3	4	5	6	7	8	9	10	11	12	13	14	15	Hole Pat.
	Bit Pat.		00				01				10				11				
			00	01	10	11	00	01	10	11	00	01	10	11	00	01	10	11	
		Hole Pat.									9	9	9	9	9	9	9	9	
			T	T			T		T		T	T			T		T		
			E	E				E		E	E	E				E		E	
			Z		Z				Z	Z	Z		Z				Z	Z	
			1	2	3	4	5	6	7	8	9	10	11	12	13	14	15	16	
0	0000																		8–1
1	0001	1																	1
2	0010	2																	2
3	0011	3																	3
4	0100	4																	4
5	0101	5																	5
6	0110	6																	6
7	0111	7																	7
8	1000	8																	8
9	1001	9																	8–1
10	1010	8–2																	8–2
11	1011	8–3																	8–3
12	1100	8–4																	8–4
13	1101	8–5																	8–5
14	1110	8–6																	8–6
15	1111	8–7																	8–7
		Hole Pat.																	

Hole Patterns:

1, 2, 3, 4, 5, 6, 7, 8, 9, 10, 11, 12, 13, 14, 15, 16

1	2

Block	Hole Patterns at:
1	Top and Left
2	Top and Right

Fig. 16.27 Decimal ASCII-256, Prime

16.21 TRANSLATION EQUATIONS

The translation equations for Decimal ASCII Prime, the original Plomondon proposal, Versions 1 and 2, and Versions 3 and 4, for both 128 characters and 256 characters are set down in Figs. 16.28 through 16.35.

Using the three simplifying assumptions previously noted, Boolean operators are counted for these equations. The results are summarized in Fig. 16.36. Results for EBCDIC are also shown for purposes of comparison.

Common expressions

$A = 2 \veebar 3$	$F = A \veebar J \veebar C$
$J = 4 \veebar 5$	$H = T \veebar E$
$C = 6 \veebar 7$	$P = (\bar{9} \wedge \bar{1}) \veebar [(\bar{F} \wedge \bar{8}) \wedge (9 \veebar 1)]$

Equations

$A7 = H \wedge P$

$A6 = [(\bar{T} \wedge \bar{E}) \veebar (H \wedge Z)] \wedge P$

$A5 = \{(\bar{T} \wedge \bar{Z}) \veebar [(T \veebar Z) \wedge E]\} \wedge P$

$A4 = \bar{1} \wedge [(\bar{9} \wedge 8) \wedge (9 \wedge \bar{8} \wedge \bar{F})]$

$A3 = J \veebar C$

$A2 = A \veebar C$

$A1 = (3 \veebar 5 \veebar 7) \veebar [\bar{8} \wedge (9 \veebar 1)]$

Fig. 16.28 Decimal ASCII-128, Prime

Common expressions

$A = 2 \veebar 3$	$K = T \veebar E$
$J = 4 \veebar 5$	$R = 9 \wedge (8 \equiv 1) \wedge \bar{F}$
$C = 6 \veebar 7$	$U = \bar{1} \wedge F$
$F = A \veebar J \veebar C$	$W = \bar{F} \veebar U$

Equations

$A8 = [(1 \wedge \bar{F}) \wedge (9 \vee 8)] \veebar [\bar{1} \wedge 9 \wedge (8 \vee F)]$

$A7 = K \wedge W$

$A6 = [(\bar{T} \wedge \bar{E}) \veebar (K \wedge Z)] \wedge W$

$A5 = \{[\bar{T} \wedge \bar{Z}] \veebar [(T \veebar Z) \wedge E]\} \wedge W$

$A4 = (8 \wedge \bar{1}) \veebar R$

$A3 = J \veebar C$

$A2 = A \veebar C$

$A1 = [3 \veebar 5 \veebar 7] \veebar \{\bar{F} \wedge \{[\bar{8} \wedge (1 \veebar 9)]\} \veebar \{1 \wedge 9\}\}$

Fig. 16.29 Decimal ASCII-256, Prime

Common expressions

$A = 2 \vee 3$

$J = 4 \vee 5$

$C = 6 \vee 7$

$D = \bar{9} \wedge \bar{8} \wedge \bar{1}$

$F = A \vee J \vee C$

$H = T \vee E$

$P = (\bar{9} \wedge \bar{1}) \vee [(\bar{F} \wedge \bar{8}) \wedge (9 \vee 1)]$

$S = \bar{F} \wedge D$

$U = \bar{1} \wedge F$

$K = \bar{T} \wedge \bar{E}$

$X = K \vee (\bar{9} \wedge H)$

Equations

$\text{A7} = H \wedge P$

$\text{A6} = [K \vee (H \wedge Z)] \wedge \{\{\bar{F} \wedge [\bar{9} \wedge (8 \vee 1)] \vee (9 \wedge \bar{8} \wedge \bar{1})\} \vee (F \wedge 9)\}\} \vee \{[K \vee (H \wedge \bar{Z})] \wedge S\}$

$\text{A5} = \{K \wedge \{[(Z \equiv D) \wedge \bar{F}] \vee [\bar{Z} \wedge U]\}\} \vee \{\bar{T} \wedge E \wedge P\}$

$\text{A4} = (K \wedge 9 \wedge 8) \vee \{\overline{(T \wedge E)} \wedge \bar{1} \wedge [(\bar{9} \wedge 8) \vee (9 \wedge \bar{8} \wedge \bar{F})]\}$

$\text{A3} = X \wedge (K \vee C)$

$\text{A2} = X \wedge (A \vee C)$

$\text{A1} = \{X \wedge [(1 \wedge \bar{8}) \vee (3 \vee 5 \vee 7)]\} \vee \{[(9 \wedge \bar{F}) \wedge (1 \wedge 8 \wedge K)] \vee [(\bar{1} \wedge \bar{8}) \wedge \overline{(T \wedge E)}]\}$

Fig. 16.30 Decimal ASCII-128, Plomondon Proposal

Common expressions

$A = 2 \vee 3$

$J = 4 \vee 5$

$C = 6 \vee 7$

$F = A \vee J \vee C$

$K = T \vee E$

$L = [(9 \vee 8) \wedge 1] \vee (9 \wedge 8)$

$N = [(9 \vee 8) \wedge \bar{1}] \vee (\bar{9} \wedge \bar{8})$

$R = 9 \wedge (8 \equiv 1) \wedge \bar{F}$

$U = \bar{1} \wedge F$

$W = \bar{F} \vee U$

$D = \bar{9} \wedge \bar{8} \wedge \bar{1}$

Equations

$\text{A8} = (T \wedge E \wedge W) \vee \{K \wedge [(\bar{F} \wedge L) \vee (9 \wedge U)]\}$

$\text{A7} = K \wedge W$

$\text{A6} = \{\bar{K} \wedge [(\bar{F} \wedge N) \vee (\bar{9} \wedge U)]\} \vee \{[K \wedge \{[\bar{F} \wedge (Z \vee D)] \vee [Z \wedge U]\}\}$

$\text{A5} = (\bar{T} \wedge \bar{E}) \wedge \{[\bar{F} \wedge (Z \equiv D)] \vee [\bar{Z} \wedge U]\} \vee \{E \wedge \overline{(T \wedge Z)} \wedge W\}$

$\text{A4} = (8 \wedge \bar{1}) \vee R$

$\text{A3} = J \vee C$

$\text{A2} = A \vee C$

$\text{A1} = [(3 \vee 5 \vee 7) \vee R] \vee (\bar{8} \wedge 1 \wedge \bar{F})$

Fig. 16.31 Decimal ASCII-256, Plomondon Proposal

Common expressions

$A = 2 \veebar 3$

$J = 4 \veebar 5$

$C = 6 \veebar 7$

$F = A \veebar J \veebar C$

$H = T \veebar E$

$Q = \bar{9} \wedge \bar{1}$

$B = (\bar{F} \wedge \bar{8}) \wedge (9 \veebar 1)$

$P = Q \veebar B$

$G = [Q \wedge (F \vee 8)] \veebar B$

$S = \bar{F} \wedge \bar{9} \wedge \bar{8} \wedge \bar{1}$

Equations

$\mathrm{A7} = H \wedge P$

$A6 = [(\bar{T} \wedge \bar{E}) \veebar (H \wedge Z)] \wedge P$

$\mathrm{A5} = \{[(\bar{T} \wedge Z) \veebar (E \wedge \bar{Z})] \wedge S\} \veebar \{\{(\bar{T} \wedge \bar{Z}) \veebar [(T \veebar Z) \wedge E]\} \wedge G\}$

$\mathrm{A4} = \bar{1} \wedge [(\bar{9} \wedge 8) \wedge (9 \wedge \bar{8} \wedge \bar{F})]$

$\mathrm{A3} = J \veebar C$

$\mathrm{A2} = A \veebar C$

$\mathrm{A1} = (3 \veebar 5 \veebar 7) \veebar [\bar{8} \wedge (9 \veebar 1)]$

Fig. 16.32 Decimal ASCII-128, Version 1

Common expressions

$A = 2 \veebar 3$

$J = 4 \veebar 5$

$C = 6 \veebar 7$

$F = A \veebar J \veebar C$

$D = \bar{9} \wedge \bar{8} \wedge \bar{1}$

$R = 9 \wedge (8 \equiv 1) \wedge \bar{F}$

$U = \bar{1} \wedge F$

$W = \bar{F} \veebar U$

$K = T \veebar E$

Equations

$\mathrm{A8} = [(1 \wedge \bar{F}) \wedge (9 \vee 8)] \veebar [\bar{1} \wedge 9 \wedge (8 \vee F)]$

$\mathrm{A7} = K \wedge W$

$\mathrm{A6} = [(\bar{T} \wedge \bar{E}) \veebar (K \wedge Z)] \wedge W$

$\mathrm{A5} = [E \wedge \overline{(T \wedge Z)} \wedge (\bar{F} \veebar U)] \veebar \{\bar{T} \wedge \bar{E}\} \wedge \{[\bar{F} \wedge (Z \equiv D)] \vee (\bar{Z} \wedge U)\}$

$\mathrm{A4} = (8 \wedge \bar{1}) \veebar R$

$\mathrm{A3} = J \veebar C$

$\mathrm{A2} = A \veebar C$

$\mathrm{A1} = [3 \veebar 5 \veebar 7] \veebar \{\bar{F} \wedge \{[\bar{8} \wedge (1 \veebar 9)]\} \veebar \{1 \wedge 9\}\}$

Fig. 16.33 Decimal ASCII-256, Version 2

Common Expressions

$A = 2 \veebar 3$

$J = 4 \veebar 5$

$C = 6 \veebar 7$

$F = A \veebar J \veebar C$

$H = T \veebar E$

$Q = \bar{9} \wedge \bar{1}$

$B = (\bar{F} \wedge \bar{8}) \wedge (9 \veebar 1)$

$P = Q \veebar B$

$G = [Q \wedge (F \vee 8)] \veebar B$

$S = \bar{F} \wedge \bar{9} \wedge \bar{8} \wedge \bar{1}$

Equations

$A7 = H \wedge P$

$A6 = (\bar{T} \wedge \bar{E} \wedge P) \veebar \{H \wedge [(\bar{Z} \wedge S) \veebar (Z \wedge G)]\}$

$A5 = \{[(\bar{T} \wedge Z) \veebar (E \wedge \bar{Z})] \wedge S\} \veebar \{\{(\bar{T} \wedge \bar{Z}) \veebar [(T \veebar Z) \wedge E]\} \wedge G\}$

$A4 = \bar{1} \wedge [(\bar{9} \wedge 8) \wedge (9 \wedge \bar{8} \wedge \bar{F})]$

$A3 = J \veebar C$

$A2 = A \veebar C$

$A1 = (3 \veebar 5 \veebar 7) \veebar [\bar{8} \wedge (9 \veebar 1)]$

Fig. 16.34 Decimal ASCII-128, Version 3

Common expressions

$A = 2 \veebar 3$

$J = 4 \veebar 5$

$C = 6 \veebar 7$

$F = A \veebar J \veebar C$

$D = \bar{9} \wedge \bar{8} \wedge \bar{1}$

$R = 9 \wedge (8 \equiv 1) \wedge \bar{F}$

$U = \bar{1} \wedge F$

$W = \bar{F} \veebar U$

$H = T \veebar E$

Equations

$A8 = [(1 \wedge \bar{F}) \wedge (9 \vee 8)] \veebar [\bar{1} \wedge 9 \wedge (8 \vee F)]$

$A7 = K \wedge W$

$A6 = (\bar{T} \wedge \bar{E} \wedge W) \veebar \{K \wedge \{[Z \wedge U] \veebar [\bar{F} \wedge (Z \veebar D)]\}\}$

$A5 = [E \wedge \overline{(T \wedge Z)} \wedge (\bar{F} \veebar U)] \veebar \{\bar{T} \wedge \bar{E}\} \wedge [\bar{F} \wedge (Z \equiv D)] \vee (\bar{Z} \wedge U)]\}$

$A4 = (8 \wedge \bar{1}) \veebar R$

$A3 = J \veebar C$

$A2 = A \veebar C$

$A1 = [3 \veebar 5 \veebar 7] \veebar \{\bar{F} \wedge \{[\bar{8} \wedge (1 \veebar 9)]\} \veebar \{1 \wedge 9\}\}$

Fig. 16.35 Decimal ASCII-256, Version 4

Proposal	Size	Common expressions	Equations	Total
Prime	128	11	22	32
	256	11	29	40
Plomondon Proposal	128	18	52	70
	256	21	38	59
Version 1	128	17	27	44
Version 2	256	13	35	48
Version 3	128	17	30	47
Version 4	256	13	39	52
EBCDIC	256	43	110	153

Fig. 16.36 Counts of Boolean operators

16.22 ANOMALY OF BOOLEAN EQUATIONS

Before discussing the comparative complexities, what seems to be an anomaly should be explained. For the Plomondon proposals, for Versions 1 and 2, and for Versions 3 and 4, the count of Boolean operators for the 256-character version is *less than* the count for the 128-character version, whereas the opposite might have been expected. One aspect of the optimization of Boolean expressions is that very often the more terms there are initially, the more combinations and condensations will result. And there are more terms initially in the 256-character cases than in the 128-character cases.

In the routine work of simplifying Boolean expressions, it is quite valid to

a) treat $A \wedge B$ as AB,

b) treat $A \vee B$ as $A + B$,

c) manipulate the Boolean variables as if they were algebraic variables with algebraic operations.

Thus $A \vee (B \wedge C)$ can be treated as if it were $A + BC$.

Example

In the derivation of Version 1 and Version 2, certain terms are found in conjunction with $T\ E\ \bar{Z}$ and $\bar{T}\ E\ \bar{Z}$.

Version 1

$$(TE\bar{Z} + \bar{T}E\bar{Z})(\bar{F}\bar{9}\bar{8}\bar{1} + \bar{F}\bar{9}\bar{8}1 + \bar{F}\bar{9}8\bar{1} + F\bar{9}\bar{8}\bar{1} + F981 + F9\bar{8}\bar{1})$$
$$= (T + \bar{T})E\bar{Z}[\bar{F}(\bar{9}\bar{8}\bar{1} + \bar{9}\bar{8}1 + \bar{9}8\bar{1} + 9\bar{8}\bar{1}) + F(\bar{9}\bar{8}\bar{1} + 9\bar{8}\bar{1})]$$
$$= E\bar{Z}[(\bar{F} + F)(\bar{9}\bar{8}\bar{1} + 9\bar{8}\bar{1}) + \bar{F}(\bar{9}\bar{8}1 + \bar{9}8\bar{1})]$$
$$= E\bar{Z}[(\bar{9} + 9)\bar{8}\bar{1} + \bar{F}\bar{9}(\bar{8}1 + 8\bar{1})]$$
$$= E\bar{Z}[\bar{8}\bar{1} + \bar{F}\bar{9}(8 + 1)]$$
$$= (E \wedge \bar{Z}) \wedge \{[\bar{8} \wedge \bar{1}] \veebar [(\bar{F} \wedge \bar{9}) \wedge (8 \vee 1)]\}$$

to put it back into Boolean form.

Version 2

$$(TE\bar{Z} + \bar{T}E\bar{Z})(\bar{F}\bar{9}\bar{8}\bar{1} + \bar{F}\bar{9}81 + \bar{F}\bar{9}\bar{8}1 + \bar{F}\bar{9}8\bar{1} + \bar{F}9\bar{8}\bar{1} + F\bar{9}\bar{8}\bar{1}$$
$$+ F\bar{9}81 + \bar{F}9\bar{8}1 + \bar{F}98\bar{1} + \bar{F}981 + F9\bar{8}\bar{1} + F98\bar{1})$$

Inspection reveals that of the 16 possible terms involving F, 9, 8, 1, four are absent:

$$F981 + F9\bar{8}1 + F\bar{9}81 + F\bar{9}\bar{8}1$$

We have

$$(T + \bar{T})E\bar{Z}\overline{(F981 + F9\bar{8}1 + F\bar{9}81 + F\bar{9}\bar{8}1)}$$
$$= E\bar{Z}\overline{[F1(98 + 9\bar{8} + \bar{9}8 + \bar{9}\bar{8})]}$$
$$= E\bar{Z}(\overline{F1})$$
$$= (E \wedge \bar{Z}) \wedge (\overline{F \wedge 1}) \text{ to put it back into Boolean form.}$$

We see therefore that, although we started with more terms in Version 2 than in Version 1, after combination and condensation, this part of Version 2 requires only three Boolean operators, whereas Version 1 requires six.

It is clear that Versions 1, 2, 3, and 4 are less complex than the initial Plomondon proposals, and therefore preferable.

The increments from Decimal ASCII Prime are revealing:

128 Characters		256 Characters	
Prime	32	Prime	40
Version 1	44	Version 2	48
Version 3	47	Version 4	52

For the 128-character versions, the perturbation from Prime to solve the Null/Blank/Zero Problem, an increment of 12, was greater than the perturbation to solve the Plus and Minus Zero Problem, an increment of 4. For the 256-character versions, the perturbation to solve the Null/Blank/Zero Problem was an increment of 7, while the perturbation to solve the Plus and Minus Zero Problem was an increment of 5. And, of course, compared to EBCDIC with a Boolean count of 153, these increments were really negligible.

However, the positions on the standards committees hardened; the issue being between minimum complexity versus provision for Plus and Minus Zero.

Plus and minus zero proponents. The increase in complexity to provide for Plus and Minus Zero is very small.

Minimum complexity proponents. Since positive and negative numeric fields on punched cards can be provided in other ways than overpunching (namely, carry the algebraic sign in a separate card column), *no* increase in complexity is justified, however small.

Technical issues on standards committees are resolved by the democratic process of a majority vote. In this case, the minimum-complexity group had more votes, and Version 1 and 2 became the draft American National Standard. Version 1 became an approved ECMA Standard.

16.23 SIC TRANSIT GLORIA DECIMAL ASCII

As the draft American National Standard moved through the various committee levels, users became very concerned. As they saw it, the consequences of Decimal ASCII becoming an approved American National Standard were that

- existing card files would have to be converted to the new card code;
- existing card equipment would have to be modified or replaced with new Decimal ASCII card equipment.

These two consequences would be immensely costly to users and they rose in opposition. IBM felt it must support its customers in this matter, reversed its position, and also came out in opposition.

At the X3 level, Decimal ASCII failed to obtain a majority, and was deemed to have failed. Ultimately, the ECMA Standard was withdrawn. The standards committee turned back to a consideration of the Hollerith card code, as will be related in Chapters 17, 18, and 20.

17
Which Hollerith?

As described in Chapter 16, the Decimal ASCII Card Code was proposed for study at the end of 1963 to ASA Subcommittee X3.2 (now ANSI X3L2). It was initially very successful in the standards committees, but technical controversies arose which delayed its final acceptance. In April 1964, opposition to the draft standard arose in Subcommittee because of its substantial incompatibility with the Hollerith card code in common use. Support for a standard based on the Hollerith card code increased, and in September 1964, Subcommittee X3L2 voted to prepare a draft American Standard Hollerith card code.

While it is correct to say that "the" Hollerith card code was in common use, in fact there were many versions in actual use, versions different between different manufacturer's equipment, and even different versions on different equipments of the same manufacturers. *Which* Hollerith card code to incorporate into the draft American Standard became the question which vexed Subcommittee X3.2. It took four years and many proposals, submitted by members of Subcommittee X3.2, to resolve this question.

Since there were many versions in common use, it was clear that the final "standard" version, whatever it was, would necessarily be different from most versions in common use, very possibly different from all of them. It was realized, therefore, that the final standard version would imply economic impact both to users and manufacturers of punched card equipment. One or another of three economic principles was considered by the members of the standards committee:

1. To minimize the impact across all users and manufacturers.
2. To equalize the impact between all users and manufacturers.

3. To minimize the impact on the users of equipment of a particular manufacturer.

It was clear to the standards committee that no single solution could satisfy all principles.

It should be realized that these economic principles, although they undoubtedly influenced the judgments of individual members, were not a subject of discussion at the meetings of the committee. Technical factors were the subject of discussion.

17.1 TECHNICAL CRITERIA

During the earlier committee discussions on candidate card code standards, which considered various binary representations as well as versions of Hollerith, technical criteria emerged and were formalized by the committee. Since some of these criteria were conflicting, no candidate card code could satisfy all of them. The criteria that are grouped below accordingly as Binary Representation, Decimal ASCII, or Hollerith did or did not satisfy the criteria. The word "Hollerith," in the discussion below, is used generically, and covers any or all versions of Hollerith then is use.

17.1.1 Satisfied by Binary Representation, Decimal ASCII, and Hollerith

Criterion 1. The code should represent the full ASCII character set. (Note: *Some* of the Hollerith proposals put before the standards committee did not, in fact, satisfy this criterion.)

Criterion 2. The code should provide for logical and orderly expansion to larger sets.

Comment. Eventually the standards committee realized that until "logical and orderly expansion" was defined, this criterion was not useful. It was claimed for all candidates that they *did* satisfy this criterion, but they clearly satisfied it in different ways, and according to some particular interpretation of the criterion.

Criterion 3. The code should not decrease the present character storage capacity of the card.

Comment. In fact, no candidate was proposed which violated this criterion. This criterion was a carryover from codes for other kinds of media, where what were called shifted or precedence codes required more than one consecutive bit pattern per character. Such a code would decrease the character storage capacity of a card, but none such were proposed.

This criterion would have ruled out the UNIVAC card code which had 45 card columns, but actually two tiers per card column, giving a card capacity of 90 characters. However, this code was a six-row code, and could accommodate a maximum of 64 characters. To extend it to 128 characters (for ASCII) would have resulted in a twelve-row code, but then it would have a capacity of only 45 characters per card.

Criterion 4. No more than one card column should be used to represent one character.

Comment. This was a criterion intended to rule out shifted or precedence codes.

Criterion 5. Character representation should be independent of card column locations.

Comment. All proposals satisfied this criterion.

Criterion 6. All hole patterns in the set should require the same number of punchable *positions.*

Comment. Again, this was a criterion intended to rule out a shifted or precedence code.

Criterion 7. The code must be capable of being implemented in the standard card.

Comment. The "standard card" was (nominally) $3\frac{1}{4}$ inches by $7\frac{3}{8}$ inches. A standards proposal at that time under study by a different standards committee implied a card of $3\frac{1}{4}$ inches by $8\frac{1}{2}$ inches, a size which would not have satisfied this criterion.

17.1.2 Satisfied by Decimal ASCII and Hollerith; Not Satisfied by Binary Representation

Criterion 8. The code, when punched in a card, should not appreciably weaken the card; that is, the code should cause a minimum number of holes to be punched. Another way of stating this is that the code should be designed for

a) minimum hole density per unit area of the card,
b) minimum hole density per column, and
c) minimum hole density per row.

Comment. This is a relative criterion, not an absolute criterion. That is to say, it is always possible to consider two candidate card codes and decide which satisfies the criterion better. For example, Decimal ASCII and Hollerith certainly satisfy it better than a Binary Representation. As is discussed in Chapter 16, the Modified Binary Representation satisfied it better than the Direct Binary Representation, with respect to the special, numeric, and alphabetic characters in columns 2, 3, 4, and 5 of ASCII.

Criterion 9. The code should be capable of being used with existing equipment.

Comment. "Existing equipment," of course, accommodated the Hollerith card code. The set of 64 hole patterns assigned to columns 2, 3, 4, and 5 of ASCII (the so-called graphic subset) for Decimal ASCII were the same set of hole patterns accommodated by much punched card equipment of the time, albeit with different graphic meanings. Thus if care was exercised within a punched card application to bear in mind the differently mapped graphic *meanings* of Decimal ASCII and Hollerith, it was contended that Decimal ASCII could "use" some of the punched card equipment of the time.

Criterion 10. The codes for the numerics should be readily sight readable.

The phrase "readily sight readable" in the above criterion is an example of jargon, with a well-understood meaning to members of the X3.2 Subcommittee. The phrase "sight readable" conveys the meaning of readability by human beings, as contrasted with readability by input/output card readers. The adverb "readily" conveys a qualification, as covered in the two examples below:

Example 1

The hole patterns assigned to numerics in the Decimal ASCII card code were the same as those in the Hollerith card code; that is, punches in card rows 0, 1, 2, . . . , 9 for numerics 0, 1, 2, . . . , 9. These would be held to be "readily sight readable".

Example 2

The hole patterns for numerics in the Direct Binary Representation card code were as follows:

Numeric	Hole Pattern
0	No punches
1	1
2	2
3	2-1
4	3
5	3-1
6	3-2
7	3-2-1
8	4
9	4-1

These hole patterns, while certainly "sight readable," would require either training or mental calculation on the part of the human to associate them with the numerics, so they were held not to be "readily" sight readable.

17.1.3 Satisfied by Decimal ASCII and Binary Representation; Not Satisfied by Hollerith

Criterion 11. The code should require minimum translation to and from ASCII.

Comment. This also was a relative criterion, not an absolute criterion. The essential design feature of Decimal ASCII was minimum translation to/from ASCII, but in the sense of being less than the translation of Hollerith to/from ASCII. Clearly the Direct Binary Representation would require even less translation than Decimal ASCII to/from ASCII.

17.1.4 Satisfied by Binary Representation; Not Satisfied by Decimal ASCII or Hollerith

Criterion 12. The code should provide for error detection (parity).

Comment. In the concept of the Direct Binary Representation where bits 1 through 7 of ASCII would be punched in card-rows 1 through 7 of the card, card-rows 12, 11, 0, 8, and 9 would then be available, if needed, for parity-row schemes. With Decimal ASCII and Hollerith, since all 12 card rows of the card are required for hole patterns of the code, no card rows are available for parity schemes.

17.1.5 Satisfied by Hollerith; Not Satisfied by Decimal ASCII or Binary Representation

Criterion 13. The code should be compatible with the common existing standard domestic code (Hollerith).

Criterion 14. The code should be such as to require the minimum number of passes in mechanical sorting.

Comment. By "mechanical sorting" was meant the mechanical sorters of the day without logic circuitry. Schemes were devised, involving multiple passes per card column, to sort Decimcal ASCII and to sort Binary Representation, but such schemes would clearly require more than the minimum number of passes required by Hollerith.

Criterion 15. The code should be compatible with international card standards.

Comment. This criterion was not really applicable because, at the time, there were no international card standards.

Criterion 16. The code should preserve the logical arrangement of the ASCII columns.

Comment. The standards committee was never able to agree what, if anything, this criterion meant.

The 16 criteria above, while meaningful in inter-code discussions on Hollerith, Decimal ASCII, and Binary Representation, were of no use in trying to decide "which Hollerith?" A survey conducted in November 1964 of various card equipments provided by eight manufacturers (Burroughs, CDC, GE, Honeywell, IBM, NCR, RCA, and UNIVAC) showed there was complete unanimity on the hole patterns for the alphabetics, numerics, the Space character, and six specials . , * / – $ but, for other special graphics, there were 21 versions of Hollerith, different to a greater or lesser degree.

The time frame in which the Hollerith discussion began and continued is significant. In April 1964, the IBM System/360 computing systems were announced, with an 8-bit architecture. Up to that time, computing systems had prevailingly been of 6-bit (or homomorphically 6-bit) architecture. Card-code sets that had consisted of up to 64 characters would need to be extended to 128 characters for ASCII, and had been extended to 256 characters by the System/360's code, EBCDIC.

As well as the problem of different versions of Hollerith, there was also the problem that there were no "common existing standard Hollerith codes" (Criterion 13) for the control characters of ASCII, and for the lower-case alphabetic characters of ASCII. Indeed, ASCII as then published (ASA X3.4-1963) did not have the lower-case alphabetics assigned to columns 6 and 7, and many of the control characters were not defined specifically.

However, when the first proposed American Standard Hollerith Representation of ASCII was drafted in September 1964, ASA Subcommittee X3.2 had agreed internally on specific definitions for all 32 control characters of ASCII, and had assigned the lower-case alphabetics and five special graphics to columns 6 and 7 of ASCII.

17.2 PROBLEMS OF DECISION

At this time, or before final approval in 1968, there were eight problems (apart from the many extant versions of Hollerith) that made consensus on "which Hollerith?" difficult.

Problem 1

No commonly used card hole patterns for lower-case alphabetics (although assignments had been made in EBCDIC for the System/360).

Problem 2

No commonly used card hole patterns for the control characters of ASCII (although about half of these control characters had been assigned in EBCDIC).

Problem 3

Two special graphics, @ (Commercial At) and , (Grave Accent) seesawed back and forth between code positions 4/0 and 6/0 of ASCII at successive meetings of ISO/TC97/SC2. A hole pattern for Commercial At was in common use. The question was whether this hole pattern should be assigned to code position 4/0 or 6/0.

Problem 4

Graphics for code positions 5/12, 7/12, and 7/14 changed and interchanged. While none of the various graphics had commonly used card hole patterns, two of them were assigned in EBCDIC.

Problem 5

There was a continuing debate on whether the final Hollerith card code and the EBCDIC card code should or should not be compatible. This was complicated by the fact that ASCII had graphics not in EBCDIC, and EBCDIC had graphics not in ASCII.

Problem 6

Two graphics, ¬ (Logical NOT) and | (Logical OR), were in and out of ASCII, and in different code positions of ASCII, at different times between 1963 and 1967. Both these graphics had assigned hole patterns in EBCDIC.

Problem 7

Code position 1/10 at the inception of the Hollerith debate was SS (Start of Special), but was subsequently changed to SUB (Substitute). This was really an administrative problem, not a code problem, but it did lead to different *looking* code charts.

Problem 8

As described in Chapters 4, and 9, the so-called A- and H- duals were broadly implemented in different punched card equipment as shown below:

Hole pattern	8-4	8-3	12	12-8-2	0-8-4
A-graphic	@	#	&	¤	%
H-graphic	′	=	+	)	(

In EBCDIC, the decision had been made to provide unique bit patterns and hole patterns for all ten of these graphics, and to replace the ¤ (lozenge) with the < (less than), as follows:

Graphic	Hole pattern
@	8-4
#	8-3
&	12
<	12-8-4
%	0-8-4
'	8-5
=	8-6
+	12-8-6
)	11-8-5
(	12-8-5

That is to say, the A-graphics (but replacing ¤ with <) were assigned their existing hole patterns, but the H-graphics were assigned new hole patterns. On the standards committee, the same question arose:

> Should the A-graphics retain existing hole patterns and the H-graphics receive new hole patterns, or should the H-graphics retain existing hole patterns and the A-graphics receive new hole patterns?

On the standards committee, there were protagonists for the former, and protagonists for the latter. Problem 1 was soon resolved (hole patterns for lower-case alphabetics), but the other problems were resolved only after many discussions and ballots, and were the source of many different proposals for a standard Hollerith card code.

Resolution of Problem 1. In deciding on hole patterns for the lower-case alphabetics, two principles were applied:

A) Each lower-case alphabetic hole pattern should bear some logical relationship to the corresponding upper-class alphabetic hole pattern.
B) The number of holes in lower-case alphabetic hole patterns should be minimum.

The obvious way to apply Principle A was to include the hole pattern for the upper-case alphabetic in the hole pattern for the lower-case alphabetic, and then to distinguish between them by adding a zone punch.

Indeed, there is no other solution than the addition of a zone punch either 0, 11, or 12. In the full set of 256 hole patterns, both the 8-punch and 9-punch act as zone punches in some hole patterns. But neither they nor indeed any numeric punch 1 through 9 could act as zone punches for the alphabetics, since they act as digit punches for the alphabetics. Ideally, it would be nice if the additional zone punch could be the *same* additional zone punch for all letters. But this was not possible. We know that

upper-case alphabetics A to I had zone punch 12,

upper-case alphabetics J to R had zone punch 11,

upper-case alphabetics S to Z had zone punch 0.

Available as new zone-punch hole patterns were 12-0, 12-11, 11-0, and 12-11-0. There were four possible hole patterns, from which three had to be chosen. No choice of three would satisfy the ideal condition.

However, Principle B clearly implied that the three choices should be 12-0, 12-11, 11-0, and not 12-11-0. The possible choices were

a to i	12-11	or	12-0,
j to r	11-0	or	12-11,
s to z	12-0	or	11-0.

Between these two sets of choices, the actual choice appeared to be quite arbitrary—with no technical reasons for or against either choice.

It was observed on the standards committee that the same choice must have been available when designing the card code for EBCDIC. The choice for EBCDIC had had to be made, and it was made, admittedly arbitrarily, for

a to i	12-0,
j to r	12-11,
s to z	11-0.

The standards committee decided that, since there was no technical reason against this choice for the Hollerith card code, there was no reason not to accept the same decision that had been made for EBCDIC. The decision was so made by the committee.

17.3 PROPOSALS

During the deliberations of the committee, seventeen proposals were submitted by various committee members. These proposals were submitted in the form of committee documents.

Proposal 1

On September 11, 1964, the first Proposed American Standard Hollerith Representation of ASCII was drafted (document X3.2.3/53). It specified hole patterns for all 128 characters* (see Fig. 17.1). What solutions did this proposal provide for the eight problems?

Lower-case alphabetics

Hole patterns matched EBCDIC hole patterns, as previously described. (This problem will not be referred to subsequently in this chapter.)

Control characters

The draft standard says

> The de facto Hollerith had not contained the ASCII control characters. Since new hole patterns had to be devised for all characters in ASCII columns 0 and 1, the hole patterns for these two columns were developed with a logical relationship to the ASCII Code.

Examination of the hole patterns for columns 0 and 1 shows this to be true:

i) Zone-punches 9-12 apply to all of column 0.
ii) Zone-punches 9-11 apply to all of column 1.
iii) With the exception of row 0 of columns 0 and 1, all digit-punch hole patterns translate to the ASCII low-order four bits on a precise and exact BCD basis.

There *was* a little problem for row 0 of columns 0 and 1. The "logical" hole patterns to correspond to part (iii) above would have been 9-12 and 9-11. But these hole patterns were already preempted for graphics I and R. As is observed in other sections of this book, this kind of preemption (for example 0-9, 12-0-9, 12-11-9, 11-0-9 are also preempted) led to the

* For a reason that will be given later, some subsequent Hollerith proposals specified fewer than 128 hole patterns. One, for example, specified only 43 hole patterns!

	b7 b6 b5	0 0 0	0 0 1	0 1 0	0 1 1	1 0 0	1 0 1	1 1 0	1 1 1
b4 b3 b2 b1	Col / Row	0	1	2	3	4	5	6	7
0 0 0 0	0	NUL 9-12-0 8-1	DLE 9-12-11 8-1	SP No Pch	0 0	\ 8-1	P 11-7	@ 8-4	p 12-11-7
0 0 0 1	1	SOH 9-12-1	DC1 9-11-1	! 11-8-2	1 1	A 12-1	Q 11-8	a 12-0-1	q 12-11-8
0 0 1 0	2	STX 9-12-2	DC2 9-11-2	" 0-8-2	2 2	B 12-2	R 11-9	b 12-0-2	r 12-11-9
0 0 1 1	3	ETX 9-12-3	DC3 9-11-3	# 8-3	3 3	C 12-3	S 0-2	c 12-0-3	s 11-0-2
0 1 0 0	4	EOT 9-12-4	DC4 9-11-4	$ 11-8-3	4 4	D 12-4	T 0-3	d 12-0-4	t 11-0-3
0 1 0 1	5	ENQ 9-12-5	NAK 9-11-5	% 0-8-4	5 5	E 12-5	U 0-4	e 12-0-5	u 11-0-4
0 1 1 0	6	ACK 9-12-6	SYN 9-11-6	& 12	6 6	F 12-6	V 0-5	f 12-0-6	v 11-0-5
0 1 1 1	7	BEL 9-12-7	ETB 9-11-7	' 8-5	7 7	G 12-7	W 0-6	g 12-0-7	w 11-0-6
1 0 0 0	8	BS 9-12 8	CAN 9-11 8	(12-8-5	8 8	H 12-8	X 0-7	h 12-0-8	x 11-0-7
1 0 0 1	9	HT 9-12 8-1	EM 9-11 8-1	) 11-8-5	9 9	I 12-9	Y 0-8	i 12-0-9	y 11-0-8
1 0 1 0	10	LF 9-12 8-2	SS 9-11 8-2	* 11-8-4	: 8-2	J 11-1	Z 0-9	j 12-11-1	z 11-0-9
1 0 1 1	11	VT 9-12 8-3	ESC 9-11 8-3	+ 12-8-6	; 11-8-6	K 11-2	[12-8-7	k 12-11-2	{ 12-8-2
1 1 0 0	12	FF 9-12 8-4	FS 9-11 8-4	, 0-8-3	< 12-8-4	L 11-3	~ 12-8-1	l 12-11-3	| 0-8-7
1 1 0 1	13	CR 9-12 8-5	GS 9-11 8-5	- 11	= 8-6	M 11-4	] 11-8-7	m 12-11-4	} 11-8-2
1 1 1 0	14	SO 9-12 8-6	RS 9-11 8-6	. 12-8-3	> 0-8-6	N 11-5	^ 11-8-1	n 12-11-5	¬ 0-8-5
1 1 1 1	15	SI 9-12 8-7	US 9-11 8-7	/ 0-1	? 8-7	O 11-6	_ 0-8-1	o 12-11-6	DEL 12-11

Fig. 17.1 Hollerith, Version 1

hole pattern 8-1 in combination with zone-punch hole patterns also being displaced, and these (in both EBCDIC and Hollerith) usually ended up in row 0 because they were the hole patterns left over to fill up the code positions in row 0. Following this line of reasoning, 9-12-0-8-1 and 9-12-11-8-1 were chosen for row 0, columns 0 and 1.

@ and '

At this time, in ASCII, ` (Grave Accent) was in code position 4/0 and @ (Commercial At) in 6/0. It is to be noted that @ received its de facto 8-4 hole pattern.

5/12, 7/12, 7/14

At this time graphics ˜ | and ¬ were in code positions 5/12, 7/12, 7/14, respectively.

EBCDIC/Hollerith compatibility

This proposal was evidently drafted by a proponent of EBCDIC Hollerith compatibility. Except for columns 0 and 1 (see above) all hole patterns were compatible, except those shown shaded in Fig. 17.1. The graphics [˜] ˆ { } were not incorporated into EBCDIC at that time. Looking back, it is not clear why the hole patterns of graphics " — ¬ were not chosen to be compatible with those of EBCDIC.

Logical OR, Logical NOT

The Logical OR, Logical NOT problem (to be described later) had not yet surfaced.

Position 1/10

Control character SS (Start of Special) was at that time in code position 1/10 in ASCII. (This problem will not be discussed again until the problem actually surfaces.)

A versus H

Since the drafter was evidently a proponent for EBCDIC/Hollerith compatibility, and since EBCDIC had chosen existing hole patterns for the A-graphics, this proposal also did so.

Comment. At this time, only two criteria were being applied:

i) Simple translation relationship, Hollerith to/from ASCII, for the control characters.

ii) EBCDIC/Hollerith compatibility as much as possible.

Proposal 2

On November 10, 1964, the second proposal was made (document X3.2.3/69) by Mr. J. L. Tobin. The proposer chose not to make any suggestions with respect to control characters, so he suggested hole patterns only for the 94 graphics, and Space. The proposer had analyzed the different versions of Hollerith previously referred to, and had counted up the number of companies (out of 8) who agreed on a particular hole pattern. He had then proposed a "consensus" approach as follows:

Unanimous	8 companies
Overwhelming	6 or 7 companies
Substantial	4 or 5 companies
Little or none	3 or less

Based on this analysis, the proposer chose the hole patterns shown in Fig. 17.2.

Comments. As might be supposed from the selection scheme, there was considerable incompatibility with EBCDIC among the specials.

This proposal did not receive support in the standards committee.

At the January 28, 1965 meeting, ASA Task Group X3.2.3 formally voted to accept the existing Hollerith hole patterns for Space, the alphabetics, 10 numerics, and 6 specials:

. , * / − $

All manufacturers' equipments provided these. It was at this meeting, therefore, that the concept of the "hard-core 43 graphics" emerged and was never subsequently objected to.

Proposal 3

On November 23, 1964, another proposal was made. The proposer was, as in the previous case, wrestling with the problem of criteria. This proposer restricted himself to 64 hole patterns, since the maximum existing implementation (except for EBCDIC on the System/360) had 64 hole patterns. The proposer, Mr. E. H. Clamons, presented a rather pragmatic set of criteria, as follows:

1. Old established codes, IBM 407.
2. New established codes, IBM BCD.
3. New established codes, UNIVAC 1004.
4. Suggested for adoption.

The proposal is shown in Fig. 17.3. It did not receive support in the standards committee.

	b7 b6 b5	0 0 0	0 0 1	0 1 0	0 1 1	1 0 0	1 0 1	1 1 0	1 1 1
b4 b3 b2 b1	Row \ Col	0	1	2	3	4	5	6	7
0 0 0 0	0	NUL	DLE	SP No Pch	0 0	@ 8-4	P 11-7	\ 0-8-1	p 12-11-7
0 0 0 1	1	SOH	DC1	! 11-8-2	1 1	A 12-1	Q 11-8	a 12-0-1	q 12-11-8
0 0 1 0	2	STX	DC2	" 0-8-7	2 2	B 12-2	R 11-9	b 12-0-2	r 12-11-9
0 0 1 1	3	ETX	DC3	# 8-3	3 3	C 12-3	S 0-2	c 12-0-3	s 11-0-2
0 1 0 0	4	EOT	DC4	$ 11-8-3	4 4	D 12-4	T 0-3	d 12-0-4	t 11-0-3
0 1 0 1	5	ENQ	NAK	% 0-8-4	5 5	E 12-5	U 0-4	e 12-0-5	u 11-0-4
0 1 1 0	6	ACK	SYN	& 12	6 6	F 12-6	V 0-5	f 12-0-6	v 11-0-5
0 1 1 1	7	BEL	ETB	' 8-2	7 7	G 12-7	W 0-6	g 12-0-7	w 11-0-6
1 0 0 0	8	BS	CAN	(12-8-5	8 8	H 12-8	X 0-7	h 12-0-8	x 11-0-7
1 0 0 1	9	HT	EM	) 11-8-5	9 9	I 12-9	Y 0-8	i 12-0-9	y 11-0-8
1 0 1 0	10	LF	SS	* 11-8-4	: 8-5	J 11-1	Z 0-9	j 12-11-1	z 11-0-9
1 0 1 1	11	VT	ESC	+ 12-8-2	; 11-8-6	K 11-2	[12-8-7	k 12-11-2	{ 12-0
1 1 0 0	12	FF	FS	, 0-8-3	< 12-8-6	L 11-3	~ 12-8-4	l 12-11-3	| 12-8-1
1 1 0 1	13	CR	GS	- 11	= 0-8-6	M 11-4	] 11-8-7	m 12-11-4	} 11-0
1 1 1 0	14	SO	RS	. 12-8-3	> 8-6	N 11-5	^ 11-8-1	n 12-11-5	¬ 8-7
1 1 1 1	15	SI	US	/ 0-1	? 0-8-2	O 11-6	_ 0-8-5	o 12-11-6	DEL

Fig. 17.2 Hollerith, Version 2

	b7	0	0	0	0	1	1	1	1
	b6	0	0	1	1	0	0	1	1
	b5	0	1	0	1	0	1	0	1
b4 b3 b2 b1	Row \ Col	0	1	2	3	4	5	6	7
0 0 0 0	0	NUL	DLE	SP No Pch	0 0	@ 8-4	P 11-7	\	p
0 0 0 1	1	SOH	DC1	! 11-0	1 1	A 12-1	Q 11-8	a	q
0 0 1 0	2	STX	DC2	" 12-8-4	2 2	B 12-2	R 11-9	b	r
0 0 1 1	3	ETX	DC3	# 8-3	3 3	C 12-3	S 0-2	c	s
0 1 0 0	4	EOT	DC4	$ 11-8-3	4 4	D 12-4	T 0-3	d	t
0 1 0 1	5	ENQ	NAK	% 0-8-4	5 5	E 12-5	U 0-4	e	u
0 1 1 0	6	ACK	SYN	& 12	6 6	F 12-6	V 0-5	f	v
0 1 1 1	7	BEL	ETB	' 8-7	7 7	G 12-7	W 0-6	g	w
1 0 0 0	8	BS	CAN	(0-8-5	8 8	H 12-8	X 0-7	h	x
1 0 0 1	9	HT	EM	) 0-8-7	9 9	I 12-9	Y 0-8	i	y
1 0 1 0	10	LF	SS	* 11-8-4	: 8-5	J 11-1	Z 0-9	j	z
1 0 1 1	11	VT	ESC	+ 8-2	; 11-8-6	K 11-2	[12-8-5	k	{
1 1 0 0	12	FF	FS	, 0-8-3	< 12-8-6	L 11-3	~ 0-8-6	l	|
1 1 0 1	13	CR	GS	- 11	= 12-8-7	M 11-4	] 11-8-5	m	}
1 1 1 0	14	SO	RS	. 12-8-3	> 8-6	N 11-5	^ 11-8-7	n	¬
1 1 1 1	15	SI	US	/ 0-1	? 12-0	O 11-6	_ 0-8-2	o	DEL

Fig. 17.3 Hollerith, Version 3

	b7 b6 b5	0 0 0	0 0 1	0 1 0	0 1 1	1 0 0	1 0 1	1 1 0	1 1 1
b4 b3 b2 b1	Col / Row	0	1	2	3	4	5	6	7
0 0 0 0	0	NUL 9-12-0 8-1	DLE 9-12-11 8-1	SP No Pch	0 0		P 11-7	@	p 12-11-7
0 0 0 1	1	SOH 9-12-1	DC1 9-11-1	! 11-8-2	1 1	A 12-1	Q 11-8	a 12-0-1	q 12-11-8
0 0 1 0	2	STX 9-12-2	DC2 9-11-2	" 8-7	2 2	B 12-2	R 11-9	b 12-0-2	r 12-11-9
0 0 1 1	3	ETX 9-12-3	DC3 9-11-3	# 8-3	3 3	C 12-3	S 0-2	c 12-0-3	s 11-0-2
0 1 0 0	4	EOT 9-12-4	DC4 9-11-4	$ 11-8-3	4 4	D 12-4	T 0-3	d 12-0-4	t 11-0-3
0 1 0 1	5	ENQ 9-12-5	NAK 9-11-5	% 0-8-4	5 5	E 12-5	U 0-4	e 12-0-5	u 11-0-4
0 1 1 0	6	ACK 9-12-6	SYN 9-11-6	& 12	6 6	F 12-6	V 0-5	f 12-0-6	v 11-0-5
0 1 1 1	7	BEL 9-12-7	ETB 9-11-7	' 8-5	7 7	G 12-7	W 0-6	g 12-0-7	w 11-0-6
1 0 0 0	8	BS 9-12 8	CAN 9-11 8	(12-8-5	8 8	H 12-8	X 0-7	h 12-0-8	x 11-0-7
1 0 0 1	9	HT 9-12 8-1	EM 9-11 8-1	) 11-8-5	9 9	I 12-9	Y 0-8	i 12-0-9	y 11-0-8
1 0 1 0	10	LF 9-12 8-2	SS 9-11 8-2	* 11-8-4	: 8-2	J 11-1	Z 0-9	j 12-11-1	z 11-0-9
1 0 1 1	11	VT 9-12 8-3	ESC 9-11 8-3	+ 12-8-6	; 11-8-6	K 11-2	[12-8-1	k 12-11-2	{ 11-0
1 1 0 0	12	FF 9-12 8-4	FS 9-11 8-4	, 0-8-3	< 12-8-4	L 11-3	~ 12-8-2	l 12-11-3	¬ 11-8-7
1 1 0 1	13	CR 9-12 8-5	GS 9-11 8-5	- 11	= 8-6	M 11-4	] 8-1	m 12-11-4	} 12-0
1 1 1 0	14	SO 9-12 8-6	RS 9-11 8-6	. 12-8-3	> 0-8-6	N 11-5	^ 0-8-2	n 12-11-5	| 12-8-7
1 1 1 1	15	SI 9-12 8-7	US 9-11 8-7	/ 0-1	? 0-8-7	O 11-6	_ 0-8-5	o 12-11-6	DEL 12-11-0 7-8-9

Fig. 17.4 Hollerith, Version 4

Proposal 4

In January, 1965 the fourth proposal was made. It was made by the IBM representative, and specified 256 hole patterns. It is shown in Fig. 17.4 in ASCII format, and in Fig. 17.5 in EBCDIC format. It was essentially

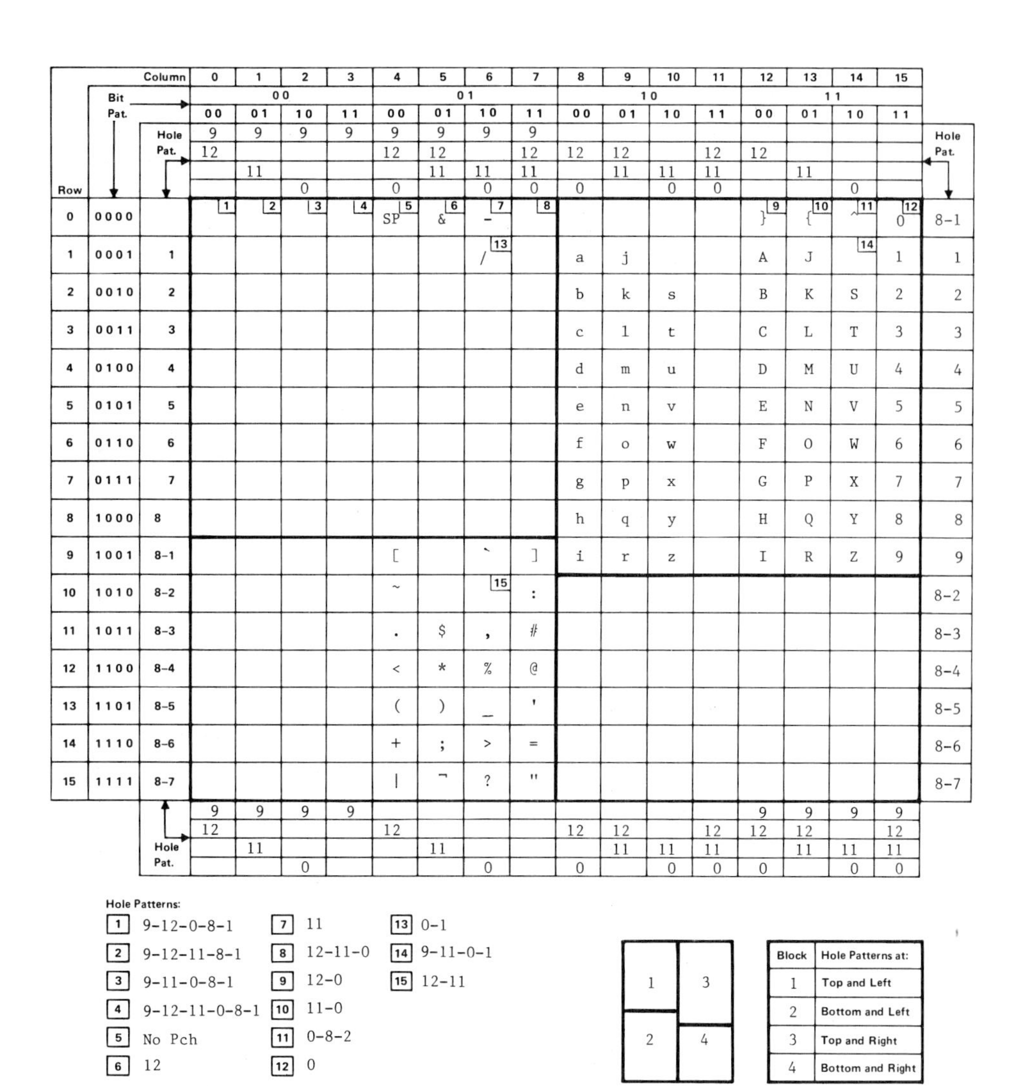

Row	Bit Pat.	Hole Pat.	0	1	2	3	4	5	6	7	8	9	10	11	12	13	14	15	Hole Pat.
	Bit Pat.		00 00	00 01	00 10	00 11	01 00	01 01	01 10	01 11	10 00	10 01	10 10	10 11	11 00	11 01	11 10	11 11	
		Hole Pat.	9 12	9 11	9 0	9	9 12 0	9 12 11	9 11 0	9 12 11 0	12 0	12 11	11 0	12 11 0	12	11	0		
0	0000		[1]	[2]	[3]	[4]	[5] SP	[6] &	[7] -	[8]					[9] }	[10] {	[11] ~	[12] 0	8-1
1	0001	1							[13] /		a	j			A	J	[14]	1	1
2	0010	2									b	k	s		B	K	S	2	2
3	0011	3									c	l	t		C	L	T	3	3
4	0100	4									d	m	u		D	M	U	4	4
5	0101	5									e	n	v		E	N	V	5	5
6	0110	6									f	o	w		F	O	W	6	6
7	0111	7									g	p	x		G	P	X	7	7
8	1000	8									h	q	y		H	Q	Y	8	8
9	1001	8-1					[		`	]	i	r	z		I	R	Z	9	9
10	1010	8-2					~		[15]	:									8-2
11	1011	8-3					.	$	,	#									8-3
12	1100	8-4					<	*	%	@									8-4
13	1101	8-5					(	)	_	'									8-5
14	1110	8-6					+	;	>	=									8-6
15	1111	8-7					|	¬	?	"									8-7
		Hole Pat.	9 12	9 11	9 0	9	12	11	0		12 0	12 11	11 0	12 11 0	9 12 0	9 12 11	9 11 0	9 12 11 0	

Hole Patterns:

1	9-12-0-8-1	7	11	13	0-1
2	9-12-11-8-1	8	12-11-0	14	9-11-0-1
3	9-11-0-8-1	9	12-0	15	12-11
4	9-12-11-0-8-1	10	11-0		
5	No Pch	11	0-8-2		
6	12	12	0		

Block	Hole Patterns at:
1	Top and Left
2	Bottom and Left
3	Top and Right
4	Bottom and Right

Fig. 17.5 EBCDIC (from Hollerith, Version 4)

EBCDIC, but with the 32 control characters of ASCII assigned to columns 0 and 1 of the 8-bit code table. Incidentally, this assignment of control characters agreed with that of Proposal 1 (Fig. 17.1).

EBCDIC at that time had 88 assigned graphics, while ASCII had 94. This proposal substituted the ASCII ˜ for the EBCDIC ¢ in position 4/10,* and positioned the six additional ASCII graphics [`] } { ˜ in positions 4/9, 6/9, 7/9, 12/0, 13/0, 14/0, respectively, of Fig. 17.5.

This proposal was accepted (at the time) by ASA Task Group X3.2.3, and drafted into a Proposed American Standard Hollerith Representation of ASCII (document X3.2.3/85, 1965 April 14).

Although ASA Task Group X3.2.3 was by now committed to development of a Hollerith standard, the parent ASA Subcommittee X3.2 had not relinquished its support for Decimal ASCII, and, at its meeting on March 4, 1965, the following motion was passed (10 yes, 2 no, 6 abstain):

> X3.2 directs X3.2.3 to develop a Decimal ASCII standard, which is to be *the* punched card code for information interchange, and to recommend a method of accommodating a 64-character Hollerith card code.

The instruction to find a method to accommodate 64 Hollerith characters led to a variety of proposals in Task Group X3.2.3. At the subsequent meeting on April 27, 28, 1965, Task Group X3.2.3, wrestling with the instruction given by X3.2, passed two motions; one motion stating X3.2.3's judgment that *two* standards, Decimal ASCII and Hollerith, are required; the other motion stating X3.2.3's judgment that *both* these standards should encompass all 128 ASCII characters. Both these motions were felt by X3.2.3 to be in consonance with the X3.2 motion.

However, at the subsequent X3.2 meeting, on April 29, 1965, a majority of X3.2 members were either opposed to two standards, or opposed to an "extended" Hollerith standard (128 characters) per se.

Proposal 5

On July 1, 1965, a fifth Hollerith proposal was made by Mr. R. H. Brown (see Fig. 17.6). The proposer restricted the set to 64 hole patterns (in the spirit of the above-mentioned X3.2 motion). The proposer was a proponent of assigning existing hole patterns to the H-graphics.

---

* The standards committee had not adopted (and still has not adopted) the hexadecimal notation for naming the sixteen rows and columns of an 8-bit code table. Instead, rows and columns were numbers 0, 1, 2, 3, . . . , 13, 14, 15 and the columns/row notation was used for designating code-table positions. In this chapter, since in most instances the 8-bit code tables are copies from actual standards committee documents, the author also uses the column/row notation, instead of the hexadecimal notation.

	b7 b6 b5	0 0 0	0 0 1	0 1 0	0 1 1	1 0 0	1 0 1	1 1 0	1 1 1
b4 b3 b2 b1	Row \ Col	0	1	2	3	4	5	6	7
0 0 0 0	0	NUL	DLE	SP No Pch	0 0	` 0-8-7	P 11-7	@	p
0 0 0 1	1	SOH	DC1	! 11-8-2	1 1	A 12-1	Q 11-8	a	q
0 0 1 0	2	STX	DC2	" 8-7	2 2	B 12-2	R 11-9	b	r
0 0 1 1	3	ETX	DC3	# 12-8-7	3 3	C 12-3	S 0-2	c	s
0 1 0 0	4	EOT	DC4	$ 11-8-3	4 4	D 12-4	T 0-3	d	t
0 1 0 1	5	ENQ	NAK	% 0-8-5	5 5	E 12-5	U 0-4	e	u
0 1 1 0	6	ACK	SYN	& 8-2	6 6	F 12-6	V 0-5	f	v
0 1 1 1	7	BEL	ETB	' 8-4	7 7	G 12-7	W 0-6	g	w
1 0 0 0	8	BS	CAN	(0-8-4	8 8	H 12-8	X 0-7	h	x
1 0 0 1	9	HT	EM	) 12-8-4	9 9	I 12-9	Y 0-8	i	y
1 0 1 0	10	LF	SS	* 11-8-4	: 8-5	J 11-1	Z 0-9	j	z
1 0 1 1	11	VT	ESC	+ 12	; 11-8-6	K 11-2	[12-8-5	k	{
1 1 0 0	12	FF	FS	, 0-8-3	< 12-8-6	L 11-3	~ 0-8-6	l	— over- line
1 1 0 1	13	CR	GS	- 11	= 8-3	M 11-4	] 11-8-5	m	}
1 1 1 0	14	SO	RS	. 12-8-3	> 8-6	N 11-5	^ 11-8-7	n	|
1 1 1 1	15	SI	US	/ 0-1	? 12-8-2	O 11-6	_ 0-8-7	o	DEL

Fig. 17.6 Hollerith, Version 5

@ and

The graphic for grave accent was at that time in position 4/0.

5/7, 7/12, 7/14

The graphics for tilde, overline, and vertical line were at that time in positions 5/7, 7/12, 7/14, respectively.

EBCDIC/Hollerith compatibility

Because the proposal assigned existing hole patterns to the H-graphics, there were many incompatibilities with EBCDIC, as shown by the shaded areas in Fig. 17.6.

Logical OR Logical NOT

The operations Logical OR and Logical NOT, previously in positions 7/12 and 7/14, had been replaced by the graphics for overline and vertical line.

1/10

Control character SS (Start of Special) was still in vogue for position 1/10.

A versus H

Existing hole patterns were assigned in this proposal to the H-graphics.

This proposal was neither rejected nor accepted by Task Group X3.2.3 at this time, but was kept under study. It was at this time that the A versus H controversy surfaced.

Proposal 6

On September 7, 1965, Proposal 6 was made by Mr. L. L. Griffin, Chairman of Subcommittee X3.2.

The proposal embodied the Decimal ASCII Card Code but additionally proposed a "Translation Table" to the hard-core 43 Hollerith characters (shown in Fig. 17.7).

Proposal 7

On September 14, 1965, Proposal 7 was made by Mr. R. M. Brown. It was substantially the same as Proposal 5, but with a difference considered to be important by the proposer. The hole patterns of 12-8-2 and 11-8-2 of Proposal 5 had been replaced by hole patterns 12-0 and 11-0 in Proposal 7 (see Fig. 17.8). The widespread practice of overpunching numerics by a 12-punch or 11-punch to indicate positive or negative numeric fields naturally required that the hole patterns 12-0 and 11-0 be included in the set of 64 hole patterns. The proposer pointed out that this particular card code was, at that time, a Draft Military Standard.

As with Proposal 5, Proposal 7 was kept for study by Task Group X3.2.3.

Proposal 8

On September 15, 1965, Proposal 8 was submitted by Mr. J. L. Tobin. Proposal 8 was, in fact, the same as Proposal 7.

b7 b6 b5		0 0 0	0 0 1	0 1 0	0 1 1	1 0 0	1 0 1	1 1 0	1 1 1
b4 b3 b2 b1	Row \ Col	0	1	2	3	4	5	6	7
0 0 0 0	0	NUL	DLE	SP No Pch	0 0	`	P 11-7	@	p
0 0 0 1	1	SOH	DC1	!	1 1	A 12-1	Q 11-8	a	q
0 0 1 0	2	STX	DC2	"	2 2	B 12-2	R 11-9	b	r
0 0 1 1	3	ETX	DC3	#	3 3	C 12-3	S 0-2	c	s
0 1 0 0	4	EOT	DC4	$ 11-8-3	4 4	D 12-4	T 0-3	d	t
0 1 0 1	5	ENQ	NAK	%	5 5	E 12-5	U 0-4	e	u
0 1 1 0	6	ACK	SYN	&	6 6	F 12-6	V 0-5	f	v
0 1 1 1	7	BEL	ETB	'	7 7	G 12-7	W 0-6	g	w
1 0 0 0	8	BS	CAN	(	8 8	H 12-8	X 0-7	h	x
1 0 0 1	9	HT	EM	)	9 9	I 12-9	Y 0-8	i	y
1 0 1 0	10	LF	SS	* 11-8-4	:	J 11-1	Z 0-9	j	z
1 0 1 1	11	VT	ESC	+	;	K 11-2	[	k	{
1 1 0 0	12	FF	FS	, 0-8-3	<	L 11-3	~	l	— over-line
1 1 0 1	13	CR	GS	- 11	=	M 11-4	]	m	}
1 1 1 0	14	SO	RS	. 12-8-3	>	N 11-5	^	n	|
1 1 1 1	15	SI	US	/ 0-1	?	O 11-6	_	o	DEL

Fig. 17.7 Hollerith, Version 6

At the September 14, 15, 1965 meeting of Task Group X3.2.3, Decimal ASCII was forwarded to Subcommittee X3.2 as a recommended American Standard. Task Group X3.2.3's opinion was about evenly divided between an H-based Hollerith and an A-based Hollerith.

	b7	0	0	0	0	1	1	1	1
	b6	0	0	1	1	0	0	1	1
	b5	0	1	0	1	0	1	0	1
b4 b3 b2 b1	Row \ Col	0	1	2	3	4	5	6	7
0 0 0 0	0	NUL	DLE	SP No Pch	0 0	` 0-8-7	P 11-7	@	p
0 0 0 1	1	SOH	DC1	! 11-0	1 1	A 12-1	Q 11-8	a	q
0 0 1 0	2	STX	DC2	" 8-7	2 2	B 12-2	R 11-9	b	r
0 0 1 1	3	ETX	DC3	# 12-8-7	3 3	C 12-3	S 0-2	c	s
0 1 0 0	4	EOT	DC4	$ 11-8-3	4 4	D 12-4	T 0-3	d	t
0 1 0 1	5	ENQ	NAK	% 0-8-5	5 5	E 12-5	U 0-4	e	u
0 1 1 0	6	ACK	SYN	& 8-2	6 6	F 12-6	V 0-5	f	v
0 1 1 1	7	BEL	ETB	' 8-4	7 7	G 12-7	W 0-6	g	w
1 0 0 0	8	BS	CAN	(0-8-4	8 8	H 12-8	X 0-7	h	x
1 0 0 1	9	HT	EM	) 12-8-4	9 9	I 12-9	Y 0-8	i	y
1 0 1 0	10	LF	SS	* 11-8-4	: 8-5	J 11-1	Z 0-9	j	z
1 0 1 1	11	VT	ESC	+ 12	; 11-8-6	K 11-2	[12-8-5	k	{
1 1 0 0	12	FF	FS	, 0-8-3	< 12-8-6	L 11-3	~ 0-8-6	l	— over-line
1 1 0 1	13	CR	GS	- 11	= 8-3	M 11-4	] 11-8-5	m	}
1 1 1 0	14	SO	RS	. 12-8-3	> 8-6	N 11-5	^ 11-8-7	n	|
1 1 1 1	15	SI	US	/ 0-1	? 12-0	O 11-6	_ 0-8-2	o	DEL

Fig. 17.8 Hollerith, Versions 7 and 8

Proposal 9

At the November 3, 4, 5, 1965 meeting of Task Group X3.2.3, an attempt was made to appease Decimal ASCII proponents (Hollerith-H proponents and Hollerith-A proponents) by incorporating *all three* card codes into a draft American Standards.

b7 b6 b5		0 0 0	0 0 1	0 1 0	0 1 1	1 0 0	1 0 1	1 1 0	1 1 1
b4 b3 b2 b1	Row \ Col	0	1	2	3	4	5	6	7
0 0 0 0	0	NUL	DLE	SP No Pch	0 0	` 0-8-7	P 11-7	@ 8-4	p
0 0 0 1	1	SOH	DC1	! 12-8-7 11-0	1 1	A 12-1	Q 11-8	a	q
0 0 1 0	2	STX	DC2	" 8-7	2 2	B 12-2	R 11-9	b	r
0 0 1 1	3	ETX	DC3	# 8-3 12-8-7	3 3	C 12-3	S 0-2	c	s
0 1 0 0	4	EOT	DC4	$ 11-8-3	4 4	D 12-4	T 0-3	d	t
0 1 0 1	5	ENQ	NAK	% 0-8-4 0-8-5	5 5	E 12-5	U 0-4	e	u
0 1 1 0	6	ACK	SYN	& 12 8-2	6 6	F 12-6	V 0-5	f	v
0 1 1 1	7	BEL	ETB	' 8-5 8-4	7 7	G 12-7	W 0-6	g	w
1 0 0 0	8	BS	CAN	(12-8-5 0-8-4	8 8	H 12-8	X 0-7	h	x
1 0 0 1	9	HT	EM	) 11-8-5 12-8-4	9 9	I 12-9	Y 0-8	i	y
1 0 1 0	10	LF		* 11-8-4	: 8-2 8-5	J 11-1	Z 0-9	j	z
1 0 1 1	11	VT	ESC	+ 12-8-6 12	; 11-8-6	K 11-2	[11-8-2 12-8-5	k	{
1 1 0 0	12	FF	FS	, 0-8-3	< 12-8-4 12-8-6	L 11-3	~ 12-8-2 0-8-6	l	— over- line
1 1 0 1	13	CR	GS	- 11	= 8-6 8-3	M 11-4	] 0-8-2 11-8-5	m	}
1 1 1 0	14	SO	RS	. 12-8-3	> 0-8-6 12-0	N 11-5	^ 11-8-7	n	|
1 1 1 1	15	SI	US	/ 0-1	? 0-8-7 12-0	O 11-6	_ 0-8-5 0-8-6	o	DEL

A
H

Fig. 17.9 Hollerith, Version 9

The two Hollerith proposals, both specifying 64 hole patterns to satisfy the dictate of Subcommittee X3.2, are shown in Fig. 17.9. The A-Hollerith version was substantially compatible with EBCDIC.

The philosophy of appeasing Decimal ASCII, Hollerith-A, and Hollerith-H proponents by combining all three card codes into a single draft American Standard did not survive. As a result of an X3.2 letter ballot on the draft American Standard, Subcommittee X3.2 pared the two 64-character Hollerith card codes into a single Hollerith card code consisting of the "hard core 43" (Fig. 17.7). This Hollerith card code was then combined with the Decimal ASCII card code into a single draft American Standard. The Hollerith Card Code Table was qualified as an "interim representation," although no time limits were expressed with respect to the interim. At the December 7, 8, 9, 1965 meeting of Subcommittee X3.2, a recorded vote was taken to forward this draft American Standard to Committee X3 for further processing.

In 1965, Decimal ASCII had been approved as an ECMA Standard, and in May, 1965, ISO/TC97/SC2 prepared a draft ISO Proposal on Decimal ASCII, which was circulated for review and comment.

In December 1965, Committee X3 issued an X3 letter ballot on Decimal ASCII. In June 1966, the X3 ballot result was reported:

15 affirmative,
13 negative,
17 not yet responded.

X3 declared that it appeared there would not be a consensus for approval of Decimal ASCII. An ad hoc committee was established to recommend a course of action. The ad hoc committee met on July 28, 1966, and, after discussion, recommended the preparation of an American Standard "BCD Card Code" (their nomenclature) based on existing Hollerith practices, and that 128 hole patterns be assigned. Task Group X3.2.3, therefore, once again approached the problem of "which Hollerith?"

Proposal 10

On August 15, 1966, Proposal 10 (Fig. 17.10) shows the Hollerith Card Code prepared in X3.2.3 in response to the X3 directive. It was a Proposed American Standard BCD Card Code (document X3.2.3/141).

Control characters

For the first time, Hollerith hole patterns compatible with EBCDIC hole patterns were assigned in columns 0 and 1.

@ and `

The graphic @ (Commercial At) was now firmly in position 4/0, and remained there thereafter. This problem will not be referred to again.

	b7	0	0	0	0	1	1	1	1
	b6	0	0	1	1	0	0	1	1
	b5	0	1	0	1	0	1	0	1
b4 b3 b2 b1	Row \ Col	0	1	2	3	4	5	6	7
0 0 0 0	0	NUL 9-12-0 8-1	DLE 9-12-11 8-1	SP No Pch	0 0	@ 8-4	P 11-7	` 8-1	p 12-11-7
0 0 0 1	1	SOH 9-12-1	DC1 9-11-1	! 11-8-2	1 1	A 12-1	Q 11-8	a 12-0-1	q 12-11-8
0 0 1 0	2	STX 9-12-2	DC2 9-11-2	" 8-7	2 2	B 12-2	R 11-9	b 12-0-2	r 12-11-9
0 0 1 1	3	ETX 9-12-3	DC3 9-11-3	# 8-3	3 3	C 12-3	S 0-2	c 12-0-3	s 11-0-2
0 1 0 0	4	EOT 9-7	DC4 9-8-4	$ 11-8-3	4 4	D 12-4	T 0-3	d 12-0-4	t 11-0-3
0 1 0 1	5	ENQ 9-0 8-5	NAK 9-8-5	% 0-8-4	5 5	E 12-5	U 0-4	e 12-0-5	u 11-0-4
0 1 1 0	6	ACK 9-0 8-6	SYN 9-2	& 12	6 6	F 12-6	V 0-5	f 12-0-6	v 11-0-5
0 1 1 1	7	BEL 9-0 8-7	ETB 9-0-6	' 8-5	7 7	G 12-7	W 0-6	g 12-0-7	w 11-0-6
1 0 0 0	8	BS 9-11-6	CAN 9-11-8	(12-8-5	8 8	H 12-8	X 0-7	h 12-0-8	x 11-0-7
1 0 0 1	9	HT 9-12-5	EM 9-11 8-1	) 11-8-5	9 9	I 12-9	Y 0-8	i 12-0-9	y 11-0-8
1 0 1 0	10	LF 9-0-5	SUB 9-8-7	* 11-8-4	: 8-2	J 11-1	Z 0-9	j 12-11-1	z 11-0-9
1 0 1 1	11	VT 9-12 8-3	ESC 9-0-7	+ 12-8-6	; 11-8-6	K 11-2	[11-0- 8-5	k 12-11-2	{ 12-0- 8-3
1 1 0 0	12	FF 9-12 8-4	FS 9-11 8-4	, 0-8-3	< 12-8-4	L 11-3	\ 12-0	l 12-11-3	| 12-8-7
1 1 0 1	13	CR 9-12 8-5	GS 9-11 8-5	- 11	= 8-6	M 11-4	] 12-11-0 8-5	m 12-11-4	} 12-11- 8-3
1 1 1 0	14	SO 9-12 8-6	RS 9-11 8-6	. 12-8-3	> 0-8-6	N 11-5	^ 11-8-7	n 12-11-5	~ 11-0
1 1 1 1	15	SI 9-12 8-7	US 9-11 8-7	/ 0-1	? 0-8-7	O 11-6	_ 0-8-5	o 12-11-6	DEL 9-12-7

Fig. 17.10 Hollerith, Version 10

5/12, 7/12, 7/14

Graphics \ | and ˜ were currently in code positions 5/12, 7/12, 7/14, respectively.

Row	Bit Pat.	Hole Pat.	0	1	2	3	4	5	6	7	8	9	10	11	12	13	14	15	Hole Pat.
	Bit Pat.		00				01				10				11				
			00	01	10	11	00	01	10	11	00	01	10	11	00	01	10	11	
		Hole Pat.	9	9	9	9	9	9	9	9									Hole Pat.
			12				12	12		12	12	12		12	12				
				11				11	11	11		11	11	11		11			
					0		0		0	0	0		0	0			0		
0	0000		[1]	[2]	[3]	[4]	SP [5]	& [6]	– [7]	[8]					\ [9]	~ [10]	[11]	0 [12]	8–1
1	0001	1							/ [13]		a	j			A	J	[14]	1	1
2	0010	2									b	k	s		B	K	S	2	2
3	0011	3									c	l	t		C	L	T	3	3
4	0100	4									d	m	u		D	M	U	4	4
5	0101	5									e	n	v		E	N	V	5	5
6	0110	6									f	o	w		F	O	W	6	6
7	0111	7									g	p	x		G	P	X	7	7
8	1000	8									h	q	y		H	Q	Y	8	8
9	1001	8–1								`	i	r	z		I	R	Z	9	9
10	1010	8–2						!	[15]	:									8–2
11	1011	8–3					.	$	,	#	{	}							8–3
12	1100	8–4					<	*	%	@									8–4
13	1101	8–5					(	)	_	'			[	]					8–5
14	1110	8–6					+	;	>	=									8–6
15	1111	8–7					|	¬	?	"									8–7
		Hole Pat.	9	9	9	9									9	9	9	9	
			12				12				12	12		12	12	12		12	
				11				11				11	11	11		11	11	11	
					0				0		0		0	0	0		0	0	

Hole Patterns:

[1]	9–12–0–8–1	[7]	11	[13]	0–1
[2]	9–12–11–8–1	[8]	12–11–0	[14]	9–11–0–1
[3]	9–11–0–8–1	[9]	12–0	[15]	12–11
[4]	9–12–11–0–8–1	[10]	11–0		
[5]	No Pch	[11]	0–8–2		
[6]	12	[12]	0		

1	3
2	4

Block	Hole Patterns at:
1	Top and Left
2	Bottom and Left
3	Top and Right
4	Bottom and Right

Fig. 17.11 EBCDIC (from Hollerith, Version 10)

EBCDIC/Hollerith compatibility

Although a number of code positions are shown shaded in Fig. 17.10, this Hollerith proposal was in fact intended to be entirely compatible with the EBCDIC then current. In an Appendix to this Proposed American Standard, an EBCDIC code table was shown "to accommodate the requirements of 8-bit environments to provide 256 hole patterns." This chart is reproduced in Fig. 17.11. It is to be noted that the columns and rows were numbered 0, 1, 2, 3, . . . , 7, 8, 9, 10, 11, 12, 13, 14, 15, instead of the hex notation 0, 1, 2, 3, . . . , 7, 8, 9, A, B, C, D, E, F.

a) Graphics [ ] { and } were shown in Fig. 17.11 in code positions 11/13, 11/14, 8/11, 9/11, respectively, because they were at that time given those EBCDIC code positions in a System/360 programming product called Text/360, a text processing program.

b) Although not at that time actually assigned in EBCDIC, graphics \ ˜ and ` were shown assigned in Fig. 17.11 in code positions 12/0, 13/0, and 7/9, respectively.

Logical OR, Logical NOT

At this time, Logical OR was in ASCII code position 7/12, and committee members were considering that ^ might serve as, or replace, the Logical NOT graphic ¬ in PL/I.

However, a representative of SHARE had stated in a letter that this was unsatisfactory. SHARE had stated a requirement that the graphics for Logical OR and Logical NOT

a) be in columns 2, 3, 4, or 5 of ASCII;

b) not be in any National Use code position.

A versus H

The A-graphics were assigned to de facto hole patterns in Proposal 10.

Position 1/10

The Substitute character, SUB, had now replaced the Start of Special character, SS, in position 1/10 of ASCII.

Proposal 11

In August 1966, Proposal 11 was made by the UNIVAC member of X3.2.3. It is shown in Fig. 17.12. This proposal was intended to achieve EBCDIC compatibility, except for columns 0 and 1. It was proposed that the control characters of ASCII be assigned identically in columns 0 and 1 of EBCDIC as in columns 0 and 1 of ASCII.

It was recognized that the following graphics were not, at that time, assigned in EBCDIC:

[ ] { } \ ` ˜

b4 b3 b2 b1	b7 b6 b5 →	0 0 0	0 0 1	0 1 0	0 1 1	1 0 0	1 0 1	1 1 0	1 1 1
	Col / Row	0	1	2	3	4	5	6	7
0 0 0 0	0	NUL 9-12-0 8-1	DLE 9-12-11 8-1	SP No Pch	0 0	@ 8-4	P 11-7	` 12-11-0 8-4	p 12-11-7
0 0 0 1	1	SOH 9-12-1	DC1 9-11-1	! 12-8-7	1 1	A 12-1	Q 11-8	a 12-0-1	q 12-11-8
0 0 1 0	2	STX 9-12-2	DC2 9-11-2	" 8-7	2 2	B 12-2	R 11-9	b 12-0-2	r 12-11-9
0 0 1 1	3	ETX 9-12-3	DC3 9-11-3	# 8-3	3 3	C 12-3	S 0-2	c 12-0-3	s 11-0-2
0 1 0 0	4	EOT 9-12-4	DC4 9-11-4	$ 11-8-3	4 4	D 12-4	T 0-3	d 12-0-4	t 11-0-3
0 1 0 1	5	ENQ 9-12-5	NAK 9-11-5	% 0-8-4	5 5	E 12-5	U 0-4	e 12-0-5	u 11-0-4
0 1 1 0	6	ACK 9-12-6	SYN 9-11-6	& 12	6 6	F 12-6	V 0-5	f 12-0-6	v 11-0-5
0 1 1 1	7	BEL 9-12-7	ETB 9-11-7	' 8-5	7 7	G 12-7	W 0-6	g 12-0-7	w 11-0-6
1 0 0 0	8	BS 9-12 8	CAN 9-11 8	(12-8-5	8 8	H 12-8	X 0-7	h 12-0-8	x 11-0-7
1 0 0 1	9	HT 9-12 8-1	EM 9-11 8-1	) 11-8-5	9 9	I 12-9	Y 0-8	i 12-0-9	y 11-0-8
1 0 1 0	10	LF 9-12 8-2	SUB 9-11 8-2	* 11-8-4	: 8-2	J 11-1	Z 0-9	j 12-11-1	z 11-0-9
1 0 1 1	11	VT 9-12 8-3	ESC 9-11 8-3	+ 12-8-6	; 11-8-6	K 11-2	[12-8-2	k 12-11-2	{ 12-0 8-2
1 1 0 0	12	FF 9-12 8-4	FS 9-11 8-4	, 0-8-3	< 12-8-4	L 11-3	\ 12-11	l 12-11-3	| 11-0 8-2
1 1 0 1	13	CR 9-12 8-5	GS 9-11 8-5	- 11	= 8-6	M 11-4	] 11-8-2	m 12-11-4	} 12-11 8-2
1 1 1 0	14	SO 9-12 8-6	RS 9-11 8-6	. 12-8-3	> 0-8-6	N 11-5	∧ 11-8-7	n 12-11-5	~ 12-11 8-7
1 1 1 1	15	SI 9-12 8-7	US 9-11 8-7	/ 0-1	? 0-8-7	O 11-6	_ 0-8-5	o 12-11-6	DEL 12-11-0 7-8-9

Fig. 17.12 Hollerith, Version 11

The proposer proposed the following:

a) Replace ¢ and ! with [and] in EBCDIC, that is, assign them to EBCDIC code positions 4/10 and 5/10 (see Fig. 17.11). The principle proposed here was that the 64 hole patterns assigned to the graphics of columns 2, 3, 4, and 5 of ASCII (the so-called basic subset) should be the set of hole patterns implemented on IBM's 029

Keypunch. EBCDIC code positions 4/10 and 5/10 had assigned hole patterns 12-8-2 and 11-8-2. Hence [and] should be assigned to these EBCDIC code positions.

b) To resolve the Logical OR, Logical NOT problem, let them retain their existing hole patterns 12-8-7 and 11-8-7, but substitute the ASCII graphics ! and ˆ for | and ¬. Therefore, assign ! and ˆ to EBCDIC code positions 4/15 and 5/15, respectively.

c) Assign \ in EBCDIC code position 6/10 "to fill up the block of specials in Quadrant 2 of EBCDIC."

d) Since { } | ` and ˜ were "paired" (that is, in the same rows) in ASCII with [ ] \ @ and ˆ they should be similarly "paired" in EBCDIC, that is, assigned to code positions as shown in Fig. 17.13. The concept of "pairing" here was that, just as there was a single-bit difference for "paired" graphics in ASCII, so there should be a single-bit difference in EBCDIC. The practical utility of this concept was not revealed by the proposer.

Proposal 12

At the same time, the representative of the Department of Defense presented Proposal 12, which assigned the H-graphics to the existing hole patterns. It is shown in Fig. 17.14.

Also raised in this proposal was a requirement that hole patterns 12-0 and 11-0 (to allow for overpunched numerics) be assigned in columns 2, 3, 4, or 5 of ASCII, rather than hole patterns 12-8-2 and 11-8-2. This created a dilemma. The 12-0 and 11-0 had not been provided on IBM's Keypunch (the 029) because of mechanical problems. So this requirement of the Department of Defense would be in conflict with the requirement (stated by the UNIVAC representative in Proposal 11 above) that the 64 hole patterns assigned to columns 2, 3, 4, and 5 of ASCII should be those implemented on the 029.

Proposal 13

On October 14, 1966, Proposal 13 was made, a proposed American Standard BCD Card Code. It is shown in Fig. 17.15. As with Proposal 10 above, it was designed for compatibility with EBCDIC (see Fig. 17.16). Also, an attempt was made to resolve the Logical OR, Logical NOT problem, and the problem of 12-0 and 11-0 raised in Proposal 12 above.

a) As shown in Fig. 17.16, graphics \ [and] were assigned in EBCDIC code positions 12/0, 13/0, 14/0, respectively. Since hole patterns 12-0 and 11-0 are assigned in EBCDIC to positions 12/0 and 13/0,

Row	Bit Pat.	Hole Pat.	0	1	2	3	4	5	6	7	8	9	10	11	12	13	14	15	Hole Pat.
	Bit Pat.		00				01				10				11				
			00	01	10	11	00	01	10	11	00	01	10	11	00	01	10	11	
		Hole Pat.	9	9	9	9	9	9	9	9									
			12				12	12		12	12	12		12	12				
				11				11	11	11		11	11	11		11			
					0		0		0	0	0		0	0			0		
0	0000		[1]	[2]	[3]	[4]	[5] SP	[6] &	[7] -	[8]					[9]	[10]	[11]	[12] 0	8-1
1	0001	1							/ [13]		a	j			A	J	[14]	1	1
2	0010	2									b	k	s		B	K	S	2	2
3	0011	3									c	l	t		C	L	T	3	3
4	0100	4									d	m	u		D	M	U	4	4
5	0101	5									e	n	v		E	N	V	5	5
6	0110	6									f	o	w		F	O	W	6	6
7	0111	7									g	p	x		G	P	X	7	7
8	1000	8									h	q	y		H	Q	Y	8	8
9	1001	8-1									i	r	z		I	R	Z	9	9
10	1010	8-2					[	]	\ [15]	:	{	}	|						8-2
11	1011	8-3					.	$	,	#									8-3
12	1100	8-4					<	*	%	@									8-4
13	1101	8-5					(	)	_	'			DEL						8-5
14	1110	8-6					+	;	>	=									8-6
15	1111	8-7					!	^	?	"									8-7
		Hole Pat.	9	9	9	9									9	9	9	9	
			12				12				12	12		12	12	12		12	
				11				11				11	11	11		11	11	11	
					0				0		0		0	0	0		0	0	

Hole Patterns:

[1] 9-12-0-8-1	[7] 11	[13] 0-1
[2] 9-12-11-8-1	[8] 12-11-0	[14] 9-11-0-1
[3] 9-11-0-8-1	[9] 12-0	[15] 12-11
[4] 9-12-11-0-8-1	[10] 11-0	
[5] No Pch	[11] 0-8-2	
[6] 12	[12] 0	

1	3
2	4

Block	Hole Patterns at:
1	Top and Left
2	Bottom and Left
3	Top and Right
4	Bottom and Right

Fig. 17.13 EBCDIC (from Hollerith, Version 11)

this would satisfy the Department of Defense requirement (although it was not known how the problem of implementing hole patterns on the 029 Keypunch would be resolved). Code position 14/0 in EBCDIC has hole pattern 0-8-2. This was implemented on the 029 Keypunch, but with no graphic assigned at that time.

	b7	0	0	0	0	1	1	1	1
	b6	0	0	1	1	0	0	1	1
	b5	0	1	0	1	0	1	0	1
b4 b3 b2 b1	Row \ Col	0	1	2	3	4	5	6	7
0 0 0 0	0	NUL 9-12-0 8-1	DLE 9-12-11 8-1	SP No Pch	0 0	@ 8-5	P 11-7	` 8-1	p 12-11-7
0 0 0 1	1	SOH 9-12-1	DC1 9-11-1	! 0-8-2	1 1	A 12-1	Q 11-8	a 12-0-1	q 12-11-8
0 0 1 0	2	STX 9-12-2	DC2 9-11-2	" 8-7	2 2	B 12-2	R 11-9	b 12-0-2	r 12-11-9
0 0 1 1	3	ETX 9-12-3	DC3 9-11-3	# 8-6	3 3	C 12-3	S 0-2	c 12-0-3	s 11-0-2
0 1 0 0	4	EOT 9-7	DC4 9-8-4	$ 11-8-3	4 4	D 12-4	T 0-3	d 12-0-4	t 11-0-3
0 1 0 1	5	ENQ 9-0 8-5	NAK 9-8-5	% 12-8-5	5 5	E 12-5	U 0-4	e 12-0-5	u 11-0-4
0 1 1 0	6	ACK 9-0 8-6	SYN 9-2	& 12-8-6	6 6	F 12-6	V 0-5	f 12-0-6	v 11-0-5
0 1 1 1	7	BEL 9-0 8-7	ETB 9-0-6	' 8-4	7 7	G 12-7	W 0-6	g 12-0-7	w 11-0-6
1 0 0 0	8	BS 9-11-6	CAN 9-11-8	(0-8-4	8 8	H 12-8	X 0-7	h 12-0-8	x 11-0-7
1 0 0 1	9	HT 9-12-5	EM 9-11 8-1	) 12-8-4	9 9	I 12-9	Y 0-8	i 12-0-9	y 11-0-8
1 0 1 0	10	LF 9-0-5	SUB 12-9 8-2	* 11-8-4	: 8-2	J 11-1	Z 0-9	j 12-11-1	z 11-0-9
1 0 1 1	11	VT 9-12 8-3	ESC 9-0-7	+ 12	; 11-8-6	K 11-2	[11-0	k 12-11-2	{ 12-0 8-3
1 1 0 0	12	FF 9-12 8-4	FS 9-11 8-4	, 0-8-3	< 11-8-6	L 11-3	\ 12-0	l 12-11-3	| 12-11-0 8-5
1 1 0 1	13	CR 9-12 8-5	GS 9-11 8-5	- 11	= 8-3	M 11-4	] 12-8-7	m 12-11-4	} 12-11 8-3
1 1 1 0	14	SO 9-12 8-6	RS 9-11 8-6	. 12-8-3	> 0-8-6	N 11-5	^ 11-8-7	n 12-11-5	~ 11-0 8-5
1 1 1 1	15	SI 9-12 8-7	US 9-11 8-7	/ 0-1	? 0-8-7	O 11-6	_ 0-8-5	o 12-11-6	DEL 12-11-0 9-8-7

Fig. 17.14 Hollerith, Version 12

b) The "pairing" concept was then invoked for a new graphic ¦ (a broken vertical line) and for { and } to position them in EBCDIC code positions 8/0, 9/0, 10/0, respectively.

c) Similarly, ˜ and ` were "paired" in EBCDIC with ˆ and @, respectively.

<table>
<tr><td></td><td>b7
b6
b5</td><td>0
0
0</td><td>0
0
1</td><td>0
1
0</td><td>0
1
1</td><td>1
0
0</td><td>1
0
1</td><td>1
1
0</td><td>1
1
1</td></tr>
<tr><td>b4 b3 b2 b1</td><td>Col / Row</td><td>0</td><td>1</td><td>2</td><td>3</td><td>4</td><td>5</td><td>6</td><td>7</td></tr>
<tr><td>0 0 0 0</td><td>0</td><td>NUL
9-12-0
8-1</td><td>DLE
9-12-11
8-1</td><td>SP
No Pch</td><td>0
0</td><td>@
8-4</td><td>P
11-7</td><td>`
12-11-0
8-4</td><td>p
12-11-7</td></tr>
<tr><td>0 0 0 1</td><td>1</td><td>SOH
9-12-1</td><td>DC1
9-11-1</td><td>! |
12-8-7</td><td>1
1</td><td>A
12-1</td><td>Q
11-8</td><td>a
12-0-1</td><td>q
12-11-8</td></tr>
<tr><td>0 0 1 0</td><td>2</td><td>STX
9-12-2</td><td>DC2
9-11-2</td><td>"
8-7</td><td>2
2</td><td>B
12-2</td><td>R
11-9</td><td>b
12-0-2</td><td>r
12-11-9</td></tr>
<tr><td>0 0 1 1</td><td>3</td><td>ETX
9-12-3</td><td>DC3
9-11-3</td><td>#
8-3</td><td>3
3</td><td>C
12-3</td><td>S
0-2</td><td>c
12-0-3</td><td>s
11-0-2</td></tr>
<tr><td>0 1 0 0</td><td>4</td><td>EOT
9-7</td><td>DC4
9-8-4</td><td>$
11-8-3</td><td>4
4</td><td>D
12-4</td><td>T
0-3</td><td>d
12-0-4</td><td>t
11-0-3</td></tr>
<tr><td>0 1 0 1</td><td>5</td><td>ENQ
9-0
8-5</td><td>NAK
9-8-5</td><td>%
0-8-4</td><td>5
5</td><td>E
12-5</td><td>U
0-4</td><td>e
12-0-5</td><td>u
11-0-4</td></tr>
<tr><td>0 1 1 0</td><td>6</td><td>ACK
9-0
8-6</td><td>SYN
9-2</td><td>&
12</td><td>6
6</td><td>F
12-6</td><td>V
0-5</td><td>f
12-0-6</td><td>v
11-0-5</td></tr>
<tr><td>0 1 1 1</td><td>7</td><td>BEL
9-0
8-7</td><td>ETB
9-0-6</td><td>'
8-5</td><td>7
7</td><td>G
12-7</td><td>W
0-6</td><td>g
12-0-7</td><td>w
11-0-6</td></tr>
<tr><td>1 0 0 0</td><td>8</td><td>BS
9-11-6</td><td>CAN
9-11-8</td><td>(
12-8-5</td><td>8
8</td><td>H
12-8</td><td>X
0-7</td><td>h
12-0-8</td><td>x
11-0-7</td></tr>
<tr><td>1 0 0 1</td><td>9</td><td>HT
9-12-5</td><td>EM
9-11
8-1</td><td>)
11-8-5</td><td>9
9</td><td>I
12-9</td><td>Y
0-8</td><td>i
12-0-9</td><td>y
11-0-8</td></tr>
<tr><td>1 0 1 0</td><td>10</td><td>LF
9-0-5</td><td>SUB
9-8-7</td><td>*
11-8-4</td><td>:
8-2</td><td>J
11-1</td><td>Z
0-9</td><td>j
12-11-1</td><td>z
11-0-9</td></tr>
<tr><td>1 0 1 1</td><td>11</td><td>VT
9-12
8-3</td><td>ESC
9-0-7</td><td>+
12-8-6</td><td>;
11-8-6</td><td>K
11-2</td><td>[
11-0</td><td>k
12-11-2</td><td>{
12-11
8-1</td></tr>
<tr><td>1 1 0 0</td><td>12</td><td>FF
9-12
8-4</td><td>FS
9-11
8-4</td><td>,
0-8-3</td><td><
12-8-4</td><td>L
11-3</td><td>\
12-0</td><td>l
12-11-3</td><td>¦
12-0
8-1</td></tr>
<tr><td>1 1 0 1</td><td>13</td><td>CR
9-12
8-5</td><td>GS
9-11
8-5</td><td>-
11</td><td>=
8-6</td><td>M
11-4</td><td>]
0-8-2</td><td>m
12-11-4</td><td>}
11-0
8-1</td></tr>
<tr><td>1 1 1 0</td><td>14</td><td>SO
9-12
8-6</td><td>RS
9-11
8-6</td><td>.
12-8-3</td><td>>
0-8-6</td><td>N
11-5</td><td>^ ¬
11-8-7</td><td>n
12-11-5</td><td>~
12-11
8-7</td></tr>
<tr><td>1 1 1 1</td><td>15</td><td>SI
9-12
8-7</td><td>US
9-11
8-7</td><td>/
0-1</td><td>?
0-8-7</td><td>O
11-6</td><td>_
0-8-5</td><td>o
12-11-6</td><td>DEL
9-12-7</td></tr>
</table>

Fig. 17.15 Hollerith, Version 13

d) At this time, it had been proposed to resolve the Logical NOT, Logical Or problem in ASCII by assigning ! and | as duals in ASCII code position 2/1, and ^ and ¬ as duals in ASCII code position 5/14. Also, the new graphic ¦ was proposed for ASCII code position 7/12, to avoid confusion with the Logical OR graphic |.

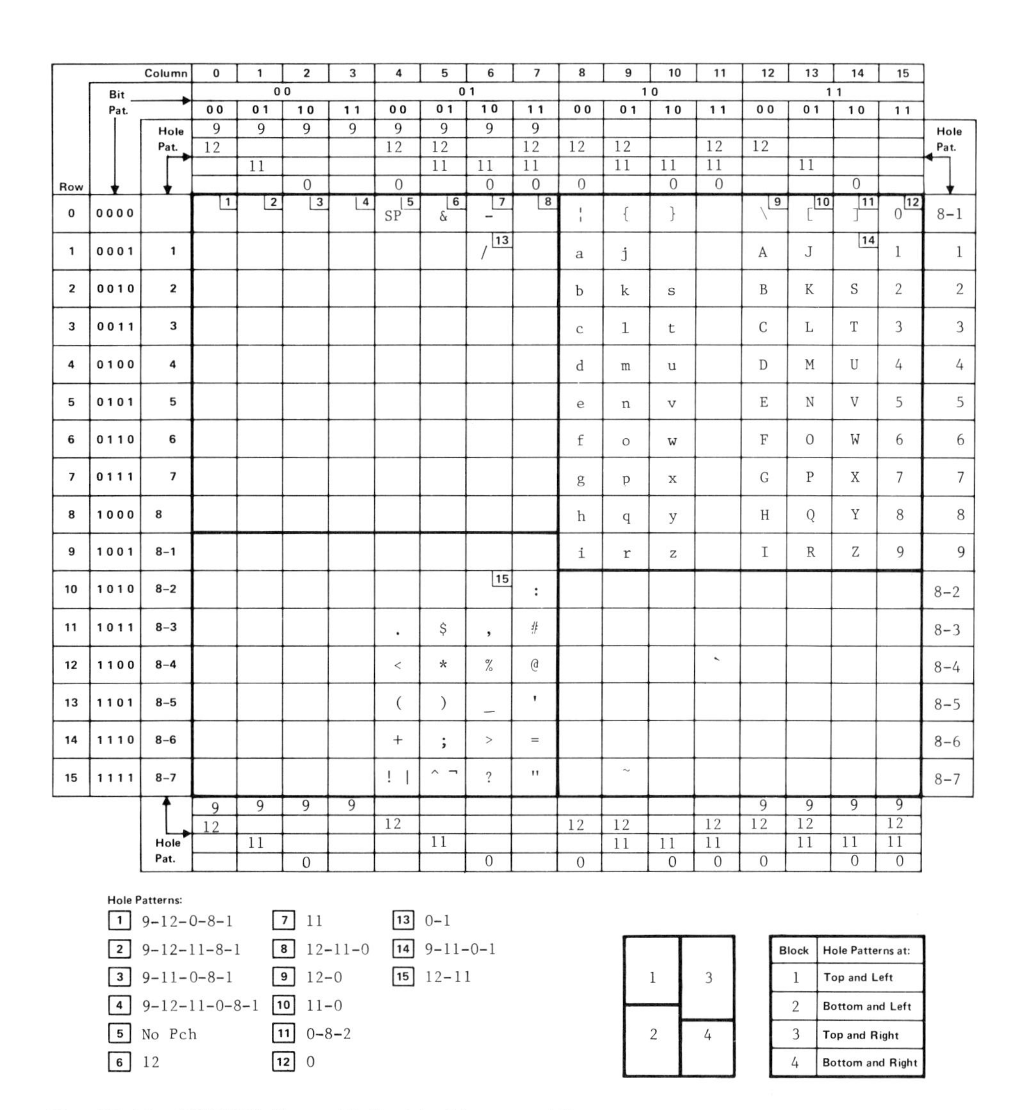

Column			0	1	2	3	4	5	6	7	8	9	10	11	12	13	14	15	
Bit Pat.			00				01				10				11				
			00	01	10	11	00	01	10	11	00	01	10	11	00	01	10	11	
		Hole Pat.	9	9	9	9	9	9	9	9									Hole Pat.
			12				12	12		12	12	12		12	12				
				11				11	11	11		11	11	11		11			
Row					0		0		0	0	0		0	0			0		
0	0000		[1]	[2]	[3]	[4]	[5] SP	[6] &	[7] -	[8]	¦	{	}		[9] \	[10] [	[11]]	[12] 0	8-1
1	0001	1							[13] /		a	j			A	J	[14]	1	1
2	0010	2									b	k	s		B	K	S	2	2
3	0011	3									c	l	t		C	L	T	3	3
4	0100	4									d	m	u		D	M	U	4	4
5	0101	5									e	n	v		E	N	V	5	5
6	0110	6									f	o	w		F	O	W	6	6
7	0111	7									g	p	x		G	P	X	7	7
8	1000	8									h	q	y		H	Q	Y	8	8
9	1001	8-1									i	r	z		I	R	Z	9	9
10	1010	8-2							[15]	:									8-2
11	1011	8-3					.	$	,	#									8-3
12	1100	8-4					<	*	%	@				`					8-4
13	1101	8-5					(	)	_	'									8-5
14	1110	8-6					+	;	>	=									8-6
15	1111	8-7					! |	^ ¬	?	"		~							8-7
		Hole Pat.	9	9	9	9									9	9	9	9	
			12				12				12	12		12	12	12		12	
				11				11				11	11	11		11	11	11	
					0				0		0		0	0	0		0	0	

Hole Patterns:

[1]	9-12-0-8-1	[7]	11	[13]	0-1
[2]	9-12-11-8-1	[8]	12-11-0	[14]	9-11-0-1
[3]	9-11-0-8-1	[9]	12-0	[15]	12-11
[4]	9-12-11-0-8-1	[10]	11-0		
[5]	No Pch	[11]	0-8-2		
[6]	12	[12]	0		

Block	Hole Patterns at:
1	Top and Left
2	Bottom and Left
3	Top and Right
4	Bottom and Right

Fig. 17.16 EBCDIC (from Hollerith, Version 13)

Proposal 14

On August 31, 1966, Task Group X3.2.3 met and considered Proposal 10. Some changes were made that resulted in Proposal 14 (Fig. 17.17),

b4 b3 b2 b1	Row \ Col	0	1	2	3	4	5	6	7
	b7 b6 b5	0 0 0	0 0 1	0 1 0	0 1 1	1 0 0	1 0 1	1 1 0	1 1 1
0 0 0 0	0	NUL 9-12-0 8-1	DLE 9-12-11 8-1	SP No Pch	0 0	@ 8-4	P 11-7	` 12-11-0 8-5	p 12-11-7
0 0 0 1	1	SOH 9-12-1	DC1 9-11-1	! 12-8-7	1 1	A 12-1	Q 11-8	a 12-0-1	q 12-11-8
0 0 1 0	2	STX 9-12-2	DC2 9-11-2	" 8-7	2 2	B 12-2	R 11-9	b 12-0-2	r 12-11-9
0 0 1 1	3	ETX 9-12-3	DC3 9-11-3	# 8-3	3 3	C 12-3	S 0-2	c 12-0-3	s 11-0-2
0 1 0 0	4	EOT 9-7	DC4 9-8-4	$ 11-8-3	4 4	D 12-4	T 0-3	d 12-0-4	t 11-0-3
0 1 0 1	5	ENQ 9-0 8-5	NAK 9-8-5	% 0-8-4	5 5	E 12-5	U 0-4	e 12-0-5	u 11-0-4
0 1 1 0	6	ACK 9-0 8-6	SYN 9-2	& 12	6 6	F 12-6	V 0-5	f 12-0-6	v 11-0-5
0 1 1 1	7	BEL 9-0 8-7	ETB 9-0-6	' 8-5	7 7	G 12-7	W 0-6	g 12-0-7	w 11-0-6
1 0 0 0	8	BS 9-11-6	CAN 9-11-8	(12-8-5	8 8	H 12-8	X 0-7	h 12-0-8	x 11-0-7
1 0 0 1	9	HT 9-12-5	EM 9-11 8-1	) 11-8-5	9 9	I 12-9	Y 0-8	i 12-0-9	y 11-0-8
1 0 1 0	10	LF 9-0-5	SUB 9-8-7	* 11-8-4	: 8-2	J 11-1	Z 0-9	j 12-11-1	z 11-0-9
1 0 1 1	11	VT 9-12 8-3	ESC 9-0-7	+ 12-8-6	; 11-8-6	K 11-2	[12-8-2	k 12-11-2	{ 12-0
1 1 0 0	12	FF 9-12 8-4	FS 9-11 8-4	, 0-8-3	< 12-8-4	L 11-3	\ 12-11	l 12-11-3	¦ 0-8-2
1 1 0 1	13	CR 9-12 8-5	GS 9-11 8-5	- 11	= 8-6	M 11-4	] 11-8-2	m 12-11-4	} 11-0
1 1 1 0	14	SO 9-12 8-6	RS 9-11 8-6	. 12-8-3	> 0-8-6	N 11-5	^ 11-8-7	n 12-11-5	~ 12-11 8-7
1 1 1 1	15	SI 9-12 8-7	US 9-11 8-7	/ 0-1	? 0-8-7	O 11-6	_ 0-8-5	o 12-11-6	DEL 9-12-7

Fig. 17.17 Hollerith, Version 14

which became a proposed USA Standard.*

a) Task Group X3.2.3 decided that hole patterns 12-8-2 and 11-8-2 (implemented on the 029 Keypunch) should be assigned in columns

* ASA had changed its name to United States of America Standards Institute (USASI), and standards were now called "USA Standard . . ."

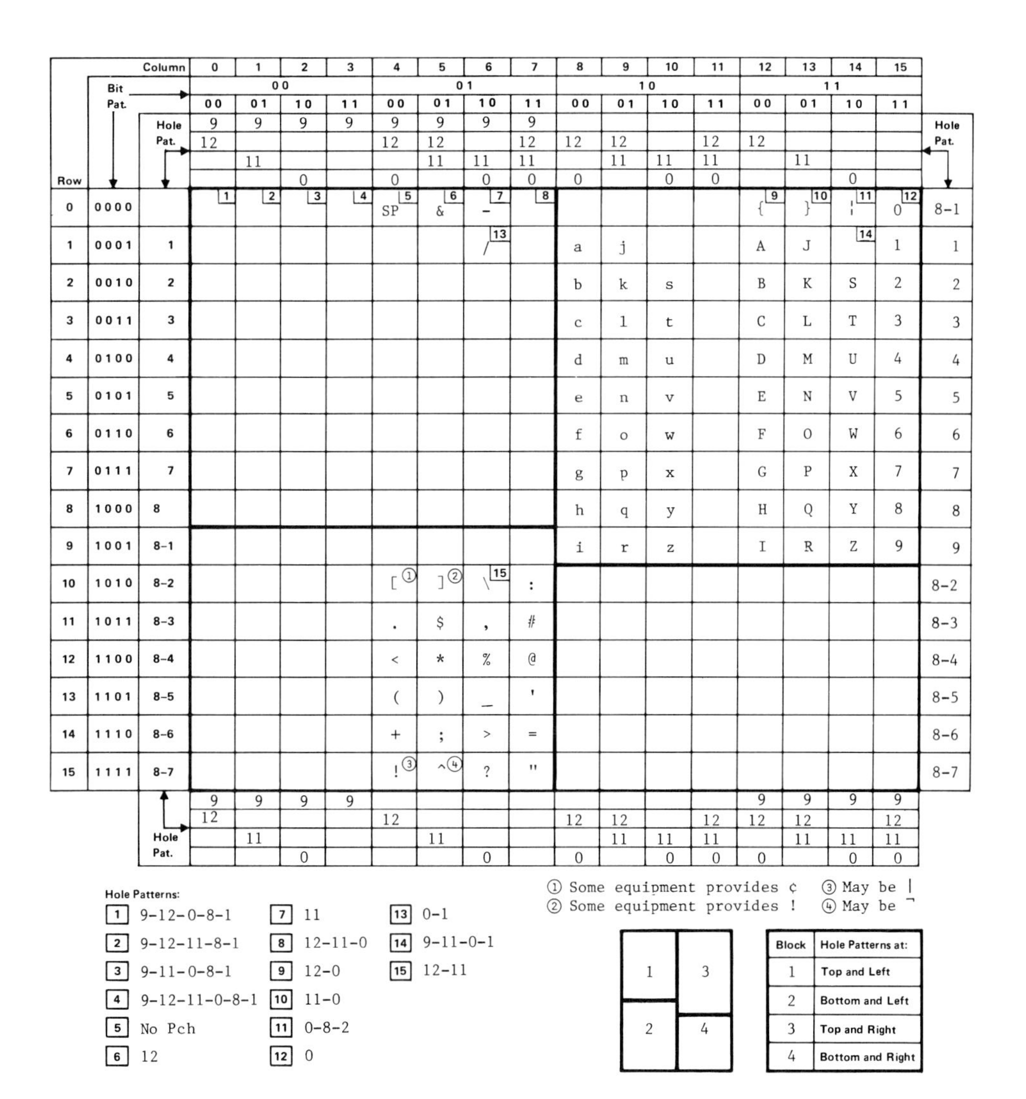

Row	Bit Pat.	Hole Pat.	0	1	2	3	4	5	6	7	8	9	10	11	12	13	14	15	Hole Pat.
		Bit Pat.	00				01				10				11				
			00	01	10	11	00	01	10	11	00	01	10	11	00	01	10	11	
		Hole Pat.	9-12	9-11	9-0	9	9-12-0	9-12-11	9-11-0	9-12-11-0	12-0	12-11	11-0	12-11-0	12	11	0		
0	0000		[1]	[2]	[3]	[4]	SP [5]	& [6]	- [7]	[8]					{ [9]	} [10]	¦ [11]	0 [12]	8-1
1	0001	1							/ [13]		a	j			A	J	[14]	1	1
2	0010	2									b	k	s		B	K	S	2	2
3	0011	3									c	l	t		C	L	T	3	3
4	0100	4									d	m	u		D	M	U	4	4
5	0101	5									e	n	v		E	N	V	5	5
6	0110	6									f	o	w		F	O	W	6	6
7	0111	7									g	p	x		G	P	X	7	7
8	1000	8									h	q	y		H	Q	Y	8	8
9	1001	8-1									i	r	z		I	R	Z	9	9
10	1010	8-2					[①	] ②	\ [15]	:									8-2
11	1011	8-3					.	$	,	#									8-3
12	1100	8-4					<	*	%	@									8-4
13	1101	8-5					(	)	_	'									8-5
14	1110	8-6					+	;	>	=									8-6
15	1111	8-7					! ③	^ ④	?	"									8-7
		Hole Pat.	9-12	9-11	9-0	9	12	11	0		12-0	12-11	11-0	12-11-0	9-12-0	9-12-11	9-11-0	9-12-11-0	

Hole Patterns:

[1]	9-12-0-8-1	[7]	11	[13]	0-1
[2]	9-12-11-8-1	[8]	12-11-0	[14]	9-11-0-1
[3]	9-11-0-8-1	[9]	12-0	[15]	12-11
[4]	9-12-11-0-8-1	[10]	11-0		
[5]	No Pch	[11]	0-8-2		
[6]	12	[12]	0		

① Some equipment provides ¢ ③ May be |
② Some equipment provides ! ④ May be ¬

Block	Hole Patterns at:
1	Top and Left
2	Bottom and Left
3	Top and Right
4	Bottom and Right

Fig. 17.18 EBCDIC (from Hollerith, Version 14)

2, 3, 4, or 5 of ASCII, rather than 12-0 and 11-0 (unimplementable on the 029 Keypunch). These hole patterns were therefore assigned to code positions 5/11 and 5/13, respectively. It was also proposed that they be incorporated into EBCDIC as duals for ¢ and ! (see Fig. 17.18).

b) It was proposed that EBCDIC show ! and ^ in code positions 4/15 and 5/15, respectively, as duals for | and ¬.

c) To ensure that 12-0 and 11-0 (not assigned in columns 2, 3, 4, or 5 of ASCII) would be assigned somewhere in the set of 128 hole patterns, it was proposed that they be assigned to code positions 7/11 and 7/13. This resulted in their being proposed to be assigned in EBCDIC code positions 12/0 and 13/0.

d) It was proposed that \ be assigned to EBCDIC code position 6/10, "to fill up the block of specials in Quadrant 2 of EBCDIC." This resulted in a hole pattern of 12-11, which raised a new problem. The 12-11 hole pattern was determined to be unimplementable on the 029 Keypunch. It is to be noted that the implementable hole pattern 0-8-2 was assigned to ASCII code position 7/12, which resulted in the | being proposed to be assigned in EBCDIC code position 14/0.

e) A significant step was taken in this proposed USA Standard toward resolving the Logical OR, Logical NOT problem. It was proposed to place in the ASCII standard, and in this Hollerith standard, the following wording:

   In specific applications it may be desirable to employ distinctive styling of individual graphics to facilitate their use for specific purposes, as, for example, to stylize the graphics in code-table positions 2/1 and 5/14 into those frequently associated with Logical OR (|) and Logical NOT (¬).

This wording which specifically allowed manufacturers to provide | and ¬ instead of ! and ^ was accepted as the final resolution of the Logical OR, Logical NOT problem.

Proposal 15

At the March 1967 meeting of ISO/TC97/SC2, it was reported that a ballot on Decimal ASCII had resulted in two countries voting "YES," six countries voting "NO," and three countries abstaining. Decimal ASCII was therefore officially terminated as an ISO Draft Proposal.

Three countries had submitted Hollerith card-code proposals. A review showed agreement on 124 of the 128 hole patterns, with disagreement in hole patterns for code positions 5/12, 5/13, 7/12, and 7/13 as shown in Fig. 17.19. The corresponding result for EBCDIC was as shown in Fig. 17.20.

b4 b3 b2 b1	b7 b6 b5	0 0 0	0 0 1	0 1 0	0 1 1	1 0 0	1 0 1	1 1 0	1 1 1
	Row \ Col	0	1	2	3	4	5	6	7
0 0 0 0	0	NUL 9-12-0 8-1	DLE 9-12-11 8-1	SP No Pch	0 0	@ 8-4	P 11-7	` 12-11-0 8-4	p 12-11-7
0 0 0 1	1	SOH 9-12-1	DC1 9-11-1	! 12-8-7	1 1	A 12-1	Q 11-8	a 12-0-1	q 12-11-8
0 0 1 0	2	STX 9-12-2	DC2 9-11-2	" 8-7	2 2	B 12-2	R 11-9	b 12-0-2	r 12-11-9
0 0 1 1	3	ETX 9-12-3	DC3 9-11-3	# 8-3	3 3	C 12-3	S 0-2	c 12-0-3	s 11-0-2
0 1 0 0	4	EOT 9-7	DC4 9-8-4	$ 11-8-3	4 4	D 12-4	T 0-3	d 12-0-4	t 11-0-3
0 1 0 1	5	ENQ 9-0 8-5	NAK 9-8-5	% 0-8-4	5 5	E 12-5	U 0-4	e 12-0-5	u 11-0-4
0 1 1 0	6	ACK 9-0 8-6	SYN 9-2	& 12	6 6	F 12-6	V 0-5	f 12-0-6	v 11-0-5
0 1 1 1	7	BEL 9-0 8-7	ETB 9-0-6	' 8-5	7 7	G 12-7	W 0-6	g 12-0-7	w 11-0-6
1 0 0 0	8	BS 9-11-6	CAN 9-11-8	(12-8-5	8 8	H 12-8	X 0-7	h 12-0-8	x 11-0-7
1 0 0 1	9	HT 9-12-5	EM 9-11 8-1	) 11-8-5	9 9	I 12-9	Y 0-8	i 12-0-9	y 11-0-8
1 0 1 0	10	LF 9-0-5	SUB 9-8-7	* 11-8-4	: 8-2	J 11-1	Z 0-9	j 12-11-1	z 11-0-9
1 0 1 1	11	VT 9-12 8-3	ESC 9-0-7	+ 12-8-6	; 11-8-6	K 11-2	[11-0	k 12-11-2	{ 12-0
1 1 0 0	12	FF 9-12 8-4	FS 9-11 8-4	, 0-8-3	< 12-8-4	L 11-3	\	l 12-11-3	¦
1 1 0 1	13	CR 9-12 8-5	GS 9-11 8-5	- 11	= 8-6	M 11-4	]	m 12-11-4	}
1 1 1 0	14	SO 9-12 8-6	RS 9-11 8-6	. 12-8-3	> 0-8-6	N 11-5	^ 11-8-7	n 12-11-5	~ 12-11 8-7
1 1 1 1	15	SI 9-12 8-7	US 9-11 8-7	/ 0-1	? 0-8-7	O 11-6	_ 0-8-5	o 12-11-6	DEL 9-12-7

	5/12	5/13	7/12	7/13
France	11-8-2	0-8-2	11-0	11-0-8
USA	11-0-8-2	11-8-2	0-8-2	11-0

Fig. 17.19 Hollerith, Version 15

Row	Bit Pat.	Hole Pat.	0	1	2	3	4	5	6	7	8	9	10	11	12	13	14	15	Hole Pat.
	Bit Pat.		00				01				10				11				
			00	01	10	11	00	01	10	11	00	01	10	11	00	01	10	11	
		Hole Pat.	9	9	9	9	9	9	9	9									
			12				12	12		12	12	12		12	12				
				11				11	11	11		11	11	11		11			
					0		0		0	0	0		0	0			0		
0	0000		[1]	[2]	[3]	[4]	SP [5]	& [6]	- [7]	[8]					{ [9]	[10]	[11]	0 [12]	8-1
1	0001	1							/ [13]		a	j			A	J	[14]	1	1
2	0010	2									b	k	s		B	K	S	2	2
3	0011	3									c	l	t		C	L	T	3	3
4	0100	4									d	m	u		D	M	U	4	4
5	0101	5									e	n	v		E	N	V	5	5
6	0110	6									f	o	w		F	O	W	6	6
7	0111	7									g	p	x		G	P	X	7	7
8	1000	8									h	q	y		H	Q	Y	8	8
9	1001	8-1									i	r	z		I	R	Z	9	9
10	1010	8-2					[		[15]	:									8-2
11	1011	8-3					.	$	,	#									8-3
12	1100	8-4					<	*	%	@									8-4
13	1101	8-5					(	)	_	'									8-5
14	1110	8-6					+	;	>	=									8-6
15	1111	8-7					!	^	?	"		~							8-7
		Hole Pat.	9	9	9	9									9	9	9	9	
			12				12				12	12		12	12	12		12	
				11				11				11	11	11		11	11	11	
					0				0		0		0	0	0		0	0	

Hole Patterns:

- [1] 9-12-0-8-1
- [2] 9-12-11-8-1
- [3] 9-11-0-8-1
- [4] 9-12-11-0-8-1
- [5] No Pch
- [6] 12
- [7] 11
- [8] 12-11-0
- [9] 12-0
- [10] 11-0
- [11] 0-8-2
- [12] 0
- [13] 0-1
- [14] 9-11-0-1
- [15] 12-11

	5/10	10/10	13/0	14/0
France	\	{	¦	]
USA	]	\	}	¦

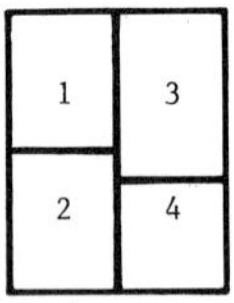

1	3
2	4

Block	Hole Patterns at:
1	Top and Left
2	Bottom and Left
3	Top and Right
4	Bottom and Right

Fig. 17.20 EBCDIC (from Hollerith, Version 15)

Proposal 16

In March 1967, Task Group X3.2.3 prepared another Proposed USA Standard Hollerith Punched Card Code (Fig. 17.21).

a) The problem of 12-11 versus 0-8-2 for code position 5/12, referred to under Proposal 14 above, was resolved by assigning 0-8-2 to position 5/12.

	b7 b6 b5	0 0 0	0 0 1	0 1 0	0 1 1	1 0 0	1 0 1	1 1 0	1 1 1
b4 b3 b2 b1	Col / Row	0	1	2	3	4	5	6	7
0 0 0 0	0	NUL 9-12-0 8-1	DLE 9-12-11 8-1	SP No Pch	0 0	@ 8-4	P 11-7	` 11-8-1	p 12-11-7
0 0 0 1	1	SOH 9-12-1	DC1 9-11-1	! 12-8-7	1 1	A 12-1	Q 11-8	a 12-0-1	q 12-11-8
0 0 1 0	2	STX 9-12-2	DC2 9-11-2	" 8-7	2 2	B 12-2	R 11-9	b 12-0-2	r 12-11-9
0 0 1 1	3	ETX 9-12-3	DC3 9-11-3	# 8-3	3 3	C 12-3	S 0-2	c 12-0-3	s 11-0-2
0 1 0 0	4	EOT 9-7	DC4 9-8-4	$ 11-8-3	4 4	D 12-4	T 0-3	d 12-0-4	t 11-0-3
0 1 0 1	5	ENQ 9-0 8-5	NAK 9-8-5	% 0-8-4	5 5	E 12-5	U 0-4	e 12-0-5	u 11-0-4
0 1 1 0	6	ACK 9-0 8-6	SYN 9-2	& 12	6 6	F 12-6	V 0-5	f 12-0-6	v 11-0-5
0 1 1 1	7	BEL 9-0 8-7	ETB 9-0-6	' 8-5	7 7	G 12-7	W 0-6	g 12-0-7	w 11-0-6
1 0 0 0	8	BS 9-11-6	CAN 9-11-8	(12-8-5	8 8	H 12-8	X 0-7	h 12-0-8	x 11-0-7
1 0 0 1	9	HT 9-12-5	EM 9-11 8-1	) 11-8-5	9 9	I 12-9	Y 0-8	i 12-0-9	y 11-0-8
1 0 1 0	10	LF 9-0-5	SUB 9-8-7	* 11-8-4	: 8-2	J 11-1	Z 0-9	j 12-11-1	z 11-0-9
1 0 1 1	11	VT 9-12 8-3	ESC 9-0-7	+ 12-8-6	; 11-8-6	K 11-2	[12-8-2	k 12-11-2	{ 12-0
1 1 0 0	12	FF 9-12 8-4	FS 9-11 8-4	, 0-8-3	< 12-8-4	L 11-3	\ 0-8-2	l 12-11-3	¦ 0-8-1
1 1 0 1	13	CR 9-12 8-5	GS 9-11 8-5	- 11	= 8-6	M 11-4	] 11-8-2	m 12-11-4	} 11-0
1 1 1 0	14	SO 9-12 8-6	RS 9-11 8-6	. 12-8-3	> 0-8-6	N 11-5	^ 11-8-7	n 12-11-5	~ 12-8-1
1 1 1 1	15	SI 9-12 8-7	US 9-11 8-7	/ 0-1	? 0-8-7	O 11-6	_ 0-8-5	o 12-11-6	DEL 9-12-7

Fig. 17.21 Hollerith, Version 16

b) All that remained was to assign hole patterns to code positions 7/12 and 7/14. Task Group X3.2.3 chose (for not very strong reasons) hole patterns 0-8-1 and 12-8-1. This resulted in ˜ and ` being proposed to be assigned to EBCDIC code positions 4/9 and 5/9, respectively, as shown in Fig. 17.22.

		Column	0	1	2	3	4	5	6	7	8	9	10	11	12	13	14	15	
	Bit Pat.		00				01				10				11				
			00	01	10	11	00	01	10	11	00	01	10	11	00	01	10	11	Hole Pat.
		Hole Pat.	9	9	9	9	9	9	9	9									
			12				12	12		12	12	12		12	12				
				11				11	11	11		11	11	11		11			
Row					0		0		0	0	0		0	0			0		
0	0000		[1]	[2]	[3]	[4]	SP [5]	& [6]	- [7]	[8]					{ [9]	} [10]	\ [11]	0 [12]	8-1
1	0001	1							/ [13]		a	j			A	J	[14]	1	1
2	0010	2									b	k	s		B	K	S	2	2
3	0011	3									c	l	t		C	L	T	3	3
4	0100	4									d	m	u		D	M	U	4	4
5	0101	5									e	n	v		E	N	V	5	5
6	0110	6									f	o	w		F	O	W	6	6
7	0111	7									g	p	x		G	P	X	7	7
8	1000	8									h	q	y		H	Q	Y	8	8
9	1001	8-1					~	`	¦		i	r	z		I	R	Z	9	9
10	1010	8-2					[	]	[15]	:									8-2
11	1011	8-3					.	$	,	#									8-3
12	1100	8-4					<	*	%	@									8-4
13	1101	8-5					(	)	_	'									8-5
14	1110	8-6					+	;	>	=									8-6
15	1111	8-7					!	^	?	"									8-7
		Hole Pat.	9	9	9	9									9	9	9	9	
			12				12				12	12		12	12	12		12	
				11				11				11	11	11		11	11	11	
					0				0		0		0	0	0		0	0	

Hole Patterns:

1	9-12-0-8-1	7	11	13	0-1
2	9-12-11-8-1	8	12-11-0	14	9-11-0-1
3	9-11-0-8-1	9	12-0	15	12-11
4	9-12-11-0-8-1	10	11-0		
5	No Pch	11	0-8-2		
6	12	12	0		

1	3
2	4

Block	Hole Patterns at:
1	Top and Left
2	Bottom and Left
3	Top and Right
4	Bottom and Right

Fig. 17.22 EBCDIC (from Hollerith, Version 16)

Proposal 17

As a result of ISO/TC97/SC2 ballots on Proposal 15, final agreement was reached, as shown in Fig. 17.23. This proposal, subsequently incorporated into a draft USA Standard Hollerith Card Code, was finally approved as a USA Standard. The results for EBCDIC are shown in Fig. 17.24.

	b7	0	0	0	0	1	1	1	1
	b6	0	0	1	1	0	0	1	1
	b5	0	1	0	1	0	1	0	1
b4 b3 b2 b1	Row \ Col	0	1	2	3	4	5	6	7
0 0 0 0	0	NUL 9-12-0 8-1	DLE 9-12-11 8-1	SP No Pch	0 0	@ 8-4	P 11-7	` 8-1	p 12-11-7
0 0 0 1	1	SOH 9-12-1	DC1 9-11-1	! [1] 12-8-7	1 1	A 12-1	Q 11-8	a 12-0-1	q 12-11-8
0 0 1 0	2	STX 9-12-2	DC2 9-11-2	" 8-7	2 2	B 12-2	R 11-9	b 12-0-2	r 12-11-9
0 0 1 1	3	ETX 9-12-3	DC3 9-11-3	# 8-3	3 3	C 12-3	S 0-2	c 12-0-3	s 11-0-2
0 1 0 0	4	EOT 9-7	DC4 9-8-4	$ 11-8-3	4 4	D 12-4	T 0-3	d 12-0-4	t 11-0-3
0 1 0 1	5	ENQ 9-0 8-5	NAK 9-8-5	% 0-8-4	5 5	E 12-5	U 0-4	e 12-0-5	u 11-0-4
0 1 1 0	6	ACK 9-0 8-6	SYN 9-2	& 12	6 6	F 12-6	V 0-5	f 12-0-6	v 11-0-5
0 1 1 1	7	BEL 9-0 8-7	ETB 9-0-6	' 8-5	7 7	G 12-7	W 0-6	g 12-0-7	w 11-0-6
1 0 0 0	8	BS 9-11-6	CAN 9-11-8	(12-8-5	8 8	H 12-8	X 0-7	h 12-0-8	x 11-0-7
1 0 0 1	9	HT 9-12-5	EM 9-11 8-1	) 11-8-5	9 9	I 12-9	Y 0-8	i 12-0-9	y 11-0-8
1 0 1 0	10	LF 9-0-5	SUB 9-8-7	* 11-8-4	: 8-2	J 11-1	Z 0-9	j 12-11-1	z 11-0-9
1 0 1 1	11	VT 9-12 8-3	ESC 9-0-7	+ 12-8-6	; 11-8-6	K 11-2	[12-8-2	k 12-11-2	{ 12-0
1 1 0 0	12	FF 9-12 8-4	FS 9-11 8-4	, 0-8-3	< 12-8-4	L 11-3	\ 0-8-2	l 12-11-3	¦ 12-11
1 1 0 1	13	CR 9-12 8-5	GS 9-11 8-5	- 11	= 8-6	M 11-4	] 11-8-2	m 12-11-4	} 11-0
1 1 1 0	14	SO 9-12 8-6	RS 9-11 8-6	. 12-8-3	> 0-8-6	N 11-5	^ [2] 11-8-7	n 12-11-5	~ 11-0-1
1 1 1 1	15	SI 9-12 8-7	US 9-11 8-7	/ 0-1	? 0-8-7	O 11-6	_ 0-8-5	o 12-11-6	DEL 9-12-7

Hole Patterns:

[1] May be |

[2] May be ¬

Fig. 17.23 Hollerith, Version 17

The Hollerith card code had finally been resolved. It subsequently became both an ISO Recommendation and an ECMA Standard. The ECMA Standard on Decimal ASCII was withdrawn.

Row	Bit Pat.	Hole Pat.	0	1	2	3	4	5	6	7	8	9	10	11	12	13	14	15	Hole Pat.
	Bit Pat.		00				01				10				11				
			00	01	10	11	00	01	10	11	00	01	10	11	00	01	10	11	
		Hole Pat.	9	9	9	9	9	9	9	9									
			12				12	12		12	12	12		12	12				
				11				11	11	11		11	11	11		11			
					0		0		0	0	0		0	0			0		
0	0000		[1]	[2]	[3]	[4]	SP [5]	& [6]	- [7]	[8]					{ [9]	} [10]	\ [11]	0 [12]	8-1
1	0001	1							/ [13]		a	j	~		A	J	[14]	1	1
2	0010	2									b	k	s		B	K	S	2	2
3	0011	3									c	l	t		C	L	T	3	3
4	0100	4									d	m	u		D	M	U	4	4
5	0101	5									e	n	v		E	N	V	5	5
6	0110	6									f	o	w		F	O	W	6	6
7	0111	7									g	p	x		G	P	X	7	7
8	1000	8									h	q	y		H	Q	Y	8	8
9	1001	8-1									i	r	z		I	R	Z	9	9
10	1010	8-2					[	]	¦ [15]	:									8-2
11	1011	8-3					.	$	,	#									8-3
12	1100	8-4					<	*	%	@									8-4
13	1101	8-5					(	)	_	'									8-5
14	1110	8-6					+	;	>	=									8-6
15	1111	8-7					! ①	^ ②	?	"									8-7
		Hole Pat.	9	9	9	9									9	9	9	9	
			12				12				12	12		12	12	12		12	
				11				11				11	11	11		11	11	11	
					0				0		0		0	0	0		0	0	

Hole Patterns:

[1]	9-12-0-8-1	[7]	11	[13]	0-1
[2]	9-12-11-8-1	[8]	12-11-0	[14]	9-11-0-1
[3]	9-11-0-8-1	[9]	12-0	[15]	12-11
[4]	9-12-11-0-8-1	[10]	11-0		
[5]	No Pch	[11]	0-8-2		
[6]	12	[12]	0		

① May be |
② May be ¬

1	3
2	4

Block	Hole Patterns at:
1	Top and Left
2	Bottom and Left
3	Top and Right
4	Bottom and Right

Fig. 17.24 EBCDIC (from Hollerith, Version 17)

Two further problems arose in the Hollerith Card Code with respect to the assignment of the Katakana graphics and with respect to the Alphabetic Extenders. These problems are discussed in Chapters 18 and 21, respectively.

18
Katakana and the Hollerith Card Code

The Japanese written language, like the Chinese written language, is ideographic; that is to say, each word is represented by an ideograph. There are thousands (estimates run to 50,000, and higher) of ideographs. In the early days of data processing in Japan, it was quite impractical to provide these thousands of symbols on either a printing or display device.

18.1 KATAKANA SYMBOLS

Instead, a set of 47 phonetic symbols was used. These symbols are called Katakana symbols (long used in Japan). The Katakana symbols, and the sounds* they represent, are shown in Fig. 18.1.

These 47 basic Katakana symbols were assigned in EBCDIC in 1964 as shown in Fig. 18.2.

18.2 KATAKANA IN PTTC

These 47 basic Katakana were implemented on IBM products using the 88-graphic PTTC code, as identified below and shown in Fig. 18.3.

Alphabetics	26
Numerics	10
Katakana	47
Specials . , - * ¥	5
	88

*This book is not the place to go into the use of the "voiced-sound symbol," the "semi-voiced sound symbol," or "small Katakana."

Shape	Name	Shape	Name
ｱ	A	ﾊ	HA
ｲ	I	ﾋ	HI
ｳ	U	ﾌ	FU
ｴ	E	ﾍ	HE
ｵ	O	ﾎ	HO
ｶ	KA	ﾏ	MA
ｷ	KI	ﾐ	MI
ｸ	KU	ﾑ	MU
ｹ	KE	ﾒ	ME
ｺ	KO	ﾓ	MO
ｻ	SA	ﾔ	YA
ｼ	SHI		
ｽ	SU	ﾕ	YU
ｾ	SE		
ｿ	SO	ﾖ	YO
ﾀ	TA	ﾗ	RA
ﾁ	CHI	ﾘ	RI
ﾂ	TSU	ﾙ	RU
ﾃ	TE	ﾚ	RE
ﾄ	TO	ﾛ	RO
ﾅ	NA	ﾜ	WA
ﾆ	NI	ﾝ	N
ﾇ	NU		
ﾈ	NE	ﾞ	Voiced Sound Symbol
ﾉ	NO	ﾟ	Semi-voiced Sound Symbol

Fig. 18.1 Basic Katakana-47

Row	Column	0	1	2	3	4	5	6	7	8	9	A	B	C	D	E	F
	Bit Pat. →	00				01				10				11			
	↓	00	01	10	11	00	01	10	11	00	01	10	11	00	01	10	11
0	0000										ｿ						
1	0001									ｱ	ﾀ						
2	0010									ｲ	ﾁ	ﾍ					
3	0011									ｳ	ﾂ	ﾎ					
4	0100									ｴ	ﾃ	ﾏ					
5	0101									ｵ	ﾄ	ﾐ					
6	0110									ｶ	ﾅ	ﾑ					
7	0111									ｷ	ﾆ	ﾒ					
8	1000									ｸ	ﾇ	ﾓ					
9	1001									ｹ	ﾈ	ﾔ					
A	1010									ｺ	ﾉ	ﾕ	ﾚ				
B	1011												ﾛ				
C	1100									ｻ		ﾖ	ﾜ				
D	1101									ｼ	ﾊ	ﾗ	ﾝ				
E	1110									ｽ	ﾋ	ﾘ	ﾞ				
F	1111									ｾ	ﾌ	ﾙ	ﾟ				

Fig. 18.2 EBCDIC basic Katakana-47

Bit Pattern		Lower Case				Upper Case			
			A	B	BA		A	B	BA
		SP	゛	ホ	ヘ	SP	°	-	*
1		ヌ	メ	マ	チ	1	¥	J	A
2		フ	ト	ノ	コ	2	S	K	B
2 1		ア	カ	リ	ソ	3	T	L	C
4		ウ	ナ	モ	シ	4	U	M	D
4 1		エ	ヒ	ミ	イ	5	V	N	E
4 2		オ	テ	ラ	ハ	6	W	O	F
4 2 1		ヤ	サ	セ	キ	7	X	P	G
8		ユ	ン	タ	ク	8	Y	Q	H
8 1		ヨ	ツ	ス	ニ	9	Z	R	I
8 2		ワ				0			
8 2 1		ケ	ネ	レ	ル	□	,	ム	.
8 4									
8 4 1									
8 4 2									
8 4 2 1									

1	3
2	4

Block	Hole Patterns at:
1	Top And Left
2	Bottom and Left
3	Top and Left
4	Bottom and Left

Fig. 18.3 Katakana-88

18.3 KATAKANA IN EBCDIC

Subsequently, 16 more Katakana symbols were introduced (see Fig. 18.4). They are called small Katakana and Katakana punctuation symbols. These 16 Katakana symbols, and the 47 described previously, are shown coded in EBCDIC in Fig. 18.5: 16 Katakana symbols in columns 4 and 5, and 47 Katakana symbols in columns 8, 9, A, and B.

Shown in Fig. 18.6 are the 88 EBCDIC symbols assigned at that time. It is to be observed that 26 of the Katakana symbols co-map into

Shape	Name
。	Katakana full stop
「	Katakana opening bracket
」	Katakana closing bracket
、	Katakana comma
・	Conjunctive symbol
ヲ	Katakana particle
ァ	a
ィ	i
ゥ	u
ェ	e
ォ	o
ャ	ya
ュ	yu
ョ	yo
ッ	tsu
ー	Prolonged sound symbol

Fig. 18.4 Small Katakana and Katakana Punctuation Symbols

Row	Column	0	1	2	3	4	5	6	7	8	9	A	B	C	D	E	F
	Bit Pat. →	00				01				10				11			
	↓	00	01	10	11	00	01	10	11	00	01	10	11	00	01	10	11
0	0000										ソ						
1	0001					。	ェ			ア	タ						
2	0010					「	ォ			イ	チ	ヘ					
3	0011					」	ャ			ウ	ツ	ホ					
4	0100					、	ュ			エ	テ	マ					
5	0101					・	ョ			オ	ト	ミ					
6	0110					ヲ	ッ			カ	ナ	ム					
7	0111					ァ				キ	ニ	メ					
8	1000					ィ	ー			ク	ヌ	モ					
9	1001					ゥ				ケ	ネ	ヤ					
A	1010									コ	ノ	ユ	レ				
B	1011												ロ				
C	1100									サ		ヨ	ワ				
D	1101									シ	ハ	ラ	ン				
E	1110									ス	ヒ	リ	゛				
F	1111									セ	フ	ル	゜				

Fig. 18.5 Katakana-63 in EBCDIC

Row	Column	0	1	2	3	4	5	6	7	8	9	A	B	C	D	E	F
	Bit Pat. →	00				01				10				11			
	↓	00	01	10	11	00	01	10	11	00	01	10	11	00	01	10	11
0	0000					SP	&	-									0
1	0001							/		a	j			A	J		1
2	0010									b	k	s		B	K	S	2
3	0011									c	l	t		C	L	T	3
4	0100									d	m	u		D	M	U	4
5	0101									e	n	v		E	N	V	5
6	0110									f	o	w		F	O	W	6
7	0111									g	p	x		G	P	X	7
8	1000									h	q	y		H	Q	Y	8
9	1001									i	r	z		I	R	Z	9
A	1010					¢	!		:								
B	1011					.	$	,	#								
C	1100					<	*	%	@								
D	1101					(	)	_	'								
E	1110					+	:	>	=								
F	1111					|	¬	?	"								

Fig. 18.6 EBCDIC-88

the same EBCDIC code positions as the small Latin alphabetics. This was not a problem at the time, since there were no data processing applications calling for the use of *both* Katakana symbols *and* small Latin alphabetics. Subsequently, a problem arose, which will be described.

18.4 JISCII

In 1968, the Code Standardization Committee of the Information Processing Society of Japan was preparing a draft Japanese Industrial Standard Code for Information Interchange (JISCII). JISCII was to be based on the ISO 7-Bit Code, but would be an 8-bit code. It is shown in Fig. 18.7, with the 94 graphics of the ISO 7-Bit Code in columns 2 through 7 and the 63 Katakana symbols in columns 10 through 13. The control character KS in code position 10/0 stands for Katakana Space.

It was observed, that in JISCII, the small Latin alphabetics and the Katakana symbols had unique code positions, whereas, as has been noted earlier, they co-map in EBCDIC. Therefore a one-to-one translation relationship between the 256 JISCII code positions and the 256 EBCDIC code positions was not possible.

Row	Column	0	1	2	3	4	5	6	7	8	9	10	11	12	13	14	15
	Bit Pat. →	00				01				10				11			
	↓	00	01	10	11	00	01	10	11	00	01	10	11	00	01	10	11
0	0000			SP	0	@	P	`	p			KS	ｰ	ﾀ	ﾐ		
1	0001			!	1	A	Q	a	q			｡	ｱ	ﾁ	ﾑ		
2	0010			"	2	B	R	b	r			｢	ｲ	ﾂ	ﾒ		
3	0011			#	3	C	S	c	s			｣	ｳ	ﾃ	ﾓ		
4	0100			$	4	D	T	d	t			､	ｴ	ﾄ	ﾔ		
5	0101			%	5	E	U	e	u			･	ｵ	ﾅ	ﾕ		
6	0110			&	6	F	V	f	v			ｦ	ｶ	ﾆ	ﾖ		
7	0111			'	7	G	W	g	w			ｧ	ｷ	ﾇ	ﾗ		
8	1000			(	8	H	X	h	x			ｨ	ｸ	ﾈ	ﾘ		
9	1001			)	9	I	Y	i	y			ｩ	ｹ	ﾉ	ﾙ		
10	1010			*	:	J	Z	j	z			ｪ	ｺ	ﾊ	ﾚ		
11	1011			+	;	K	[	k	{			ｫ	ｻ	ﾋ	ﾛ		
12	1100			,	<	L	¥	l	¦			ｬ	ｼ	ﾌ	ﾜ		
13	1101			-	=	M	]	m	}			ｭ	ｽ	ﾍ	ﾝ		
14	1110			.	>	N	^	n	~			ｮ	ｾ	ﾎ	ﾞ		
15	1111			/	?	O	_	o	DEL			ｯ	ｿ	ﾏ	ﾟ		

Fig. 18.7 JISCII

It was foreseen that the requirement would come to translate EBCDIC to/from JISCII. Clearly, there were only two possibilities; either change EBCDIC or change JISCII, so that a one-to-one relationship was possible. The probability seemed low that, working through IBM representatives to the Japanese Code Standardization Committee, the Committee would change JISCII. Therefore a decision was made to change EBCDIC with respect to the Katakana symbols.

18.5 JISCII, HOLLERITH, AND EBCDIC

At this time, the International Code Standards Committee, ISO/TC97/SC2, was working on the standardization of the Hollerith Card Code (which they called the Twelve-Row Card Code), and had accepted the requirement to standardize 256 hole patterns. Since it seemed likely that the 256 hole patterns of the EBCDIC card code would be the 256 hole patterns selected for the 256-character Hollerith Card Code, relations between three codes, JISCII, Hollerith, and EBCDIC, occupied the attention of ISO/TC97/SC2.

18.6 OBJECTIVES FOR THE HOLLERITH CARD CODE

Objectives were set for the standardization of the Hollerith Card Code.

Objective 1. 256 hole-patterns should be provided, to meet the needs of 8-bit computer manufacturers.

Objective 2. The assignment of hole patterns to control and graphic meanings should be as compatible as possible with existing assignments on 6-bit and on 8-bit computers.

Objective 3. The needs of countries using non-Latin alphabets, as well as the needs of countries using Latin alphabets, should be given consideration.

Objective 4. The translation of the Hollerith hole patterns to the EBCDIC bit patterns should be as simple as possible.

Objective 5. The translation of the Hollerith hole patterns to the bit patterns of ISO-8 (the 8-bit expansion of the ISO 7-Bit Code) should be as simple as possible.

Objective 6. The collating sequence of an alphabet should be code independent.

Comment. It was recognized in ISO/TC97/SC2 that it was not possible to achieve *all* these objectives. In particular, Objectives 4 and 5 are not mutually achievable.

18.7 ASSUMPTIONS FOR THE HOLLERITH CARD CODE

Some assumptions were accepted by ISO/TC97/SC2.

Assumption 1. The set of 256 Hollerith hole patterns shown in Fig. 18.8 should be used.*

Assumption 2. The structure of ISO-8 would be as follows (see Fig. 18.9):

a) The embedment algorithm would be E8 = 0; that is, the 128 characters of ISO-7 would be embedded contiguously in the first 8 columns of the 16-by-16 code table.
b) Columns 8 and 9 would be reserved for future assignment of control characters. For purposes of reference, these code positions are designated K0 through K31.

*This set of 256 hole patterns was also the starting point for the EBCDIC card code (Fig. 11.3).

Hole Pat.																
	9	9	9	9	9	9	9	9								
	12	12	12	12					12	12	12	12				
	11	11			11	11			11	11			11	11		
	0		0		0		0		0		0		0		0	
1																
2																
3																
4																
5																
6																
7																
8																
8-1																
8-2																
8-3																
8-4																
8-5																
8-6																
8-7																

Fig. 18.8 256 hole patterns

Row	Column	0	1	2	3	4	5	6	7	8	9	10	11	12	13	14	15
	Bit Pat.	00				01				10				11			
		00	01	10	11	00	01	10	11	00	01	10	11	00	01	10	11
0	0000	NUL	DLE	SP	0	@	P		p	K0	K16	N0	N16	N32	N48	G0	G16
1	0001	SOH	DC1	!	1	A	Q	a	q	K1	K17	N1	N17	N33	N49	G1	G17
2	0010	STX	DC2	"	2	B	R	b	r	K2	K18	N2	N18	N34	N50	G2	G18
3	0011	ETX	DC3	#	3	C	S	c	s	K3	K19	N3	N19	N35	N51	G3	G19
4	0100	EOT	DC4	$	4	D	T	d	t	K4	K20	N4	N20	N36	N52	G4	G20
5	0101	ENQ	NAK	%	5	E	U	e	u	K5	K21	N5	N21	N37	N53	G5	G21
6	0110	ACK	SYN	&	6	F	V	f	v	K6	K22	N6	N22	N38	N54	G6	G22
7	0111	BEL	ETB	'	7	G	W	g	w	K7	K23	N7	N23	N39	N55	G7	G23
8	1000	BS	CAN	(	8	H	X	h	x	K8	K24	N8	N24	N40	N56	G8	G24
9	1001	HT	EM	)	9	I	Y	i	y	K9	K25	N9	N25	N41	N57	G9	G25
10	1010	LF	SUB	*	:	J	Z	j	z	K10	K26	N10	N26	N42	N58	G10	G26
11	1011	VT	ESC	+	;	K	[	k	{	K11	K27	N11	N27	N43	N59	G11	G27
12	1100	FF	FS	,	<	L	\	l	¦	K12	K28	N12	N28	N44	N60	G12	G28
13	1101	CR	GS	-	=	M	]	m	}	K13	K29	N13	N29	N45	N61	G13	G29
14	1110	SO	RS	.	>	N	^	n	~	K14	K30	N14	N30	N46	N62	G14	G30
15	1111	SI	US	/	?	O	_	o	DEL	K15	K31	N15	N31	N47	N63	G15	G31

Fig. 18.9 Structure of ISO-8

c) Columns 10 through 13 would be reserved for future assignment of non-Latin alphabets. For purposes of reference, these code positions are designated N0 through N63.

d) Columns 14 and 15 would be reserved for future assignment of special graphics. For purposes of reference, these code positions are designated G0 through G31.

Assumption 3. In countries with non-Latin alphabetic (Katakana, Cyrillic, etc.), programming language source statements would use capital Latin alphabetics, but normal data processing applications could use the non-Latin alphabetics.

Assumption 4. If small non-Latin alphabetics are required (as in Cyrillic, for example), they can be co-mapped into the same code positions as the small Latin alphabetics, if necessary.

Comment. What was assumed here was that there would be no data processing application requiring four alphabets—the small and capital Latin alphabets and small and capital non-Latin alphabets.

Assumption 5. The gross collating sequence for the Katakana symbols, small Katakana and Katakana punctuation symbols collating low to basic Katakana symbols, could be reversed in the future, if necessary.

Comment. One of the proposals described later did invoke this assumption.

Assumption 6. The collating sequence of non-Latin alphabetics should be the same in ISO-8 and in EBCDIC.

Assumption 7. Non-Latin alphabetics should be self-contiguous in ISO-8, but need not be so in EBCDIC.

Some criteria, arising from Hollerith Card Code practices and implementations of that time, were agreed to by ISO/TC97/SC2.

Criterion 1. The hole patterns long associated with Space, numerics, capital alphabetics, and many specials, should be used.

Criterion 2. The hole patterns already associated with small alphabetics in some manufacturers' card equipments should be used.

Criterion 3. The hole patterns associated with certain control characters (NUL, HT, DEL, BS, DC3, LF, ETB, ESC, EOT, SUB) in some manufacturers' card equipments should be used.

Comment. EBCDIC is structured so that the first 4 columns are for control characters and the last 12 columns are for graphic characters. ISO-8 (Fig. 18.9) is structured so that columns 0, 1, 8, and 9 are for control characters; columns 2 through 7 and 10 through 15 are for graphic characters.

Criterion 4. The same set of 64 hole patterns should be used for control characters in both ISO-8 and EBCDIC. The same set of 192 hole patterns should be used for graphic characters in both ISO-8 and EBCDIC.

Comment. The folding characteristics of printer control units for EBCDIC-based computing systems should be incorporated because it would facilitate the provision of Latin, Katakana, and Cyrillic subsets on printers.

Comment. Due to an anomaly, the simple 64-for-controls/192-for-graphics correspondence between ISO-8 and EBCDIC described in Criterion 4 above cannot exist precisely. The Delete character is in the control character section of EBCDIC, but in column 7 of ISO-8 (therefore not in the control columns of ISO-8). To put it another way, only 63 of the control positions in columns 0, 1, 8, and 9 of ISO-8 can correspond to the control positions in columns 0 through 3 of EBCDIC. The 64th control position of ISO-8 (whatever it may be) must correspond to a position in the graphic columns, 4 through F, of EBCDIC. The consequence of this realization is described later.

Criterion 5. Given that Katakana (and Cyrillic) were to be reassigned in EBCDIC, folding* capability should be available in the revised EBCDIC.

Comment. Following accepted conventions, the 16 columns and 16 rows of the EBCDIC code table are numbered according to the hexadecimal convention, 0, 1, 2, 3, 4, 5, 6, 7, 8, 9, A, B, C, D, E, F, whereas the 16 columns and rows of ISO-8 are numbered 0, 1, 2, 3, 4, 5, 6, 7, 8, 9, 10, 11, 12, 13, 14, 15.

18.8 DEVELOPMENT OF THE HOLLERITH CARD CODE

Given the Objectives, Assumptions, and Criteria above, development of a 256-character Hollerith Card Code proceeded in ISO/TC97/SC2.

Criteria 1, 2, and 3 essentially prescribed the Hollerith hole patterns for the 94 graphics, for Space, for Delete, and for 9 control characters (NUL, HT, BS, DC3, LF, ETB, ESC, EOT, SUB) of columns 0 through 7 of ISO-8.

---

*As will be described later, this "folding" criterion came into conflict with the collating sequence of Katakana, and conflicting proposals were made to ISO/TC97/SC2.

Considering the remaining 23 control characters in columns 0 and 1 of the ISO 7-Bit code, it was observed, under Criterion 4, that zone-punch combinations of 9, 9-0, 9-11, and 9-12 were used in columns 0 through 3 of EBCDIC, but with zone-punch combinations of 9-12-0, 9-12-11, 9-11-0, and 9-12-11-0 in row 0 of these columns. It was realized that the closest approach to Objective 5 (translation simplicity, Hollerith to/from ISO-8) could be achieved if the digit punches were associated within rows with the BCD low-order four bits of the 8-bit bit patterns. Accordingly, assignments were made as shown in Fig. 18.10.

Observe that, with the exception of hole patterns for DLE and SYN, the BCD relationship is fairly good. The hole pattern 9-12-11-8-1 was one of the four in row 0, columns 0 through 3 of EBCDIC, and had to go somewhere in ISO-8. It is not obvious, in retrospect, why 9-2 was assigned to SYN.

This now left the following 31 hole patterns from columns 0 through 3 of EBCDIC, to be assigned to columns 8 and 9 of ISO-8 under Criterion 4. (Of course, 32 hole patterns were needed, but the anomaly

Low-order 4 bits	Column→ Row ↓	0		1	
0000	0			DLE	9-12-11-8-1
0001	1	SOH	9-12-1	DC1	9-11-1
0010	2	STX	9-12-2	DC2	9-11-2
0011	3	ETX	9-12-3		
0100	4			DC4	9-11-4
0101	5	ENQ	9-0-8-5	NAK	9-8-5
0110	6	ACK	9-0-8-6	SYN	9-2
0111	7	BEL	9-0-8-7		
1000	8			CAN	9-11-8
1001	9			EM	9-11-8-1
1010	10				
1011	11	VT	9-12-8-3		
1100	12	FF	9-12-8-4	FS	9-11-8-4
1101	13	CR	9-12-8-5	GS	9-11-8-5
1110	14	SO	9-12-8-6	RS	9-11-8-6
1111	15	SI	9-12-8-7	US	9-11-8-7

Fig. 18.10 Hole patterns, columns 0 and 1

(see Comment on Criterion 4) would play a role here.)

9-11-0-8-1	9-0-8	9-12-11-0-8-1	9-8
9-0-1	9-0-8-1	9-1	9-8-1
9-0-2	9-0-8-2	9-11-8-2	9-8-2
9-0-3	9-0-8-3	9-3	9-8-3
9-0-4	9-0-8-4	9-4	9-12-4
9-11-5	9-12-8-1	9-5	9-11-4
9-12-6	9-12-8-2	9-6	9-8-6
9-11-7	9-12-8-3	9-12-8	

Following the same BCD translation rule described above, these were assigned to columns 8 and 9 of ISO-8, as shown in Fig. 18.11.

Observe that, with the exception of hole patterns assigned to K13, K14, K15, K23, K28, and K29, the BCD relationship is about as good as it can be, given the hole patterns available for columns 8 and 9 of ISO-8.

Low-order 4 bits	Column→ Row ↓	8		9	
0000	0	K0	9-11-0-8-1	K16	9-12-11-0-8-1
0001	1	K1	9-0-1	K17	9-1
0010	2	K2	9-0-2	K18	9-11-8-2
0011	3	K3	9-0-3	K19	9-3
0100	4	K4	9-0-4	K20	9-4
0101	5	K5	9-11-5	K21	9-5
0110	6	K6	9-12-6	K22	9-6
0111	7	K7	9-11-7	K23	9-12-8
1000	8	K8	9-0-8	K24	9-8
1001	9	K9	9-0-8-1	K25	9-8-1
1010	10	K10	9-0-8-2	K26	9-8-2
1011	11	K11	9-0-8-3	K27	9-8-3
1100	12	K12	9-0-8-4	K28	9-12-4
1101	13	K13	9-12-8-1	K29	9-11-4
1110	14	K14	9-12-8-2	K30	9-8-6
1111	15	K15	9-11-8-3	K31	

Fig. 18.11 Hole patterns, columns 8 and 9

18.9 THE 64th HOLE PATTERN

For a hole pattern for K31, the anomaly (see Comment to Criterion 4 above) now came into play. All 64 hole patterns from columns 0 through 3 had been assigned to the Delete character and to 63 of the 64 control positions in columns 0, 1, 8, and 9 of ISO-8. Where was the 64th hole pattern to come from?

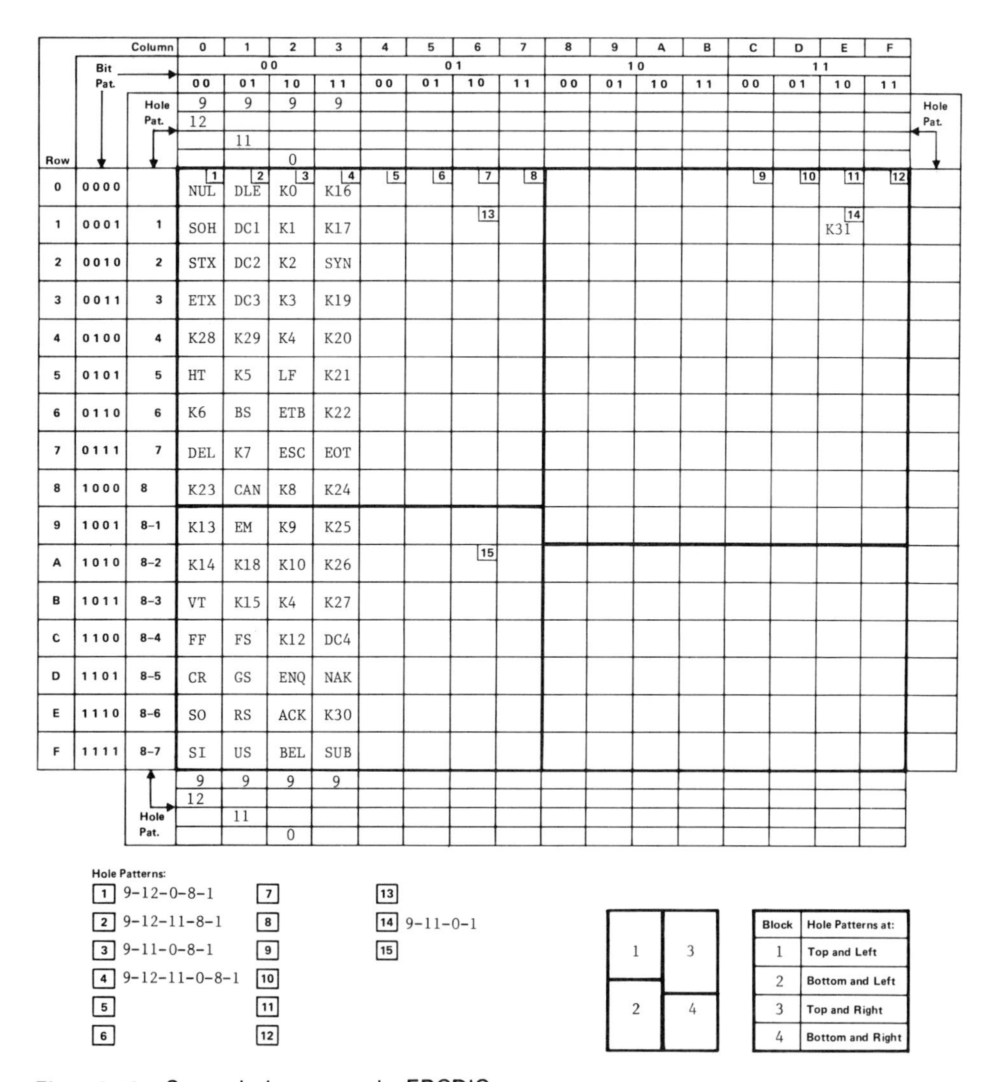

Row	Bit Pat.	Hole Pat.	0	1	2	3	4	5	6	7	8	9	A	B	C	D	E	F
	Bit Pat. (column)		00 00	00 01	00 10	00 11	01 00	01 01	01 10	01 11	10 00	10 01	10 10	10 11	11 00	11 01	11 10	11 11
		Hole Pat. (top)	9 12	9 11	9 0	9												
0	0000		[1] NUL	[2] DLE	[3] K0	[4] K16	[5]	[6]	[7]	[8]					[9]	[10]	[11]	[12]
1	0001	1	SOH	DC1	K1	K17			[13]								[14] K31	
2	0010	2	STX	DC2	K2	SYN												
3	0011	3	ETX	DC3	K3	K19												
4	0100	4	K28	K29	K4	K20												
5	0101	5	HT	K5	LF	K21												
6	0110	6	K6	BS	ETB	K22												
7	0111	7	DEL	K7	ESC	EOT												
8	1000	8	K23	CAN	K8	K24												
9	1001	8–1	K13	EM	K9	K25												
A	1010	8–2	K14	K18	K10	K26			[15]									
B	1011	8–3	VT	K15	K4	K27												
C	1100	8–4	FF	FS	K12	DC4												
D	1101	8–5	CR	GS	ENQ	NAK												
E	1110	8–6	SO	RS	ACK	K30												
F	1111	8–7	SI	US	BEL	SUB												
		Hole Pat. (bottom)	9 12	9 11	9 0	9												

Hole Patterns:

- [1] 9–12–0–8–1
- [2] 9–12–11–8–1
- [3] 9–11–0–8–1
- [4] 9–12–11–0–8–1
- [5]
- [6]
- [7]
- [8]
- [9]
- [10]
- [11]
- [12]
- [13]
- [14] 9–11–0–1
- [15]

Block	Hole Patterns at:
1	Top and Left
2	Bottom and Left
3	Top and Right
4	Bottom and Right

Fig. 18.12 Control characters in EBCDIC

Eventually after much discussion, ISO/TC97/SC2 chose the hole pattern 9-11-0-1, which comes from EBCDIC hex position E1.

Incidentally, the assignment of hole patterns for columns 0, 1, 8, and 9 in ISO-8 now dictated where the 23 control characters in columns 0 and 1 of ISO-7 (not previously assigned in EBCDIC) and the 32 control character positions (K0 through K31) in columns 8 and 9 of ISO-8 should be assigned in EBCDIC. This is shown in Fig. 18.12.

As described earlier, EBCDIC had been revised with respect to the positioning of the Katakana graphics. This was made known to the standards committees. During the discussions on the committees, it became apparent that Criterion 5, the "folding" criterion, would come into conflict with the collating sequence of Katakana.

18.10 EXAMPLES OF FOLDING

In order to appreciate the significance of the criterion on folding, four examples are given.

	Column	0	1	2	3	4	5	6	7	8	9	A	B	C	D	E	F
	Bit →	00				01				10				11			
Row	Pat. ↓	00	01	10	11	00	01	10	11	00	01	10	11	00	01	10	11
0	0000					SP	&	-			ｿ						0
1	0001							/		ｱ	ﾀ			A	J		1
2	0010									ｲ	ﾁ	ﾍ		B	K	S	2
3	0011									ｳ	ﾂ	ﾎ		C	L	T	3
4	0100									ｴ	ﾃ	ﾏ		D	M	U	4
5	0101									ｵ	ﾄ	ﾐ		E	N	V	5
6	0110									ｶ	ﾅ	ﾑ		F	O	W	6
7	0111									ｷ	ﾆ	ﾒ		G	P	X	7
8	1000									ｸ	ﾇ	ﾓ		H	Q	Y	8
9	1001									ｹ	ﾈ	ﾔ		I	R	Z	9
A	1010					¢	!		:	ｺ	ﾉ	ﾕ	ﾚ				
B	1011					.	$ ¥	,	#				ﾛ				
C	1100					<	*	%	@	ｻ		ﾖ	ﾜ				
D	1101					(	)	_	'	ｼ	ﾊ	ﾗ	ﾝ				
E	1110					+	;	>	=	ｽ	ﾋ	ﾘ	ﾞ				
F	1111					|	¬	?	"	ｾ	ﾌ	ﾙ	ﾟ				

Fig. 18.13 EBCDIC Latin and basic Katakana

Example 1

The 47 Basic Katakana symbols, the 26 capital Latin alphabetics, the 10 numerics, 26 specials, and the Space character, as then assigned in EBCDIC, are shown in Fig. 18.13.

If the two high-order bits of the 8-bit bit patterns are dropped, it will be observed, as shown in Fig. 18.14, that the 26 Latin alphabetics, the 10 numerics, the 26 specials, and the Space character "fold" into a 6-bit tableau. This dropping of the two high-order bits and the 6-bit resultant tableau to "address" printing positions on a line printer is precisely what a printer control unit can easily do. Then, if the graphic shapes shown in Fig. 18.13 are actually in the addressed printing positions of the print element (which could be a print chain, or a print train, for example), the appropriate EBCDIC graphics will be printed when the appropriate EBCDIC bit patterns are sent to the printer control unit.

Bit Pattern		0 0	0 1	1 0	1 1
0 0 0 0		SP	&	-	0
0 0 0 1		A	J	/	1
0 0 1 0		B	K	S	2
0 0 1 1		C	L	T	3
0 1 0 0		D	M	U	4
0 1 0 1		E	N	V	5
0 1 1 0		F	O	W	6
0 1 1 1		G	P	X	7
1 0 0 0		H	Q	Y	8
1 0 0 1		I	R	Z	9
1 0 1 0		¢	!		:
1 0 1 1		.	$	,	#
1 1 0 0		<	*	%	@
1 1 0 1		(	)	_	'
1 1 1 0		+	;	>	=
1 1 1 1		|	¬	?	"

Fig. 18.14 Folded Latin-63

Example 2

As a further example, if the bit patterns for the 47 Basic Katakana symbols, for the 10 numerics, for the following 6 specials

. ¥ , * – /

and for the Space character are sent to the printer control unit, the dropping of the two high-order bits yields a folded Katakana set, as shown in Fig. 18.15. Again, if these graphic symbols are in the addressed printing positions of the print element, this Katakana subset will be printed.

Observe, then, that the printer control unit has performed the identical operation on both the Latin set and the Katakana set—drop the two high-order bits, and address the resultant 6-bit bit patterns to printing positions on the print element. The two print elements are, of course,

Bit Pattern → ↓		0 0	0 1	1 0	1 1
0 0 0 0		SP	ソ	–	0
0 0 0 1		ア	タ	/	1
0 0 1 0		イ	チ	ヘ	2
0 0 1 1		ウ	ツ	ホ	3
0 1 0 0		エ	テ	マ	4
0 1 0 1		オ	ト	ミ	5
0 1 1 0		カ	ナ	ム	6
0 1 1 1		キ	ニ	メ	7
1 0 0 0		ク	リ	モ	8
1 0 0 1		ケ	ネ	ヤ	9
1 0 1 0		コ	ノ	ユ	レ
1 0 1 1		.	¥	,	ム
1 1 0 0		サ	*	ヨ	ク
1 1 0 1		シ	ハ	ラ	ン
1 1 1 0		ス	ヒ	リ	゛
1 1 1 1		セ	フ	ル	゜

Fig. 18.15 Folded Katakana-64

different, but they have a common characteristic—the appropriate graphic is in the appropriate printing position on the print element.

Example 3

A somewhat more complex folding is required to provide 48-character printing sets. The first part of the process in the printer-control unit is the same, the dropping of the two high-order bits of the 8-bit bit patterns. But, additionally, in the resultant 6-bit tableau, rows with low-order 4-bits equal to 1010, 1101, and 1111 are blocked. The three 6-bit bit patterns 011110, 101110, and 111110 are also blocked. The 48 positions then addressed to the printing positions of the print element are designated by X in Fig. 18.16.

Referring back to Fig. 18.13, it can be seen that this 48-character folding yields the 48-character folded Latin set shown in Fig. 18.17.

Bit Pattern		0 0	0 1	1 0	1 1
0 0 0 0		SP	X	X	X
0 0 0 1		X	X	X	X
0 0 1 0		X	X	X	X
0 0 1 1		X	X	X	X
0 1 0 0		X	X	X	X
0 1 0 1		X	X	X	X
0 1 1 0		X	X	X	X
0 1 1 1		X	X	X	X
1 0 0 0		X	X	X	X
1 0 0 1		X	X	X	X
1 0 1 0					
1 0 1 1		X	X	X	X
1 1 0 0		X	X	X	X
1 1 0 1					
1 1 1 0		X			
1 1 1 1					

Fig. 18.16 48-character printing positions

Bit Pattern		00	01	10	11
0000		SP	&	-	0
0001		A	J	/	1
0010		B	K	S	2
0011		C	L	T	3
0100		D	M	U	4
0101		E	N	V	5
0110		F	O	W	6
0111		G	P	X	7
1000		H	Q	Y	8
1001		I	R	Z	9
1010					
1011		.	$	,	#
1100		<	*	%	@
1101					
1110		+			
1111					

Fig. 18.17 EBCDIC folded Latin-48

Row	Column	0	1	2	3	4	5	6	7	8	9	A	B	C	D	E	F
	Bit Pat.	00				01				10				11			
		00	01	10	11	00	01	10	11	00	01	10	11	00	01	10	11
0	0000					SP		-			К						0
1	0001							/		Ю	Л						1
2	0010									А	М	У					2
3	0011									Б	Н	Ж					3
4	0100									Ц	О	В					4
5	0101									Д	П	Ь					5
6	0110									Е	Я	ы					6
7	0111									Ф	Р	З					7
8	1000									Г	С	Ш					8
9	1001									Х	Т	Э					9
A	1010																
B	1011					.	CS	,					Ч				
C	1100						*			И		Щ	Ъ				
D	1101																
E	1110									Й							
F	1111																

Fig. 18.18 EBCDIC Cyrillic-48

Example 4

The 32 large Cyrillic alphabetics, as then assigned in EBCDIC, are shown in Fig. 18.18. Applying the 48-character folding process yields the 32 large Cyrillic alphabetics, the 10 numerics, and the following 6 specials,

. CS , * – /

as shown in Fig. 18.19, where CS stands for Currency Symbol.

Bit Pattern		0 0	0 1	1 0	1 1
0 0 0 0		SP	К	–	0
0 0 0 1		Ю	Л	/	1
0 0 1 0		А	М	У	2
0 0 1 1		Б	Н	Ж	3
0 1 0 0		Ц	О	В	4
0 1 0 1		Д	П	Ь	5
0 1 1 0		Е	Я	,	6
0 1 1 1		Ф	Р	З	7
1 0 0 0		Г	С	Ш	8
1 0 0 1		Х	Т	Э	9
1 0 1 0					
1 0 1 1		.	CS	,	Ч
1 1 0 0		И	*	Щ	Ъ
1 1 0 1					
1 1 1 0		Й			
1 1 1 1					

Fig. 18.19 Folded Cyrillic-48

18.11 KATAKANA COLLATING SEQUENCE

As stated in Assumption 2(c), the non-Latin alphabetics would be assigned to columns 10 through 13 (Fig. 18.9). The proposal made to ANSI X3L2 at this time met this assumption. However, the proposal had two characteristics: (1) the 47 basic Katakana symbols would be assigned to

	Column	0	1	2	3	4	5	6	7	8	9	A	B	C	D	E	F
	Bit Pat. →	00				01				10				11			
Row	↓	00	01	10	11	00	01	10	11	00	01	10	11	00	01	10	11
0	0000								N0		N32						
1	0001					N1	N10		N48				N56				
2	0010					N2	N11	N19	N49				N57				
3	0011					N3	N12	N20	N50				N58				
4	0100					N4	N13	N21	N51				N59				
5	0101					N5	N14	N22	N52				N60				
6	0110					N6	N15	N23	N53				N61				
7	0111					N7	N16	N24	N54				N62				
8	1000					N8	N17	N25	N55				N63				
9	1001					N9	N18	N26									
A	1010									N27	N33	N37	N42				
B	1011												N43				
C	1100									N28		N38	N44				
D	1101									N29	N34	N39	N45				
E	1110									N30	N35	N40	N46				
F	1111									N31	N36	N41	N47				

Fig. 18.20 Proposed revised EBCDIC non-Latin

positions N1 through N47, and the small Katakana and Katakana punctuation symbols would be assigned to positions N48 through N63;* (2) the assignment of these non-Latin code positions into EBCDIC was not only noncontiguous (which was acceptable under Assumption 6), it was also *not* in correct collating sequence, as shown in Fig. 18.20.

---

*It is interesting that the proposal to change the gross collating sequence of the Katakana symbols in ISO-8 implied that JISCII should also be modified accordingly. As was stated earlier in this chapter, it had been reckoned that JISCII could not be changed, but nevertheless an attempt was being made here, indirectly, to make that change happen.

Row	Column	0	1	2	3	4	5	6	7	8	9	A	B	C	D	E	F
	Bit Pat. →	00				01				10				11			
	↓	00	01	10	11	00	01	10	11	00	01	10	11	00	01	10	11
0	0000					SP		-			N32						0
1	0001					N1	N10	/									1
2	0010					N2	N11	N19									2
3	0011					N3	N12	N20									3
4	0100					N4	N13	N21									4
5	0101					N5	N14	N22									5
6	0110					N6	N15	N23									6
7	0111					N7	N16	N24									7
8	1000					N8	N17	N25									8
9	1001					N9	N18	N26									9
A	1010									N27	N33	N37	N42				
B	1011					.	¥	,					N43				
C	1100						*			N28		N38	N44				
D	1101									N29	N34	N39	N45				
E	1110									N30	N35	N40	N46				
F	1111									N31	N36	N41	N47				

Fig. 18.21 Proposed revised EBCDIC Katakana

Comment. The rationale put forward to justify the proposed change in the Katakana gross collating sequence was as follows. The small Katakana and Katakana punctuation symbols had been provided on data processing equipment by few if any manufacturers, so both had been used little (if at all) in user applications. In any event, if it was necessary for some user application, to provide the "correct" collating sequence for all 63 Katakana symbols, then it was a fact that manufacturers, in their system sorting programs, provided easy methods for a user to achieve any collating sequence whatsoever, regardless of the native collating sequence of the CPU.

The intent behind the proposal was to preserve the folding capability in EBCDIC for Katakana and Cyrillic printing sets. Figure 18.21 shows how the 47 basic Katakana symbols would, under the proposal, be repositioned in EBCDIC (compare with Fig. 18.13). Figure 18.22 shows the folded 64-character Katakana set derivable from this EBCDIC positioning (compare with Fig. 18.14).

Bit Pattern		0 0	0 1	1 0	1 1
0 0 0 0		SP	N32	-	0
0 0 0 1		N1	N10	/	1
0 0 1 0		N2	N11	N19	2
0 0 1 1		N3	N12	N20	3
0 1 0 0		N4	N13	N21	4
0 1 0 1		N5	N14	N22	5
0 1 1 0		N6	N15	N23	6
0 1 1 1		N7	N16	N24	7
1 0 0 0		N8	N17	N25	8
1 0 0 1		N9	N18	N26	9
1 0 1 0		N27	N33	N37	N42
1 0 1 1		.	¥	,	N43
1 1 0 0		N28	*	N38	N44
1 1 0 1		N29	N34	N39	N45
1 1 1 0		N30	N35	N40	N46
1 1 1 1		N31	N36	N41	N47

Fig. 18.22 Proposed revised folded Katakana-64

18.12 CYRILLIC IN EBCDIC

Figure 18.23 shows how the 32 capital Cyrillic alphabetics would be repositioned in EBCDIC under the proposal (compare with Fig. 18.18). It should be noted that the first 26 capital Cyrillic alphabetics would go into EBCDIC positions N1 through N26, while the 27th through the 32nd capital Cyrillic alphabetics would go into EBCDIC code positions N28, N30, N32, N38, N43, and N44. The 48-character folded Cyrillic set resulting from the EBCDIC positioning is shown in Fig. 18.24 (compare with Fig. 18.19).

Row	Column	0	1	2	3	4	5	6	7	8	9	A	B	C	D	E	F
	Bit Pat.	00				01				10				11			
		00	01	10	11	00	01	10	11	00	01	10	11	00	01	10	11
0	0000					SP		-			N32						0
1	0001					N1	N10	/									1
2	0010					N2	N11	N19									2
3	0011					N3	N12	N20									3
4	0100					N4	N13	N21									4
5	0101					N5	N14	N22									5
6	0110					N6	N15	N23									6
7	0111					N7	N16	N24									7
8	1000					N8	N17	N25									8
9	1001					N9	N18	N26									9
A	1010																
B	1011					.	CS	,					N43				
C	1100						*			N28		N38	N44				
D	1101																
E	1110									N30							
F	1111																

Fig. 18.23 Proposed revised EBCDIC Cyrillic

Bit Pattern		0 0	0 1	1 0	1 1
0 0 0 0		SP	N32	-	0
0 0 0 1		N1	N10	/	1
0 0 1 0		N2	N11	N19	2
0 0 1 1		N3	N12	N20	3
0 1 0 0		N4	N13	N21	4
0 1 0 1		N5	N14	N22	5
0 1 1 0		N6	N15	N23	6
0 1 1 1		N7	N16	N24	7
1 0 0 0		N8	N17	N25	8
1 0 0 1		N9	N18	N26	9
1 0 1 0					
1 0 1 1		.	CS	,	N43
1 1 0 0		N28	*	N38	N44
1 1 0 1					
1 1 1 0		N30			
1 1 1 1					

Fig. 18.24 Proposed revised folded Cyrillic-48

18.13 THE U.S.A. PROPOSAL

The proposal for a revised EBCDIC had been put forward to the U.S.A. code standards committee X3L2. Subsequently, X3L2 proposed it to ISO/TC97/SC2, where it became known as the "U.S.A. Proposal."

It was realized that the "incorrect" gross collating sequence for Katakana in the proposed revised EBCDIC was not a serious "defect," since any collating sequence whatsoever could be provided by sort programs. Nevertheless, this defect disturbed people, and an alternate proposed revised EBCDIC came forward in France.

18.14 THE FRENCH PROPOSAL

The essence of the "French Proposal," as it came to be known, was that the coding positions N0 through N63 and G0 through G31 in ISO-8 (see Fig. 18.9) should be assigned consecutively (though not contiguously) in the 94 remaining code positions of EBCDIC. This is shown in Fig. 18.25.

	Column	0	1	2	3	4	5	6	7	8	9	A	B	C	D	E	F
	Bit Pat. →	00				01				10				11			
Row	↓	00	01	10	11	00	01	10	11	00	01	10	11	00	01	10	11
0	0000								N26	N35	N42	N49	N56				
1	0001					N0	N9		N27				N57				
2	0010					N1	N10	N18	N28				N58				
3	0011					N2	N11	N19	N29				N59				
4	0100					N3	N12	N20	N30				N60				
5	0101					N4	N13	N21	N31				N61				
6	0110					N5	N14	N22	N32				N62				
7	0111					N6	N15	N23	N33				N63				
8	1000					N7	N16	N24	N34				G0				
9	1001					N8	N17	N25					G1				
A	1010									N36	N43	N50	G2	G8	G14	G20	G26
B	1011									N37	N44	N51	G3	G9	G15	G21	G27
C	1100									N38	N45	N52	G4	G10	G16	G22	G28
D	1101									N39	N46	N53	G5	G11	G17	G23	G29
E	1110									N40	N47	N54	G6	G12	G18	G24	G30
F	1111									N41	N48	N55	G7	G13	G19	G25	G31

Fig. 18.25 Alternate proposed revised EBCDIC

18.15 FOLDING VERSUS COLLATING

Two points can be made with respect to the alternate proposed revised EBCDIC: (1) the collating sequence of Katakana (and, indeed, of any non-Latin alphabet correctly sequenced in ISO-8) would be maintained in EBCDIC, and (2) there would be no way whatsoever to set up a simple folding algorithm for either 48-character or 64-character sets from the N0 through N63 code positions in EBCDIC.

These two aspects—collating sequence (the French proposal) and folding capability (the U.S.A. proposal)—came to characterize the two proposals in discussions on the standards committees. Eventually, the French proposal won more adherents in ISO/TC97/SC2, and it was adopted.

It should be borne in mind that the objective of the standards committees was not to standardize a revised EBCDIC (although it might seem so from the previous discussion) but to standardize a 256-character Hollerith Card Code. There was, perhaps, a realization that, regardless of whether it was called the EBCDIC card code, or the Hollerith Card Code, or the Twelve-Row card code, it should be the same. For example, the hole pattern 12-1 should be the hole pattern for the alphabetic A and the hole pattern 12-8-5 should be the hole pattern for the "left parenthesis" in all these card codes.

18.16 THE HOLLERITH CARD CODE, FINAL VERSION

In any event, a one-to-one correspondence had now been established between the 256 bit patterns of EBCDIC and the 256 bit patterns of ISO-8. Given this correspondence, what remained to be done to specify the 256-character Hollerith Card Code was quite mechanical. The algorithm was as follows:

- Take an EBCDIC bit pattern;
- take its associated EBCDIC hole pattern;
- associate this hole pattern with the ISO-8 bit pattern corresponding to the EBCDIC bit pattern;
- Do this for all 256 EBCDIC bit patterns.

The final result is shown in Fig. 18.26. This reflects the notation in Fig. 18.9, where the control characters and graphic characters shown in columns 0 through 7 of that figure are used in Fig. 18.26, and the position-designators K0 through K31, N0 through N63, and G0 through G31 in columns 8 through 15 of that figure are also used in Fig. 18.26.

Hole Pat.																	Hole Pat.
	12				12	12		12	12				12	12		12	
		11				11	11	11		11				11	11	11	
			0		0		0	0			0		0		0	0	
	&	-	0	SP	{	¦	}	N26	N8	N17	N25	`	N35	N42	N49	N56	8-1
1	A	J	/	1	a	j	~	N57	SOH	DC1	K1	K17	N0	N9	K31	N27	1
2	B	K	S	2	b	k	s	N58	STX	DC2	K2	SYN	N1	N10	N18	N28	2
3	C	L	T	3	c	l	t	N59	ETX	DC3	K3	K19	N2	N11	N19	N29	3
4	D	M	U	4	d	m	u	N60	K28	K29	K4	K20	N3	N12	N20	N30	4
5	E	N	V	5	e	n	v	N61	HT	K5	LF	K21	N4	N13	N21	N31	5
6	F	O	W	6	f	o	w	N62	K6	BS	ETB	K22	N5	N14	N22	N32	6
7	G	P	X	7	g	p	x	N63	DEL	K7	ESC	EOT	N6	N15	N23	N33	7
8	H	Q	Y	8	h	q	y	G0	K23	CAN	K8	K24	N7	N16	N24	N34	8
9	I	R	Z	9	i	r	z	G1	K13	EM	K9	K25	NUL	DLE	K0	K16	8-1
8-2	[	]	\	:	N36	N43	N50	G2	K14	K18	K10	K26	G8	G14	G20	G26	8-2
8-3	.	$	,	#	N37	N44	N51	G3	VT	K15	K11	K27	G9	G15	G21	G27	8-3
8-4	<	*	%	@	N38	N45	N52	G4	FF	FS	K12	DC4	G10	G16	G22	G28	8-4
8-5	(	)	_	'	N39	N46	N53	G5	CR	GS	ENQ	NAK	G11	G17	G23	G29	8-5
8-6	+	;	>	=	N40	N47	N54	G6	SO	RS	ACK	K30	G12	G18	G24	G30	8-6
8-7	!	^	?	"	N41	N48	N55	G7	SI	US	BEL	SUB	G13	G19	G25	G31	8-7
Hole Pat.									9	9	9	9	9	9	9	9	
									12				12	12		12	
										11				11	11	11	
											0		0		0	0	

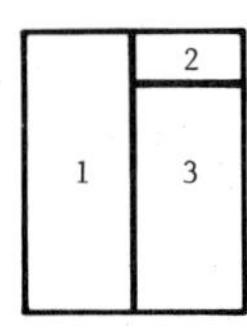

Block	Hole Patterns at:
1	Top and Left
2	Top and Right
3	Bottom and Right

Fig. 18.26 256-character Hollerith card code

18.17 REVISED KATAKANA IN EBCDIC

The final revised Katakana for EBCDIC is shown in Fig. 18.27 (compare with Fig. 18.5). The original EBCDIC Katakana (Fig. 18.5) had been implemented on the IBM System/360, and was also implemented on the IBM System/370. The revised EBCDIC Katakana of Fig. 18.27 was implemented on the IBM System/3.

	Column	0	1	2	3	4	5	6	7	8	9	A	B	C	D	E	F
	Bit Pat. →	00				01				10				11			
Row	↓	00	01	10	11	00	01	10	11	00	01	10	11	00	01	10	11
0	0000								コ	テ	ハ	ム	リ				
1	0001						ゥ		サ				ル				
2	0010					。	ェ	イ	シ				レ				
3	0011					「	ォ	ウ	ス				ロ				
4	0100					」	ャ	エ	セ				ワ				
5	0101					、	ュ	オ	ソ				ン				
6	0110					・	ョ	カ	タ				゛				
7	0111					ヲ	ッ	キ	チ				゜				
8	1000					ァ	ー	ク	ツ								
9	1001					ィ	ア	ケ									
A	1010									ト	ヒ	メ					
B	1011									ナ	フ	モ					
C	1100									ニ	ヘ	ヤ					
D	1101									ヌ	ホ	ユ					
E	1110									ネ	マ	ヨ					
F	1111									ノ	ミ	ラ					

Fig. 18.27 Final revised EBCDIC Katakana

It was considered at that time that there would be a data interchange problem if users wanted to interchange the Katakana data between a System/3 and a System/370, or if a user wanted to migrate from a System/3 to a System/370, which has greater capabilities, and still use the System/3 Katakana data bases. As it turned out, there was little or no user interchange of Katakana data between System/3's and System/370's. Additional capabilities were provided for the System/3 itself over the years, so that there was little or no user migration from System/3's to System/370's. In short, neither of the two potential problems materialized.

19
What Is a CPU Code?

19.1 INTRODUCTION

Central Processing Unit (CPU): The unit of a computing system that includes the circuits controlling the interpretation and execution of instructions.

What is a CPU Code? To answer this question by saying that a CPU code is the code used by a CPU answers the letter but not the spirit of the question. A CPU inputs, manipulates, processes, and outputs data in many shapes and forms. It is not uncommon to view a character code as being the only code form of significance to a CPU. But many other code forms—packed decimal, signed numerics, binary numbers, floating-point numbers, bit strings, and so on—are processed and manipulated by a CPU. The question would have been more meaningful if it had been

> What are the attributes of a character code, the presence or absence of which would cause the code to be categorized as a viable or nonviable CPU code?

As a preliminary to answering the question, the attributes of codes of more limited context, such as a magnetic tape code, a data transmission device code, and a punched card code are analyzed.

19.2 MAGNETIC TAPE CODE

What are the attributes that make a code suitable or desirable for magnetic tape? There are two attributes—one speaking to the format of data recorded on magnetic tape, the other speaking to control of the magnetic tape drive.

Suppose a magnetic tape has nine recording tracks, with one track dedicated to parity and the other eight tracks available for recording data, or for recording a code. Then the code should be eight bits or less in byte size. Similarly, for seven-track tape, with one track dedicated to parity and six tracks for the recording of either data or a code, the magnetic tape code should be six bits or less in byte size. The phrases "eight bits or less" and "six bits or less" were used with the realization that if, for example, a code of less than eight bits is to be recorded on eight data tracks, it is simple to fill (or pad in) zero bits to bring the byte size of the magnetic tape up to eight. By contrast, to record, for example, a seven-bit or eight-bit code on six data tracks would take a scheme which, while feasible, is complex. The hardware to implement such a recording scheme would be more complex than the hardware to record a seven-bit code on eight data tracks.

The other attribute of a magnetic-tape code is that it must contain the control characters necessary both to control the tape drive and to format or to structure the records recorded on tape. For seven-track magnetic tape and six-bit codes, as many as seven different control characters were used to control the tape movement or to implement various data formatting schemes on different CPU systems. On early nine-track tape drives, only one control character was used to control tape movement, and on recent nine-track tape drives, no control characters are used. For the latter type of tape drives, control is exercised by the execution of either computer instructions or channel commands.

19.3 DATA TRANSMISSION DEVICE CODE

When the environment of a transmission medium involves printing or display devices, another attribute is necessary besides those described above. The code for such a medium must also provide the graphic characters to meet the requirements of applications that use the medium and associated devices. For such environments, three attributes are necessary:

a) byte size commensurate with the transmission format of the medium;
b) control characters to control the terminal, printing, or display devices;
c) graphic characters to meet the application requirements that use the transmission medium and associated devices.

19.4 PUNCHED CARD CODE

A punched card code such as Hollerith has attributes that, though desirable, are conflicting. As a consequence, manufacturers of punched card equipment have to make trade-off decisions on these conflicting attributes.

In order to process on computing systems the data from punched cards, the punched card code must be translated to some other code form. In some computing systems, for example, the digit punches were translated to their binary-coded decimal equivalents, so that the system could add and subtract the data. In these computing systems, the hardware translation was implemented in electronic logic. The translation circuitry was usually located in the computer, not in the reader/punch unit. Such logic was costly. It was estimated that compared to a card reader/punch used as input/output to a computer, the cost of the translation hardware was one third of the cost of the total hardware circuitry of a reader/punch.

A desirable attribute of a card code, then, is that the translation to/from a related bit code should be as simple as possible. The translation hardware for a binary punched card code would have been substantially less complex than it would have been for the Hollerith punched card code.

The punched card code chosen by standards committees for standardization, however, was not binary, because a binary card code has two consequences that are quite undesirable. A binary card code requires more holes per character than the Hollerith card code. For example, a binary card code to represent a 64-character six-bit code would require as many as six holes per character, whereas the 64-character Hollerith Card Code requires no more than three holes per character. The additional holes per character of a binary card code have undesirable consequences: (1) the punched card itself would be structurally weak, and hence unreliable; (2) if there are more holes per character, the punch dies and plate must be of much more rugged construction (that is, higher manufacturing cost), and maintenance costs will be higher. (For a fuller discussion of these points, see Chapter 16.)

To sum up for punched card codes, translation simplicity to/from a related bit code is certainly a desirable attribute, but the simplest translation scheme, binary, has undesirable consequences—card unreliability, and higher manufacturing and equipment maintenance costs.

19.5 CPU CODE

By looking at magnetic tape codes, punched card codes, and data transmission device codes, three fundamental attributes of a media code have been discerned:

- The byte size of the code must be commensurate with the recording or transmitting format of the associated physical medium.

Bit Pattern →			A	B	BA
↓	Hole Pattern → ↓		0	11	12
		SP	ƀ [1]	-	& or +
1	1	1	/	J	A
2	2	2	S	K	B
2 1	3	3	T	L	C
4	4	4	U	M	D
4 1	5	5	V	N	E
4 2	6	6	W	O	F
4 2 1	7	7	X	P	G
8	8	8	Y	Q	H
8 1	9	9	Z	R	I
8 2	0	0	‡ [2]	!	?
8 2 1	8-3	# or =	,	$	.
8 4	8-4	@ or '	% or (	*	⌑ or)
8 4 1	8-5	:	γ	]	[
8 4 2	8-6	>	\	;	<
8 4 2 1	8-7	√	⧻	Δ	⧧

SP - Space

Hole Patterns:
[1] 8-2
[2] 0-8-2

Fig. 19.1 BCDIC

- The code must provide control characters to control associated devices.
- If there are associated printing or display devices, the code must provide graphic characters to meet the requirements of applications using the devices.

In short, a code must meet the functional requirements of the associated medium and associated product(s).

The functional requirements for CPU codes are much broader, more subtle, and more complex than they are for the media codes discussed above. For example, since a CPU may control the media and products discussed above, it must meet their functional requirements as well as its own intrinsic functional requirements.

A number of functional requirements of CPU codes will be discussed:

- control characters for associated products,
- graphic characters for associated products,
- numeric capabilities,
- collating sequence,
- translation simplicity to media codes,
- compatibility with other codes.

(Contiguity or non-contiguity of alphabetic characters will be discussed in Chapter 25.)

These aspects will be discussed for three prominent character codes:

- BCDIC, BCD Interchange Code, a 64-character, 6-bit bit code and 12-row card code (Fig. 19.1).
- EBCDIC, Extended BCD Interchange Code, a 256-character, 8-bit bit code and 12-row card code.
- ASCII, A 128-character, 7-bit bit code.

Other character codes that will be discussed in less detail are:

- Hollerith Card Code, a 256-character, 12-row card code, with 64-character and 128-character subsets,
- CCITT #2, a 58-character, 5-bit bit code,
- Fieldata, a 128-character, 7-bit code.

19.6 CONTROL CHARACTERS FOR ASSOCIATED PRODUCTS

19.6.1 BCDIC

The seven control characters in BCDIC and the graphics provided to represent them are shown below.

Graphic	Control character
b̸	Substitute Blank
γ	Word Separator
Δ	Mode Change
ǂ	Record Mark
⧧	Group Mark
⧻	Segment Mark
√	Tape Mark

Such graphics were useful in printouts for debugging programs.

These control characters controlled either the movement of the seven-track magnetic tape associated with the computing systems current at that time or the formatting and structuring of data to be recorded on magnetic tape. Not all of these control characters functioned on all computing systems, and indeed, some of them functioned differently from one system to another.

Tape Mark and *Segment Mark* were used to control the movement of tape. *Record Mark* and *Group Mark* were used for formatting and structuring of data to be recorded on tape.

Magnetic tape systems of those days were described as odd-parity systems, or as even-parity systems, according as the seven-track magnetic tape associated with the system was odd or even parity. On even-parity systems, the Space character (whose bit pattern is all zero bits), if recorded on tape, would be indistinguishable from blank tape. This situation rendered the Space character essentially unusable on such tapes. Instead, the *Substitute Blank* was used in its place.

The *Word Separator* character was necessary on 1401–1410 systems, which in a sense had 7-bit memories. One of the options available on the system was that when a bit pattern that had a one-bit as its seventh bit was recorded from memory on tape, the seventh bit would be stripped off, and a Word Separator bit pattern would be injected in the string of bit patterns being recorded on tape. On reading from magnetic tape to memory, the opposite process would ensue. The Word Separator character, then, was a means of making the 7-bit CPU code commensurate with the 6-bit byte size of magnetic tape.

The *Mode Change* character was used on magnetic tape for 7070 systems to indicate the beginning and end of numerical mode.

19.6.2 General Definitions for Control Characters

The control characters of EBCDIC (Fig. 19.2) and ASCII (Fig. 19.3) fall into seven classifications by function. Ten of these control characters are subclassified as data communication systems control, and are indicated by an asterisk (*) in the listings below.

Customer use

Characters used to designate user-assigned function, which may be realizable by user software:

	ASCII	EBCDIC
CU1 Customer Use 1		X
CU2 Customer Use 2		X
CU3 Customer Use 3		X

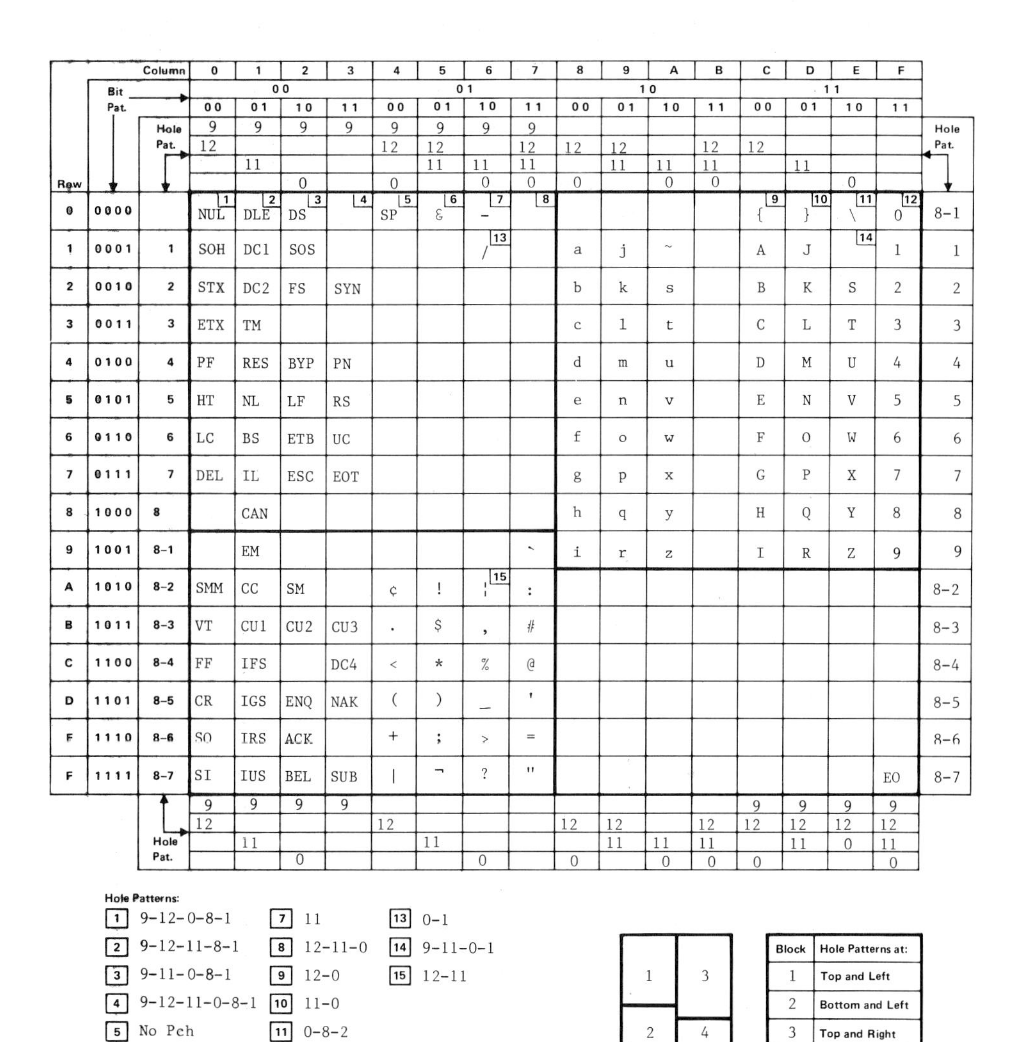

Row	Bit Pat.	Hole Pat.	0	1	2	3	4	5	6	7	8	9	A	B	C	D	E	F	Hole Pat.
	Bit Pat.		00				01				10				11				
			00	01	10	11	00	01	10	11	00	01	10	11	00	01	10	11	
		Hole Pat.	9	9	9	9	9	9	9	9									
			12				12	12		12	12	12		12	12				
				11				11	11	11		11	11	11		11			
					0		0		0	0	0		0	0			0		
0	0000		NUL [1]	DLE [2]	DS [3]	[4]	SP [5]	& [6]	- [7]	[8]					{ [9]	} [10]	\ [11]	0 [12]	8-1
1	0001	1	SOH	DC1	SOS				/ [13]		a	j	~		A	J	[14]	1	1
2	0010	2	STX	DC2	FS	SYN					b	k	s		B	K	S	2	2
3	0011	3	ETX	TM							c	l	t		C	L	T	3	3
4	0100	4	PF	RES	BYP	PN					d	m	u		D	M	U	4	4
5	0101	5	HT	NL	LF	RS					e	n	v		E	N	V	5	5
6	0110	6	LC	BS	ETB	UC					f	o	w		F	O	W	6	6
7	0111	7	DEL	IL	ESC	EOT					g	p	x		G	P	X	7	7
8	1000	8		CAN							h	q	y		H	Q	Y	8	8
9	1001	8-1		EM						`	i	r	z		I	R	Z	9	9
A	1010	8-2	SMM	CC	SM		¢	!	¦ [15]	:									8-2
B	1011	8-3	VT	CU1	CU2	CU3	.	$	,	#									8-3
C	1100	8-4	FF	IFS		DC4	<	*	%	@									8-4
D	1101	8-5	CR	IGS	ENQ	NAK	(	)	_	'									8-5
F	1110	8-6	SO	IRS	ACK		+	;	>	=									8-6
F	1111	8-7	SI	IUS	BEL	SUB	|	¬	?	"								EO	8-7
		Hole Pat.	9	9	9	9									9	9	9	9	
			12				12				12	12		12	12	12	12	12	
				11				11				11	11	11		11	0	11	
					0				0		0		0	0	0			0	

Hole Patterns:

[1]	9-12-0-8-1	[7]	11	[13]	0-1
[2]	9-12-11-8-1	[8]	12-11-0	[14]	9-11-0-1
[3]	9-11-0-8-1	[9]	12-0	[15]	12-11
[4]	9-12-11-0-8-1	[10]	11-0		
[5]	No Pch	[11]	0-8-2		
[6]	12	[12]	0		

1	3
2	4

Block	Hole Patterns at:
1	Top and Left
2	Bottom and Left
3	Top and Right
4	Bottom and Right

Fig. 19.2 EBCDIC

	b7	0	0	0	0	1	1	1	1
	b6	0	0	1	1	0	0	1	1
	b5	0	1	0	1	0	1	0	1
b4 b3 b2 b1	Row \ Col	0	1	2	3	4	5	6	7
0 0 0 0	0	NUL	DLE	SP	0	@	P		p
0 0 0 1	1	SOH	DC1	!	1	A	Q	a	q
0 0 1 0	2	STX	DC2	"	2	B	R	b	r
0 0 1 1	3	ETX	DC3	#	3	C	S	c	s
0 1 0 0	4	EOT	DC4	$	4	D	T	d	t
0 1 0 1	5	ENQ	NAK	%	5	E	U	e	u
0 1 1 0	6	ACK	SYN	&	6	F	V	f	v
0 1 1 1	7	BEL	ETB	'	7	G	W	g	w
1 0 0 0	8	BS	CAN	(	8	H	X	h	x
1 0 0 1	9	HT	EM	)	9	I	Y	i	y
1 0 1 0	10	LF	SUB	*	:	J	Z	j	z
1 0 1 1	11	VT	ESC	+	;	K	[	k	{
1 1 0 0	12	FF	FS	,	<	L	\	l	¦
1 1 0 1	13	CR	GS	-	=	M	]	m	}
1 1 1 0	14	SO	RS	.	>	N	^	n	~
1 1 1 1	15	SI	US	/	?	O	_	o	DEL

Fig. 19.3 ASCII

Device control

Characters used to control devices or to control major functions of devices:

	ASCII	EBCDIC
PF Punch Off		X
PN Punch On		X
RS Reader Stop		X
DC1 Device Control 1	X	X
DC2 Device Control 2	X	X
DC3 Device Control 3	X	X
DC4 Device Control 4	X	X

Error control

Characters used for error control, for indicating "alarms," or for identifying or requesting identification of stations in a communications system:

	ASCII	EBCDIC
DEL Delete	X	X
CAN Cancel	X	X
*ENQ Enquiry	X	X
*ACK Acknowledge	X	X
*NAK Negative Acknowledge	X	X
BEL Bell	X	X
SUB Substitute	X	X
EO Eight Ones		X

Formatting or editing control

Characters used for formatting or editing data:

		ASCII	EBCDIC
HT	Horizontal Tab	X	X
VT	Vertical Tab	X	X
FF	Form Feed	X	X
CR	Carriage Return	X	X
NL	New Line		X
BS	Backspace	X	X
LF	Line Feed	X	X
RLF	Reverse Line Feed		X
DS	Digit Select		X
SOS	Start of Significance		X
FS	Field Separator		X
SP	Space	X	X

Grouping control

Characters used to group data or information:

		ASCII	EBCDIC
*SOH	Start of Heading	X	X
*STX	Start of Text	X	X
*ETX	End of Text	X	X
SMM	Start of Manual Message		X
EM	End of Medium	X	X
CC	Cursor Control		X
IFS	Interchange File Separator	X	X
IGS	Interchange Group Separator	X	X
IRS	Interchange Record Separator	X	X
IUS	Interchange Unit Separator	X	X
*ETB	End of Transmission Block	X	X
*EOT	End of Transmission	X	X

In ASCII, the following nomenclature is used:

FS File Separator
GS Group Separator
RS Record Separator
US Unit Separator

Mode control

Characters used to set, change, or restore a mode of operation:

		ASCII	EBCDIC
LC	Lower Case		X
UC	Upper Case		X
SI	Shift In	X	X
SO	Shift Out	X	X
ESC	Escape	X	X
*DLE	Data Link Escape	X	X
GE	Graphic Escape		X
BYP	Bypass		X
RES	Restore		X
SM	Set Mode		X

Synchronization control

Characters used for synchronization of communication systems, or for synchronization of data within a format, or for synchronization of data streams with certain timing characteristics of a function of some device:

		ASCII	EBCDIC
NUL	Null	X	X
IL	Idle		X
*SYN	Synchronous Idle	X	X

ASCII lacks many of the control characters deemed essential for the CPU Code, EBCDIC, but such characters could be assigned into code extensions of ASCII.

19.7 GRAPHIC CHARACTER CAPABILITY

19.7.1 BCDIC

BCDIC has 68 graphic characters, as follows:

26 Alphabetics	A to Z
10 Numerics	0 to 9
15 Specials	. , : ; * [ ] ⟨ ⟩ $ / \ ! ? —
10 Duals	# & ¤ % @ = +) (-
7 Graphics to represent controls	⧺ γ Δ √ ǂ ⩩ ƀ

Note. There are 68 graphics, but with the five dual pairs, only 63 code positions are utilized.

The graphics used to represent control functions were considered to be useful for printouts in debugging programs.

The other graphics were sufficient for the commerical and scientific/engineering applications of the time, and for some programming languages (FORTRAN, COBOL, and various Assemblers).

The duals of BCDIC were not created or invented at the same time. On the tabulating and accounting products and systems of the early 1950s, a 48-graphic set adequate for "commercial" applications was provided:

1 Space	
10 Numerics	0 to 9
26 Alphabetics	A to Z
11 Specials	. , * / - $ % ¤ & # @

With the advent of FORTRAN, for "scientific" applications, a different 48-graphic set was required:

1 Space	
10 Numerics	0 to 9
26 Alphabetics	A to Z
11 Specials	. , * / - $ ( ) + = '

The "inventors" and users of these overlapping graphic sets thought that the application areas were separable. They were, until COBOL created a requirement for both the "commerical" and "scientific" graphics within a single application, or at least within a single computing installation. Intended application has a profound bearing on code design.

19.7.2 EBCDIC

EBCDIC has 192 code positions reserved for graphic characters. The Space character and the 94 graphic characters of ASCII are assigned.

Graphics are assigned for various non-Latin alphabets:

- Katakana (see Fig. 10.10)
- Cyrillic (see Fig. 2.34)
- Hebrew (not shown here)
- Arabic (not shown here)
- Greek (not shown here)

Graphics are assigned for various Latin alphabets which require more than the 26 letters of English-speaking countries (not shown here). Graphics for FORTRAN, COBOL, PL/I, and ALGOL (standard subset) are assigned.

For text processing applications, 120 graphics are assigned (see Chapter 26).

19.7.3 ASCII

ASCII has 94 graphic characters, sufficient for most data processing applications. It lacks others for applications such as text processing, non-Latin alphabets, but these could be assigned into code extensions of ASCII.

19.8 NUMERIC CAPABILITY

An aspect of numeric capability of a CPU code, signed numerics, will be discussed.

19.8.1 Signed numerics

BCDIC, EBCDIC, and ASCII are alike in one very important characteristic—they can be called, generically, BCD codes. The four low-order bits of the bit patterns that represent the numerics are binary coded decimal (BCD), as shown in Fig. 19.4. Note that the sequence in BCDIC—1, 2, 3, 4, 5, 6, 7, 8, 9, 0—will not affect the train of the discussion to follow. The arithmetic circuitry of BCDIC computers took into account the particular BCD bit pattern 1010, for 0.

Numeric	BCDIC	EBCDIC	ASCII
0	///////////	1111 0000	011 0000
1	00 0001	1111 0001	011 0001
2	00 0010	1111 0010	011 0010
3	00 0011	1111 0011	011 0011
4	00 0100	1111 0100	011 0100
5	00 0101	1111 0101	011 0101
6	00 0110	1111 0110	011 0110
7	00 0111	1111 0111	011 0111
8	00 1000	1111 1000	011 1000
9	00 1001	1111 1001	011 1001
0	00 1010	/////////	////////

Fig. 19.4 BCD numerics

The BCD characteristic, in fact, is the source of the names BCDIC and EBCDIC—BCD Interchange Code and Extended BCD Interchange Code. And this BCD characteristic was quite intentionally built into ASCII.

From ANSI X3.4–1963, ASCII, Criterion C2.6 reads:

> The numerals 0 through 9 shall be so coded that the four low-order bits shall be the binary coded decimal form of the particular numeral that the code represents.

The same criterion worded slightly differently, is found in ANSI X3.4–1968.

The BCD concept, as it relates to signed numerics, grew from the Hollerith Card Code (Fig. 19.5). The concept of overpunching a numeric with a 12-punch or 11-punch to indicate positive or negative numerics was, and is, common practice in punched-card applications. Thus 12-0, 12-1, 12-2, . . . , 12-9 punches represent +0, +1, +2, . . . , +9, respectively; 11-0, 11-1, 11-2, . . . , 11-9 punches represent −0, −1, −2, . . . , −9 respectively; and 0, 1, 2, . . . , 9 punches represent absolute numerics 0, 1, 2, . . . , 9, respectively.

Hole Pat.	12	11	0		12 0	12 11	11 0	12 11 0	12	11	0		12 0	12 11	11 0	12 11 0	Hole Pat.
	&	-	0	SP	{	¦	}										8-1
1	A	J	/	1	a	j	~		SOH	DC1							9-1
2	B	K	S	2	b	k	s		STX	DC2		SYN					9-2
3	C	L	T	3	c	l	t		ETX	DC3							9-3
4	D	M	U	4	d	m	u										9-4
5	E	N	V	5	e	n	v		HT		LF						9-5
6	F	O	W	6	f	o	w			BS	ETB						9-6
7	G	P	X	7	g	p	x		DEL		ESC	EOT					9-7
8	H	Q	Y	8	h	q	y			CAN							9
9	I	R	Z	9	i	r	z			EM			NUL	DLE			9-1
8-2	[	]	'	:													9-2
8-3	.	$	,	#					VT								9-3
8-4	<	*	%	@					FF	FS		DC4					9-4
8-5	(	)	_	`					CR	GS	ENQ	NAK					9-5
8-6	+	;	>	=					SO	RS	ACK						9-6
8-7	!	^	?	"					SI	US	BEL	SUB					9-7
Hole Pat.	12 8	11 8	0 8	8	12 0 8	12 11 8	11 0 8	12 11 0 8	12 8	11 8	0 8	8	12 0 8	12 11 8	11 0 8	12 11 0 8	

1	3
2	4

Block	Hole Patterns at:
1	Top and Left
2	Bottom and Left
3	Top and Right
4	Bottom and Right

Fig. 19.5 Hollerith Card Code

All three bit codes, BCDIC, EBCDIC, and ASCII, have a specified relationship to the Hollerith Card Code. In order for the signed-numerics concept to carry over into a CPU code, the bit patterns from the positive, negative, and absolute numerics of the Hollerith Card Code must exhibit the following characteristics in the bit code:

a) For all numerics, signed or absolute, the numerics 0 through 9 have the four low-order bits as BCD bit patterns.

Bit Pattern →		0 0	0 1	1 0	1 1
↓	Hole Pattern → ↓	12	11		
0 0 0 0					
0 0 0 1	1	+1	−1	1	
0 0 1 0	2	+2	−2	2	
0 0 1 1	3	+3	−3	3	
0 1 0 0	4	+4	−4	4	
0 1 0 1	5	+5	−5	5	
0 1 1 0	6	+6	−6	6	
0 1 1 1	7	+7	−7	7	
1 0 0 0	8	+8	−8	8	
1 0 0 1	9	+9	−9	9	
1 0 1 0	0	+0	−0	0	

Fig. 19.6 BCDIC signed numerics

b) For all positive numerics 0 through 9, the high-order bits* are the same.

c) For all negative numerics 0 through 9, the high-order bits* are the same.

d) For all absolute numerics 0 through 9, the high-order bits* are the same.

The code positions into which the Hollerith overpunched numeric hole patterns will translate for BCDIC, EBCDIC, and ASCII are shown in Figs. 19.6, 19.7, and 19.8, respectively. It may be seen that BCDIC and EBCDIC exhibit characteristics (a), (b), (c), and (d), but ASCII does not exhibit characteristics (a), (b), and (c).

---

*The actual high-order bits for parts (b), (c), and (d) do not matter, What matters is that within each category, (b), (c), and (d), the high-order bits are the same. The actual high-order bits will be accommodated by the arithmetic circuitry of the CPU implementing the code.

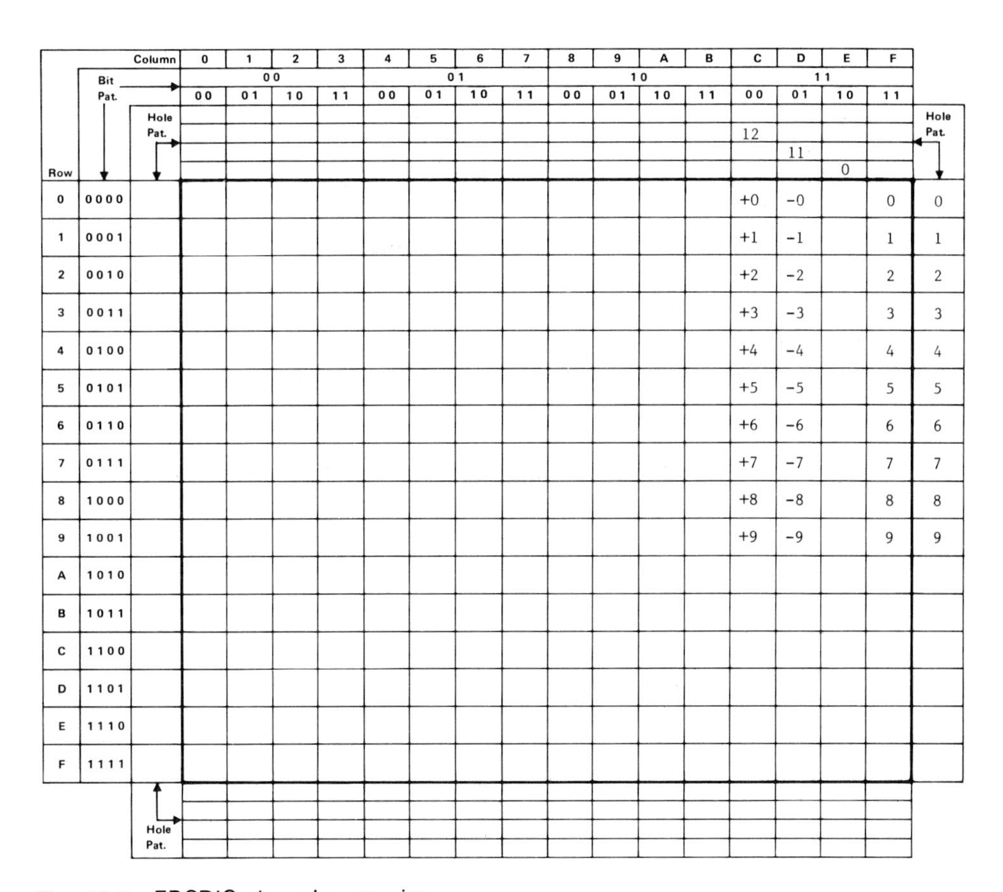

Row	Bit Pat.	0	1	2	3	4	5	6	7	8	9	A	B	C	D	E	F	Hole Pat.
		00				01				10				11				
		00	01	10	11	00	01	10	11	00	01	10	11	00	01	10	11	
	Hole Pat.													12	11	0		
0	0000													+0	-0		0	0
1	0001													+1	-1		1	1
2	0010													+2	-2		2	2
3	0011													+3	-3		3	3
4	0100													+4	-4		4	4
5	0101													+5	-5		5	5
6	0110													+6	-6		6	6
7	0111													+7	-7		7	7
8	1000													+8	-8		8	8
9	1001													+9	-9		9	9
A	1010																	
B	1011																	
C	1100																	
D	1101																	
E	1110																	
F	1111																	
	Hole Pat.																	

Fig. 19.7 EBCDIC signed numerics

Note that the American National Standard Hollerith Punched Card Code (ANSI X3.26–1970) contains the following caution about the practice of overpunching:

> *Section* 3.4
>
> Punched card systems have used the convention of overpunching digits with 12 or 11 to represent signed numbers or for other purposes. This standard does not provide a simple translation of overpunched digits to the ASCII representation of digits. Where possible, signs of numbers should be in separate columns. Overpunched digits should be used in information interchange only by specific agreement between sender and receiver.

The admonition does not state that the practice of overpunching numerics

	Column	0	1	2	3	4	5	6	7
	Bit Pattern b7	0	0	0	0	1	1	1	1
	b6	0	0	1	1	0	0	1	1
	b5	0	1	0	1	0	1	0	1
Row	b4 b3 b2 b1								
0	0 0 0 0				0		−7		
1	0 0 0 1				1	+1	−8		
2	0 0 1 0				2	+2	−9		
3	0 0 1 1				3	+3			
4	0 1 0 0				4	+4			
5	0 1 0 1				5	+5			
6	0 1 1 0				6	+6			
7	0 1 1 1				7	+7			
8	1 0 0 0				8	+8			
9	1 0 0 1				9	+9			
10	1 0 1 0					−1			
11	1 0 1 1					−2			+0
12	1 1 0 0					−3			
13	1 1 0 1					−4			−0
14	1 1 1 0					−5			
15	1 1 1 1					−6			

Fig. 19.8 ASCII "signed numerics" from Hollerith

is bad per se. It does point to a problem in translation (to be discussed in Section 19.10). Nor does the admonition say not to use the practice. It says to carry the signs in separate card columns "where possible." Indeed, the admonition does not enjoin against the use of the practice in information interchange, but says that use in information interchange should be only "by specific agreement between sender and receiver." That is a reasonable precaution. This agreement, and many others, are obvious precursors to information interchange. A receiver who does not know the card layout of the sender will not be able to process the cards.

The problem is not the use of overpunched numerics. The problem is in translation to ASCII. Doubtless, some scheme of translation of overpunched numerics (similar to that of Fig. 19.7) could be devised for ASCII but then that scheme would be different from the scheme of Fig. 19.8. While it would no doubt be feasible to build two different translation schemes—Hollerith to ASCII—into a translator, there is no way, intrinsic just to the data itself, for the translator to know when to activate one or the other translation scheme. That is to say, to build logic into the translator to recognize and respond to "the specific agreement between sender and receiver" would be quite impractical, and probably impossible. Signed numerics and ASCII are mathematically incompatible.

It has been argued that the solution to this dilemma is to forbid the use of overpunched numerics and to require that the sign for a numeric card field be carried in a separate field position. Certainly, this is a theoretical solution. It is a solution, however, contrary to a widespread and entrenched user practice. To implement such a solution would require conversion of card data fields, and a reprogramming of the user's application programs.

19.9 COLLATING SEQUENCE

The graphics, as assigned in a code, have a certain bit sequence. For reasons outside the code, the graphics may be assigned to a particular sequence, which is called the collating sequence. The bit sequence may or may not match the collating sequence.

19.9.1 BCDIC

An undesirable attribute in BCDIC as a CPU code was that the standard collating sequence for BCDIC did not match the bit sequence. In the code table of Fig. 19.1, BCDIC is shown in bit sequence. The collating sequence of BCDIC is shown in the code table of Fig. 19.9. The convention for this table is that the collating sequence, from low to high, is

space through % or (,
then γ through F,
then G through T,
then U through 9.

The disparity between bit sequence and collating sequence is evident.

		SP	γ	G	U
		.	\	H	V
		⌑ or)	⧻	I	W
		[	ƀ	!	X
		<	# or =	J	Y
		‡	@ or '	K	Z
		& or +	:	L	0
		$	>	M	1
		*	√	N	2
		]	?	O	3
		;	A	P	4
		Δ	B	Q	5
		-	C	R	6
		/	D	‡	7
		,	E	S	8
		% or (	F	T	9

Fig. 19.9 BCDIC in collating sequence

The disparity between BCDIC bit sequence and collating sequence led to "costs" for users of the 6-bit computing systems which would not have been incurred if the two sequences had matched. For the 7090 system, the cost was for computer usage time. Data fields that were to be operated on by comparison operations were, in advance, converted by a program to bit patterns that matched the collating sequence, and reconverted back to their original bit patterns (again, by a program) after the sorting or collating operations. For the 1400 and 7080 systems, the cost was additional hardware. A hardware comparator was provided which matched the bit sequence to the collating sequence during the comparison operations of sorting and collating.

19.9.2 EBCDIC

There are 256 character positions in EBCDIC, with bit patterns ranging from 0000 0000 to 1111 1111. The collating sequence of EBCDIC, from low to high, is prescribed to match the bit sequence. As a result, in sorting and collating operations, no hardware comparator is needed, and no pre- or post-conversion by software is needed.

19.9.3 ASCII

There are 128 character positions in ASCII, with bit patterns ranging from 000 0000 to 111 1111. As with EBCDIC, the collating sequence, from low to high, is prescribed to match the bit sequence.

19.10 TRANSLATION SIMPLICITY

CPU codes as related to magnetic tape and punched card codes will be discussed.

19.10.1 BCDIC

The translation from the 6-bit CPU code to the 6-bit magnetic tape code was a one-to-one bit translation. The translation to the 64-character punched card code was quite simple; digit punches 0 to 9 translated on a binary coded decimal basis, and zone punches 0, 11, 12 translated on a binary basis, as shown by the code chart of Fig. 19.1 (with the two exceptions noted, 8-2 and 0-8-2).

19.10.2 EBCDIC

The magnetic tape code for EBCDIC matches the CPU code, bit for bit and bit pattern for bit pattern. No translation is required for magnetic tape.

For punched card codes, the situation is different. The optimum theoretical EBCDIC, from a card-code-translation-simplicity point of view, would be the card code shown in Fig. 19.10. The four high-order bits of EBCDIC would translate to the four zone punches 12, 11, 0, and 9 on a pure binary basis, and the four low-order bits would translate to the digit punches, 0 through 9, on a binary coded decimal basis. However, this theoretical EBCDIC was rejected for two primary reasons—translation simplicity to BCDIC and collating sequence compatibility with BCDIC.

If the graphics of BCDIC had been positioned in theoretical EBCDIC according to the BCDIC card code, they would have been positioned as shown in Fig. 19.10. For these 63 graphics and Space, the EBCDIC card hole patterns would have matched precisely the assignments in common usage, a very desirable attribute. However, the translation from BCDIC *bit patterns* to EBCDIC *bit patterns*, under the scheme of Fig. 19.10, would have been complex and hence undesirable.

Row	Bit Pat. 4567	Hole Pat.	0	1	2	3	4	5	6	7	8	9	A	B	C	D	E	F
	Bit Pat. 01		00				01				10				11			
	Bit Pat. 23		00	01	10	11	00	01	10	11	00	01	10	11	00	01	10	11
		Hole Pat.									9	9	9	9	9	9	9	9
							12	12	12	12					12	12	12	12
					11	11			11	11			11	11			11	11
				0		0		0		0		0		0		0		0
0	0000		SP	0	-	!	& +	?			9	Z	R		I			
1	0001	1	1	/	J		A											
2	0010	2	2	S	K		B											
3	0011	3	3	T	L		C											
4	0100	4	4	U	M		D											
5	0101	5	5	V	N		E											
6	0110	6	6	W	O		F											
7	0111	7	7	X	P		G											
8	1000	8	8	Y	Q		H											
9	1001	8-1																
A	1010	8-2	b̸	‡														
B	1011	8-3	# =	,	$		.											
C	1100	8-4	@ '	% (	*		⌑)											
D	1101	8-5	:	γ	]		[											
E	1110	8-6	>	\	;		<											
F	1111	8-7	√	⧻	Δ		ǂ											
		Hole Pat.																

Fig. 19.10 Theoretical EBCDIC, based on optimum bit-pattern-to-hole-pattern relationship

More significantly, the bit sequence of the BCDIC graphics in Fig. 19.10 would have been radically different from the BCDIC collating sequence; that is, BCDIC and EBCDIC would have been incompatible from a collating-sequence point of view.

In short, four desirable attributes of a CPU code were conflicting, and not all could be achieved:

a) translation simplicity to a punched card code;
b) translation simplicity to a previous CPU code (BCDIC);
c) collating-sequence compatibility to a previous CPU code, BCDIC;
d) card-code compatibility to the card code in common usage.

	Column	0	1	2	3	4	5	6	7	8	9	A	B	C	D	E	F
	Bit Pat.	00				01				10				11			
		00	01	10	11	00	01	10	11	00	01	10	11	00	01	10	11
Row																	
0	0000					0	6	12						25	35	45	54
1	0001							13						26	36		55
2	0010													27	37	46	56
3	0011													28	38	47	57
4	0100													29	39	48	58
5	0101													30	40	49	59
6	0110													31	41	50	60
7	0111													32	42	51	61
8	1000													33	43	52	62
9	1001													34	44	53	63
A	1010								19								
B	1011					1	7	14	20								
C	1100					2	8	15	21								
D	1101					3	9	16	22								
E	1110					4	10	17	23								
F	1111					5	11	18	24								

Fig. 19.11 Embedded collating sequence

It was decided that two of these attributes, collating-sequence compatibility to BCDIC and card-code compatibility to the card code in common usage, were more important than the other two attributes.

After numbering the Space and the 63 graphics of BCDIC (Fig. 19.9), from 0 to 63, in collating sequence order, it was decided to embed these 64 characters in EBCDIC as shown in Fig. 19.11. The BCDIC collating sequence is embedded in the EBCDIC collating sequence, but not contiguously.

A consequence of these two attributes for EBCDIC is that the translation relationships, BCDIC bit patterns to EBCDIC bit patterns, are somewhat complex—more complex than for the simple scheme of Fig. 19.10.

19.10.3 ASCII

Is it feasible to translate ASCII to/from EBCDIC? Certainly, taking into account the facts that ASCII is a 128-character, 7-bit code and that EBCDIC is a 256-character, 8-bit code, the translation relationship is well known. Software for automatic translation has been provided by some manufacturers.

The straightforward translation is immensely complicated if the user intermixes pure character code forms (ASCII or EBCDIC) with other code forms, such as signed numeric, packed numeric, binary data, bit strings. The complication is both administrative and technical.

The representation of ASCII on magnetic tape is prescribed by an American National Standard. That same Standard prescribes the recording of pure ASCII character data only; that is to say, other code forms are ruled out. Magnetic tape with ASCII character data intermixed with other code forms is nonstandard, and would pose administrative problems.

Technically, the problem has an interesting aspect; it cannot be solved by manufacturers, only by users. The mixture of code forms in a user's application would vary from application to application, and even within applications. For a given application, translation is always possible, but a generalized translation program applicable to all applications, such as a manufacturer might provide, is not possible in the absence of a data descriptive language. A user who conforms ASCII data on magnetic tape to American National Standards has no translation problems. A user who chooses to intermix other code forms with ASCII data would create translation problems only he could solve.

With respect to punched card code, the situation for ASCII is not simple. ASCII is in a one-to-one correspondence to 128 characters of the American Standard Hollerith Card Code, so translation is certainly feasible. But no logical translation relationships (or almost none) exist, so the translation is on a brute-force, character-for-character basis.

It is interesting to note that in 1963 a card code called Decimal ASCII was proposed as an American National Standard, which had the characteristic of optimum translation simplicity to ASCII. Further, the concept of signed numerics could, from this card code, have been incorporated into ASCII. It had, however, a very undesirable attribute. The card hole patterns assigned to the numerics and to the alphabetics A through I matched the assignments in common usage in the data processing industry, but the assignments for the alphabetics J through Z, and for virtually all special graphics, did not match those of common usage. This mismatch implied such considerable conversion costs that users rejected Decimal ASCII when it was voted on at X3.

19.11 COMPATIBILITY

There are a number of aspects involved in compatibility between two different codes:

- The codes should be structurally similar. BCDIC and EBCDIC are structurally similar, but ASCII is structurally dissimilar to both.
- The collating sequence of the two codes should be the same. If the codes are of different size, the collating sequence of the smaller code should be embedded, not necessarily contiguously, in the collating sequence of the larger code.
- The codes should be functionally equivalent; that is, they should have the same set of control and graphic characters, although not necessarily with the same bit patterns. A smaller code is functionally equivalent upward to a larger code if the smaller code's set of graphic and control characters is contained in the larger set of characters. EBCDIC and the Hollerith Card Code are functionally equivalent. ASCII is upward functionally equivalent to EBCDIC.
- Translation relationships between the two codes should be simple.

In debates on code compatibility, it often turns out that one debater views the codes as incompatible because not *all* of the four aspects above are present, while the other debater views the codes to be compatible because at least *one* of the aspects above is present.

19.11.1 BCDIC

The magnetic tape code, punched card code, and CPU code of BCDIC are deemed to be compatible, in that they are functionally equivalent and translate simply to each other.

19.11.2 EBCDIC and BCDIC

EBCDIC is structurally similar to BCDIC. The collating sequence of BCDIC is embedded in the collating sequence of EBCDIC. All characters of BCDIC are included in EBCDIC, so there is upward functional equivalence. The translation relationship, BCDIC to EBCDIC, is not as simple as it could theoretically be, but it is certainly feasible.

19.11.3 EBCDIC and ASCII

EBCDIC and ASCII are structurally dissimilar and the collating sequences are different. However, all characters of ASCII are included in EBCDIC, so there is upward functional equivalence. The translation

Bit pattern	Letter case	Figure case	Bit pattern	Letter case	Figure case
00000	Not used	Not used	10000	E	3
00001	T	5	10001	Z	+ or "
00010	CR	CR	10010	D	(2)
00011	O	9	10011	B	?
00100	SP	SP	10100	S	'
00101	H	(1)	10101	Y	6
00110	N	,	10110	F	(1)
00111	M	.	10111	X	/
01000	LF	LF	11000	A	–
01001	L	)	11001	W	2
01010	R	4	11010	J	Bell
01011	G	(1)	11011	FS	FS
01100	I	8	11100	U	7
01101	P	0	11101	Q	1
01110	C	:	11110	K	(
01111	V	= or ;	11111	(3)LS	LS

(1) For National Use
(2) Used for Answer Back
(3) Also used for Delete

CR Carriage Return
SP Space
LF Line Feed
FS Figure Shift
LS Letter Shift

Fig. 19.12 CCITT #2

relationship, ASCII to/from EBCDIC, is quite complex but certainly feasible, in view of the upward functional equivalence.

Other codes in common use are CCITT #2 (Fig. 19.12) (sometimes called the Baudot or Teletype code) and Fieldata (Fig. 19.13).

19.11.4 CCITT 2 AND ASCII

ASCII and CCITT #2 are structurally dissimilar, have different collating sequences, and have a complex translation relationship. CCITT #2 is upward functionally equivalent to ASCII.

Row	Column	0	1	2	3	4	5	6	7
	Bit Pattern b7	0	0	0	0	1	1	1	1
	b6	0	0	1	1	0	0	1	1
	b5	0	1	0	1	0	1	0	1
Row	b4 b3 b2 b1								
0	0 0 0 0					MS	K	)	0
1	0 0 0 1					UC	L	-	1
2	0 0 1 0					LC	M	+	2
3	0 0 1 1					LF	N	<	3
4	0 1 0 0					CR	O	=	4
5	0 1 0 1					SP	P	>	5
6	0 1 1 0					A	Q	_	6
7	0 1 1 1		CONTROL (NOT DEFINED)			B	R	$	7
8	1 0 0 0					C	S	*	8
9	1 0 0 1					D	T	(	9
10	1 0 1 0					E	U	"	'
11	1 0 1 1					F	V	:	;
12	1 1 0 0					G	W	?	/
13	1 1 0 1					H	X	!	.
14	1 1 1 0					I	Y	,	SPEC
15	1 1 1 1					J	Z	STOP	IDLE

Fig. 19.13 Fieldata

19.11.5 BCDIC and ASCII

ASCII and BCDIC are structurally dissimilar, have different collating sequences, and do not have a translation relationship because they are functionally inequivalent.

19.11.6 FIELDATA

Fieldata is incompatible in all four aspects with BCDIC, EBCDIC, and ASCII, mainly because of the control functions assigned to columns 0, 1, 2, and 3 in various implementations.

19.12 SUMMARY OF FUNCTIONAL REQUIREMENTS OF A CPU CODE

- Control characters for associated media and for associated media products.
- Control characters for intrinsic CPU operations, such as editing.
- Graphics for associated printing/display products, to satisfy data processing applications, such as

  Commerical applications,
  Scientific/engineering applications,
  Applications such as meteorology, text processing, chemical abstracting, library bibliographing,
  Programming languages,
  Latin alphabetics,
  Non-Latin alphabetics,
  Graphics to represent control characters.
- Arithmetic capability, such as signed numerics.
- Collating sequence matching the bit sequence.
- Translation simplicity to media codes.
- Compatibility with other codes:

  Structural similarity,
  Functional equivalence,
  Same collating sequence,
  Translation simplicity.

An additional attribute, contiguity or noncontiguity of alphabetics, will be discussed in Chapter 25.

20 ASCII in an 8-Bit Interchange Environment

It had been decided by the standards committees in 1963 that the format for the standard magnetic tape would be 9 tracks (Fig. 20.1). One track would contain the parity bit; the other 8 tracks would contain "information bits."

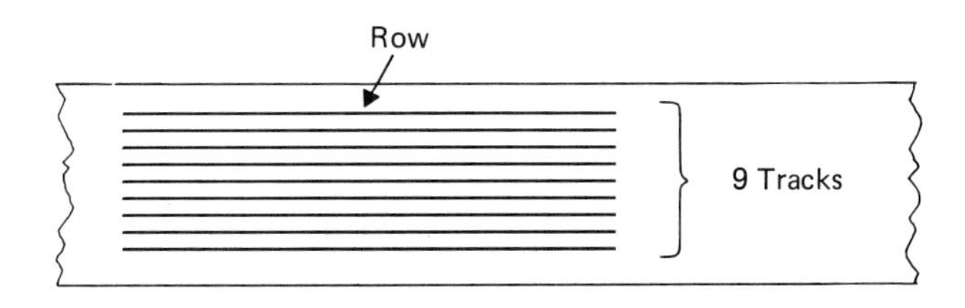

Figure 20.1

In consequence of this decision, 8 tracks were available on which to record the 7 bits of the 7-Bit Code. After some technical discussion, and some pushing and shoving, 7 specific tracks (of the 8 available tracks) and the specific track-to-bit relationship were decided upon.

20.1 ENGINEERING CONSIDERATIONS

The solution of the problem now exposed a new problem. What should be done with what came to be called "the eighth track"? Various suggestions, which might be described as being of a magnetic tape engineering nature, were presented to the standards committee with respect to the eighth track. The following list, while certainly not exhaustive, is representative of the variety of suggestions that were put forward:

1. Record all bits on the eighth track as zero-bits.
2. Record all bits on the eighth track as one-bits.
3. When recording the 7-bit bit-patterns of the 7-Bit Code, record all bits on the eighth track as zero-bits. This is not quite the same as part (1) above. In part (1) *all* bits on the eighth track would be set to zero, regardless of what information (7-Bit Code or otherwise) was being recorded. In part (3), only when the 7-bit bit patterns of the 7-Bit Code were being recorded would the bits on the eighth track be set to zero. But when other kinds of data (packed numeric data which requires 8 bits or the 8-bit bit patterns of an 8-bit code, for example) were being recorded, let the bits on the eighth track be recorded as either zero- or one-bits *as required.*

The thought here was that 7-bit code data on the tape could be distinguished from non-7-bit code data on the tape. A record of 7-Bit Code data would have the characteristic that the eighth track would be uniformly zero. Non-7-bit code data would have some one-bits in the eighth track.

This proposal came to take on a different implication. The supporters of the 7-Bit Code considered it to be "the standard code," and all other codes as "nonstandard." Hence, the above approach could be used to distinguish between "standard data" and "nonstandard data."

The weakness of this proposed facility for testing was that it is quite possible to envisage a string of 8-bit bytes containing packed numerics or a string of 8-bit bytes containing the 8-bit bit patterns of an 8-bit code, which would fortuitously exhibit the characteristic that for every byte, the bit to be recorded on the eighth track was, in fact, a zero bit. Therefore, even a string of so-called nonstandard data would pass the test for "standard data."

The supporters of this approach, while admitting the theoretical possibility of such data strings, claimed that they were very unlikely to occur in actual applications, and so the test would generally be valid in actual practice.

4. Use the 8 tracks as a clocking track to improve reliability of the tape drive.

As it turned out, none of the various suggestions was sufficiently appealing to gain a majority concensus on the standards committees. So the standards committees had arrived at that singularly frustrating situation in the drafting of a standard where every technical detail except one had achieved committee agreement. They decided, therefore, to proceed, even lacking agreement on what to do with the eighth track. The standard was drafted and approved with a specific statement that the eighth track was "undefined." Any bits in this track, however, were to be included in parity. "Undefined" meant, in the minds of the committee members, that it could be used for any purpose whatsoever. That is, any of the proposals above, or any other, could be implemented without violating the letter or spirit of the standard.

The committee, thus having sent the draft standard on its way to higher levels of standards authority, now tackled the remaining question of the "undefined" track with great vigor. But a new aspect more oriented to the aspects of coded character sets in general, than to the specific field of magnetic tape, now came on the scene. Consideration of this new aspect overshadowed all previous discussions and became the central topic of discussion in the standards committees.

20.2 8-BIT ENVIRONMENT

This aspect was the 8-bit environment which emerged in the data processing world as a result of the introduction of IBM's System/360. An 8-bit CPU code provided by the System/360 was EBCDIC (which is discussed in other chapters of this book). From the viewpoint of the standards committees, the main aspect of EBCDIC was that, structurally, it bore absolutely no relationship whatsoever with the 7-Bit Code. The most obvious structural difference was that the alphabetics were contiguous in bit sequence in the 7-Bit Code and noncontiguous in EBCDIC.

However, also provided by the System/360 was another 8-bit CPU code, called USASCII-8 (Fig. 20.2). This 8-bit code *was* structurally related to the 7-Bit Code. The eight columns of the 7-Bit Code had been distributed unaltered, albeit not contiguously, into eight of the sixteen columns of USASCII-8. This version of ASCII was slightly different from the 1963 version (see Fig. 14.11) and also slightly different from the ASCII 1967 version (see Chapter 24).

The attention of the standards committees was now focussed on the concept of an 8-bit code and, more particularly, on an 8-bit code structurally related to the 7-Bit Code. This standards development work rejoiced in the euphemistic title of "an 8-bit representation of the 7-Bit Code in an 8-bit environment."

Relating this standards effort back to the problem of what to do with the eighth track on 9-track magnetic tape, it was clear that if the structure of an 8-bit code was determined, then the recording of this 8-bit code on

Row	Column	0	1	2	3	4	5	6	7	8	9	A	B	C	D	E	F
	Bit Pat. →	00				01				10				11			
	↓	00	01	10	11	00	01	10	11	00	01	10	11	00	01	10	11
0	0000	NUL	DC0			SP	0					@	P				p
1	0001	SOM	DC1			!	1					A	Q			a	q
2	0010	EOA	DC2			"	2					B	R			b	r
3	0011	EOM	DC3			#	3					C	S			c	s
4	0100	EOT	DC4			$	4					D	T			d	t
5	0101	WRU	ERR			%	5					E	U			e	u
6	0110	RU	SYN			&	6					F	V			f	v
7	0111	BEL	LEM			'	7					G	W			g	w
8	1000	BS	S0			(	8					H	X			h	x
9	1001	HT	S1			)	9					I	Y			i	y
A	1010	LF	S2			*	:					J	Z			j	z
B	1011	VT	S3			+	;					K	[			k	
C	1100	FF	S4			,	<					L	\			l	
D	1101	CR	S5			-	=					M	]			m	
E	1110	SO	S6			.	>					N	↑			n	ESC
F	1111	SI	S7			/	?					O	←			o	DEL

Fig. 20.2 USASCII-8

the 8 data tracks of 9-track magnetic tape would necessarily define the contents of the eighth track.

The problem, which had initially been addressed as a magnetic tape engineering problem, was now addressed as a coded character sets problem. The problem was now restated. How should the 128 characters of the 7-Bit Code be embedded in the 256 code positions of an 8-bit code?

20.3 EMBEDMENT OF 7 BITS IN 8 BITS

It should be realized that, mathematically, the *number* of different possible embedments is very large. In the case of embedding 128 characters (of a 7-bit code) in the 256 code positions (of an 8-bit code), the number of different possible embedments is

$$256 \times 255 \times 254 \times \cdots 131 \times 130 \times 129 = \frac{256!}{128!} \simeq 2.2 \times 10^{291}$$

which is quite a large number indeed. However, if constraints are placed

	Column	0	1	2	3	4	5	6	7
	Bit Pattern b7	0	0	0	0	1	1	1	1
	b6	0	0	1	1	0	0	1	1
	b5	0	1	0	1	0	1	0	1
Row	b4 b3 b2 b1								
0	0 0 0 0								
1	0 0 0 1								
2	0 0 1 0								
3	0 0 1 1								
4	0 1 0 0								
5	0 1 0 1								
6	0 1 1 0								
7	0 1 1 1								
8	1 0 0 0								
9	1 0 0 1								
10	1 0 1 0								
11	1 0 1 1								
12	1 1 0 0								
13	1 1 0 1								
14	1 1 1 0								
15	1 1 1 1								

Fig. 20.3 7-bit code table

on the nature of the embedment, the number of different possible embedments reduces in size. Suppose the 7-bit code table and the 8-bit code table are exhibited in the customary columnar fashion (see Figs. 20.3 and 20.4).

Column		0	1	2	3	4	5	6	7	8	9	10	11	12	13	14	15
	→	00				01				10				11			
Row	Bit Pat. ↓	00	01	10	11	00	01	10	11	00	01	10	11	00	01	10	11
0	0000																
1	0001																
2	0010																
3	0011																
4	0100																
5	0101																
6	0110																
7	0111																
8	1000																
9	1001																
10	1010																
11	1011																
12	1100																
13	1101																
14	1110																
15	1111																

Fig. 20.4 8-bit code table

20.4 EMBEDMENT CONSTRAINTS

Columnar constraint. Suppose the constraint is to maintain columns; that is, each column from the 7-bit code table must be embedded unaltered into a column in the 8-bit code table. Then the number of different possible embedments is $16\times 15\times 14\times 13\times 12\times 11\times 10\times 9 =$ 518,918,400, which although smaller than the previous number, is still a respectably large number.

Sequence constraint. Suppose an additional constraint is applied; namely, that the eight columns of the 7-bit code table must be embedded in the sixteen column positions of the 8-bit code table in the same columnar sequence, although not necessarily contiguously. Then the number shrinks to 10,776.

Contiguous column–pair constraint. Suppose the columns of the 7-bit code table must be embedded in sequence, and in contiguous pairs, so as to maintain both the contiguous upper-case alphabet and the contiguous lower-case alphabet; the number of possible embedments reduces to 486.

Contiguous 8-column constraint. Finally, if the 8 columns of the 7-bit code table must be embedded in sequence, and the 8 columns must remain contiguous, the number of different possible embedments is 9.

These constraints—the columnar restraint, the sequence constraint, the contiguous column–pair constraint, and the contiguous 8-column constraint—are nested; that is, an embedment meeting the last constraint meets the immediately preceding constraint, which in turn meets its immediately preceding constraint, and so on.

The standards committees did not seriously consider all the vast number of possible embedments. Only six embedments received serious considerations (see Fig. 20.5). All of these embedments met the columnar constraint, the sequence constraint, and the contiguous column–pair constraint.

These three constraints, then, became the three major criteria for embedment. However, since all six candidate embedments met these three criteria, these criteria were clearly not factors for decision between the six candidates.

Two of the candidate embedments met the contiguous 8-column constraint, four candidates did not. So this criterion was a factor for decision between the six candidates.

20.5 EMBEDMENT NOTATION

While considering the embedments, the committees used a notation which helped to exhibit the embedments compactly. The eight columns of the 7-bit code table were named as follows:

Column 0	C	for Control character
Column 1	C	for Control character
Column 2	S	column of Specials
Column 3	N	column with Numerics
Column 4	A	column with upper case A
Column 5	Z	column with upper case Z
Column 6	a	column with lower case a
Column 7	z	column with lower case z

The 7-bit code table could then be compactly exhibited as follows:

Columns→	0	1	2	3	4	5	6	7
	C	C	S	N	A	Z	a	z

The problem was now restated. How should the eight columns of the 7-Bit Code be embedded in the sixteen column positions of an 8-bit code table?

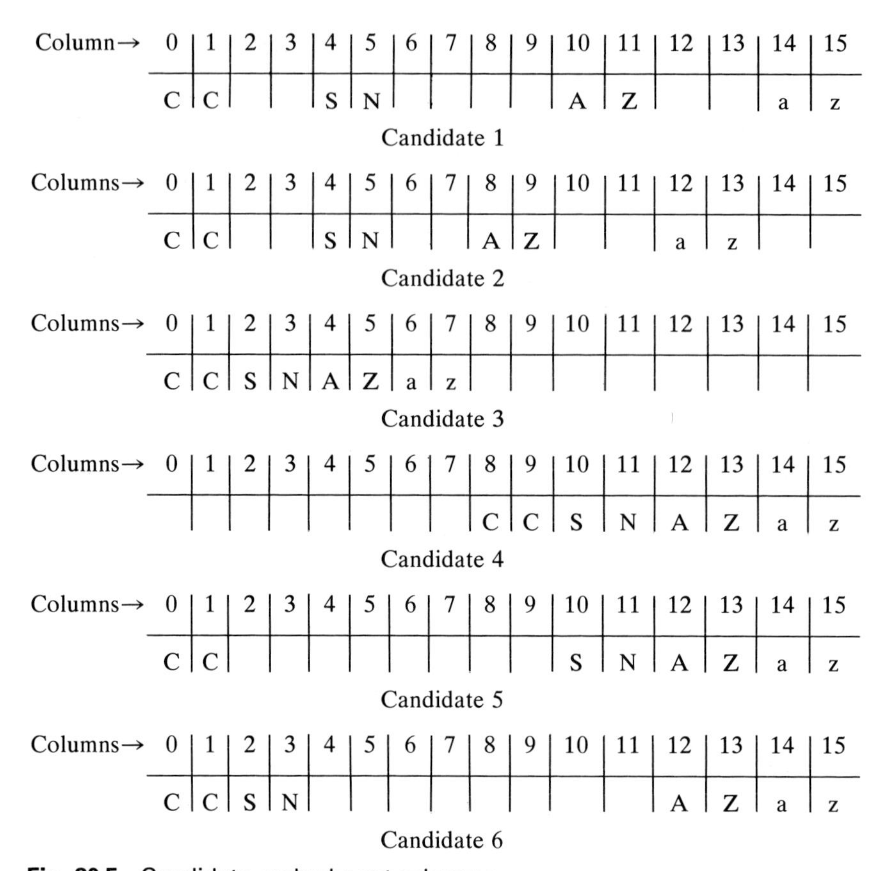

Fig. 20.5 Candidate embedment schemes

20.6 EMBEDMENT SCHEMES

Six embedment schemes were proposed to the standards committees. These are shown in Fig. 20.5. Candidate 6 is the embedment scheme first proposed (late 1963) to the standards committees. Candidate 1 is the embedment scheme actually implemented on the System/360 (see Fig. 20.2).

20.7 TRANSFORMATION ALGORITHM

It would be reasonable to suppose that, in determining the optimum 8-bit representation of the 7-bit code in an 8-bit environment, the only factor that would need to be considered would be the simplicity of the transformation algorithm, when transforming a 7-bit byte into an 8-bit byte and when transforming an 8-bit byte into a 6-bit byte. As a step in evaluating the superiority or inferiority of various transformation algorithms, it is

Column	0	1	2	3	4	5	6	7	8	9	10	11	12	13	14	15	
Bit Pat.	0 0				0 1				1 0				1 1				
	0 0	0 1	1 0	1 1	0 0	0 1	1 0	1 1	0 0	0 1	1 0	1 1	0 0	0 1	1 0	1 1	
	000	001			010	011					100	101			110	111	1
	000	001			010	011			100	101			110	111			2
	000	001	101	011	100	101	110	111									3
									000	001	010	011	100	101	110	111	4
	000	001									010	011	100	101	110	111	5
	000	001	010	011									100	101	110	111	6

Candidates

Fig. 20.6 Three high-order bits, as embedded

necessary to number or name the bits of a 7-bit byte and the bits of an 8-bit byte. The notation used by the standards committees was as follows:

7-bit byte		b7	b6	b5	b4	b3	b2	b1
8-bit byte	E8	E7	E6	E5	E4	E3	E2	E1

The six candidates evaluated by the committees are shown in Fig. 20.5. In Fig. 20.6, the three high-order bits, b7, b6, b5, of the 7-bit byte are shown in relation to the four high-order bits, E8, E7, E6, E5, of the 8-bit byte. It is to be observed that, for all candidates, b5 = E5.

The columnar restraint referred to earlier ensured that, for all candidates, the four low-order bits, b4, b3, b2, b1, of the 7-bit byte are identical to the four low-order bits, E4, E3, E2, E1, of the 8-bit byte.

In summary, for all candidates, the transformation algorithm, with respect to the five low-order bits, 7-bits to 8-bits or 8-bits to 7-bits, is

$$E5 \quad E4 \quad E3 \quad E2 \quad E1 = b5 \quad b4 \quad b3 \quad b2 \quad b1$$

The relationship between E8 E7 E6 and b7 b6 is, however, different for all candidates and the candidates came to be characterized by the transformation algorithms of the high-order bits.

Candidate	Transformation Algorithm
1	E8 = b7 E7 = b6 E6 = b7
2	E8 = b7 E7 = b6 E6 = 0
3	E8 = 0 E7 = b7 E6 = b6
4	E8 = 1 E7 = b7 E6 = b6
5	E8 = b7|b6 E7 = b7 E6 = b6
6	E8 = b7 E7 = b7 E6 = b6

Boolean analysis of Fig. 20.6 shows that the transformation algorithms for the six candidates are as follows. (The notation E8 = b7|b6 means that E8 = 1 if b7 = 1, or if b6 = 1, or if b7 and b6 = 1.)

It is to be observed that all the transformation algorithms involve fairly simple logic. They *could* be rated in degree of complexity, which could then be a factor to decide for the optimum (least complex) algorithm. However, the logic circuits to implement these algorithms would, in fact, be trivially different in complexity. The relative complexity of the transformation algorithms cannot sensibly be taken as a significant factor for decision. The standards committees recognized this, and developed other criteria.

Although the committees were ostensibly trying to determine the best 8-bit representation of the 7-bit code in an 8-bit environment, and were not supposed to be developing an 8-bit code,* all the criteria below except

*In short, the committee was indeed working to develop an 8-bit code. The rationale was that the committee was not working to develop an 8-bit code, but rather to develop an 8-bit representation of the 7-bit code in an 8-bit environment. However, some day in the future, an 8-bit code might be needed. Therefore, the 8-bit representation should be designed now so that it could serve as an 8-bit code if needed.

the first four would be considered of significance for a code. The first four criteria relate to the transformation algorithm. Of these four, three were met by all six candidates, and so were not factors for decision.

20.8 EMBEDMENT CRITERIA

The standards committee developed and considered the eighteen criteria presented below. It is to be emphasized that

a) Some of these criteria are important for environments that are mainly computer oriented.

b) Some of these criteria are important for environments that are mainly communications oriented.

c) Some of these criteria are important for both environments.

d) It is a matter of individual judgment as to which criteria fit under parts (a), (b), and (c) above.

e) It is a matter of individual judgment as to how the criteria are ranked or weighted in order of importance.

f) The set of criteria are mutually self-conflicting. In particular, Criteria 11 and 13 are conflicting, and Criteria 16 and 17 are conflicting. In consequence, *no* 8-bit representation can satisfy *all* criteria.

For purposes of reference later in this chapter, the criteria are headed by a "short form."

1. Column integrity. Each column of the 7-bit code should be embedded unaltered in a column of the 8-bit representation.

2. Column sequence. The eight columns of the 7-bit code should be embedded in the same columnar sequence in the sixteen column positions of the 8-bit representation, although not necessarily in contiguous column sequence.

3. Contiguous column pairs. Contiguous column pairs (columns 0 and 1, columns 2 and 3, columns 4 and 5, columns 6 and 7) of the 7-bit code should be embedded in contiguous column pairs in the 8-bit representation.

4. Contiguous 8-columns. The eight columns of the 7-bit code should be contiguous in the 8-bit representation.

5. Collating sequence. The 6-Bit Subset,* the 7-Bit Code, and the 8-Bit Representation should have the same relative collating sequence.

*The 6-Bit Subset referred to here consists of the 64 characters in columns 2, 3, 4, and 5 of the 7-Bit Code. The bit patterns of this subset are derivable by dropping b6 of the bit patterns of the 7-Bit Code.

6. Katakana 64. A contiguous 64-character block should be available in the 8-Bit Representation to which the 64 Japanese Katakana graphics could be assigned.

7. Latin–Katakana contiguity. A 64-character block for the Katakana alphabet and the Latin alphabet should be contiguous in the 8-Bit Representation.

8. Contiguous controls. Unassigned positions in the 8-Bit Representation should be available contiguous to the columns containing control characters from the 7-Bit Code to which new control characters could be assigned.

9. Single-bit test, Latin alphabetics. Latin alphabetic characters should be distinguishable from nonalphabetic characters in the 8-Bit Representation, as in the 7-Bit Code, by a single-bit test.

10. Symmetry. The columns of the 7-Bit Code should be distributed symmetrically in the 8-Bit Representation.

11. Space collate low. The Space character should collate low to the graphic characters from the 7-Bit Code and also to all (unassigned) graphic positions in the 8-Bit Representation.

Comment. The committee had the concept that if an 8-bit code were developed providing 128 code positions additional to those from the 7-Bit Code, 32 of these code positions, or two columns, would be assigned to new control characters, and 96 code positions, or six columns, would be assigned to new graphic characters.

12. Signed numerics greater than 9. In the 8-Bit Representation, the four high-order bits for the column containing the numerics, interpreted as a binary coded decimal, should be numerically greater than 9, to facilitate checking on arithmetic operations in computers.

13. Packed numerics. Packed numerics, if used in interchange and interpreted as 8-bit bytes, will have bit patterns from columns 0 through 9 and rows 0 through 9 of the 8-Bit Representation. Control characters may be assigned to these bit patterns without causing trouble, but the graphics from the 7-Bit Code should not be assigned to these columns.

14. Null = all zeros. The Null character from the 7-bit Code should be in the all-zeros bit-pattern position of the 8-Bit Representation; that is, in column 0, row 0.

15. Delete = all ones. The Delete character from the 7-Bit Code should be in the all-ones bit-pattern position of the 8-Bit Representation; that is, in column 15, row 15.

16. Low-order 7 bits, 7 to/from 8. The seven bits of the 7-Bit Code should become the low-order seven bits of the 8-Bit Representation.

17. Low-order 6 bits, 6 to/from 8. The six bits of the 6-Bit Subset should become the low-order six bits of the 8-Bit Representation.

18. Single-bit test, 7 versus non-7. In the 8-Bit Representation, all bit patterns from the 7-Bit Code should be distinguishable from all bit patterns not from the 7-Bit Code by a single-bit test.

20.9 ANALYSIS OF EMBEDMENTS

The six candidates of 8-Bit Representation (Fig. 20.5) were than analyzed against the 18 criteria, as shown in Fig. 20.7. (An "X" in the table means the candidate meets the criterion.)

Criterion \ Candidate	1	2	3	4	5	6
1. Column integrity	X	X	X	X	X	X
2. Column sequence	X	X	X	X	X	X
3. Contiguous column pairs	X	X	X	X	X	X
4. Contiguous 8 columns			X	X		
5. Collating sequence	X	X	X	X	X	X
6. Katakana 64	X		X	X	X	X
7. Latin–Katakana contiguity	X					X
8. Contiguous controls	X	X			X	
9. Single bit test, Latin alphabetics	X	X				X
10. Symmetry	X					X
11. Space collate low	X	X	X			X
12. Signed numerics greater than 9	X			X	X	X
13. Packed numerics				X	X	
14. Null = all zeros	X	X	X		X	X
15. Delete = all ones	X			X	X	X
16. Low-order 7 bits, 7 to/from 8			X	X	X	X
17. Low-order 6 bits, 6 to/from 8	X					
18. Single-bit test, 7 versus non-7		X	X	X		

Fig. 20.7 Candidates and criteria

20.10 COMMITTEE DECISION

It is an interesting exercise, weighting the various criteria as deemed appropriate, to judge which of the six candidates is the superior 8-Bit Representation. The standards committees, by majority but not by unanimous vote, decided in favor of candidate 3, characterized by its transformation algorithm:

$$E8 = 0$$

$$E7, 6, 5, 4, 3, 2, 1 = b7, 6, 5, 4, 3, 2, 1$$

This transformation algorithm having been decided by the coded character sets committees, the magnetic tape committees now specified in 9-track magnetic tape standards that what was previously called the "eighth" or undefined track would now be set uniformly to zero.

They might have (should have?) specified the eighth track more precisely as being set to zero for every row in which is recorded a bit pattern from the 7-Bit Code. Instead they chose to specify the eighth track as uniformly recorded with zero bits. A consequence of this particular specification is that any 9-track tape that records either an 8-bit code (such as EBCDIC) or packed numerics or binary data, such that a one-bit is recorded anywhere along the eighth track, is deemed to be nonstandard.

Since virtually all 9-track tapes recorded in computing installations will have packed numeric data or binary data, virtually all computing installation magnetic tapes are nonstandard. A curious consequence indeed.

21 The Alphabetic Extender Problem

As described in Chapter 4, accommodation of European countries with Latin alphabets of 29 letters was provided on products by duals. The three additional letters were associated with card hole patterns as shown in Fig. 21.1.

Subsequently, as described in Chapter 9, these card-code assignments were carried forward into EBCDIC. The EBCDIC code positions assigned to such graphic meanings were designated as alphabetic-extender positions.

BCDIC was a monocase alphabet code, while EBCDIC was a duocase alphabet code. The card hole patterns for the BCDIC alphabetic extenders were assigned to EBCDIC as capital alphabetic extenders, and new code positions were assigned to small alphabetic extenders, as shown in Fig. 21.2.

Hole pattern	U.S.A.	Germany	Sweden	Finland	Norway	Denmark
8-3	#	Ä	Ä	Ä	Æ	Æ
8-4	@	Ö	Ö	Ö	Ø	Ø
11-8-3	$	Ü	Å	Å	Å	Å

Fig. 21.1 Monocase and capital alphabetic extenders

Hole pattern	U.S.A.	Germany	Sweden	Finland	Norway	Denmark
8-7	"	ä	ä	ä	æ	æ
12-8-2	¢	ö	ö	ö	ø	ø
11-8-2	!	ü	å	å	å	å

Fig. 21.2 Small alphabetic extenders

21.1 THE ISO 7-BIT CODE

In 1967, the ISO 7-Bit Coded Character Set for Information Processing Interchange was standardized in an approved ISO Recommendation, R646–1967 (see Fig. 21.3).

Of particular significance were the third and fourth footnotes. They are reproduced here (in part).

	Column	0	1	2	3	4	5	6	7
	Bit Pattern b7 b6 b5	0 0 0	0 0 1	0 1 0	0 1 1	1 0 0	1 0 1	1 1 0	1 1 1
Row	b4 b3 b2 b1								
0	0 0 0 0	NUL	DLE	SP	0	@ [3]	P	` [4]	p
1	0 0 0 1	SOH	DC1	!	1	A	Q	a	q
2	0 0 1 0	STX	DC2	"	2	B	R	b	r
3	0 0 1 1	ETX	DC3	£	3	C	S	c	s
4	0 1 0 0	EOT	DC4	$	4	D	T	d	t
5	0 1 0 1	ENQ	NAK	%	5	E	U	e	u
6	0 1 1 0	ACK	SYN	&	6	F	V	f	v
7	0 1 1 1	BEL	ETB	'	7	G	W	g	w
8	1 0 0 0	BS	CAN	(	8	H	X	h	x
9	1 0 0 1	HT	EM	)	9	I	Y	i	y
10	1 0 1 0	LF	SUB	*	:	J	Z	j	z
11	1 0 1 1	VT	ESC	+	;	K	[[3]	k	[3]
12	1 1 0 0	FF	FS	,	<	L	[3]	l	[3]
13	1 1 0 1	CR	GS	-	=	M	] [3]	m	[3]
14	1 1 1 0	SO	RS	.	>	N	^ [4]	n	‾ [4]
15	1 1 1 1	SI	US	/	?	O	_	o	DEL

Fig. 21.3 ISO 7-Bit Code

[3]*Reserved for National Use* These positions are primarily intended for alphabetic extension. If they are not required for that purpose, they may be used for symbols.

[4]*Positions* 5/14, 6/0, *and* 7/14 · · · may be used for other graphical symbols when it is necessary to have 8, 9, or 10 positions for national use.

In the ISO 7-Bit Code, therefore, the requirement for alphabetic extender positions was recognized, as it had been in EBCDIC.

21.2 EBCDIC AND THE 7-BIT CODE

At this point in time, then, there was a slight mismatch between EBCDIC and the ISO 7-Bit Code, in that the former assigned 6 positions for alphabetic extenders, and the latter assigned 7 "primary" positions and 3 more "secondary" positions if needed by some country.

This mismatch could have been rectified easily by assigning additional alphabetic extender positions in EBCDIC. But a much more worrisome mismatch arose when the ISO/TC97/SC2 was working on the standardization of the Hollerith Card Code (called, by SC2, the Twelve-Row Card Code).

21.3 EBCDIC AND THE HOLLERITH CARD CODE

As described in Chapter 17, a slight mismatch existed between the American Standard Hollerith Card Code and the EBCDIC Card Code, as shown below:

Hole pattern	Hollerith	EBCDIC
12-8-7	!	|
11-8-7	^	¬
12-8-2	[	¢
11-8-2	]	!

The first pair of these, ! (exclamation point) versus | (Logical OR) and ^ (circumflex) versus ¬ (Logical NOT), was resolved, as described in Chapter 24, by text in the American Standard which specifically permitted the "stylization" of ^ as ¬, and of ! as |.

The second pair of these, [(left bracket) versus ¢ (cent sign), and] (right bracket) versus ! (exclamation point), was not resolved in the American Standard, but was resolved in EBCDIC by permitting dualization; which is to say, by permitting the left bracket and the right bracket to be provided instead of the cent sign and the exclamation point.

The more worrisome mismatch referred to above resulted from applying the following algorithm:

- Take a graphic in an ISO 7-Bit Code code position.
- Take the card hole pattern corresponding to that ISO 7-Bit Code code position.
- Take that hole pattern and the corresponding EBCDIC code position.
- Take the EBCDIC graphic in that EBCDIC code position.

21.4 THE GERMAN 7-BIT CODE

In applying this algorithm, it is necessary to apply it *not* to the ISO 7-Bit Code but rather to some national variant; the German 7-Bit Code (DIN 66003–1967, Informationsverarbeitung 7-Bit Code) is used here for illustrative purposes. The result is shown in Fig. 21.4.

It should be pointed out that this alphabetic extender mismatch did *not* exist between the American Standard Hollerith Card Code and EBCDIC, as shown in Figure 21.5; however, the slight mismatches referred to earlier in this chapter are seen.

The mismatches between EBCDIC and the Hollerith Card Code in the U.S.A., then, were with respect to specials (albeit in alphabetic

German			EBCDIC	
Graphic	7-bit code position	Hole pattern	Code position	Graphic
#	2/3	8-3	7B	Ä
@	4/0	8-4	5B	Ö
$	2/4	11-8-3	7C	Ü
Ä	5/11	12-8-2	4A	ö
Ö	5/12	0-8-2	E0	Not assigned*
Ü	5/13	11-8-2	5A	ü
ä	7/11	12-0	C0	Not assigned*
ö	7/12	12-11	6A	Not assigned*
ü	7/13	11-0	D0	Not assigned*
"	2/2	8-7	7F	ä

Fig. 21.4 DIN 66003 and EBCDIC mismatch. (Asterisk refers to year 1967.)

U.S.A.			EBCDIC	
Graphic	7-bit code position	Hole pattern	Code position	Graphic
#	2/3	8-3	7B	#
@	4/0	8-4	5B	@
$	2/4	11-8-3	7C	$
[	5/11	12-8-2	4A	¢
\	5/12	0-8-2	E0	Not assigned*
]	5/13	11-8-2	5A	!
{	7/11	12-0	C0	Not assigned*
¦	7/12	12-11	6A	Not assigned*
}	7/13	11-0	D0	Not assigned*
"	2/2	8-7	7F	"

Fig. 21.5 Hollerith and EBCDIC mismatches. (Asterisk refers to year 1967.)

extender positions), but in Germany and also in the four Scandinavian countries the mismatches were with respect to alphabetic extenders themselves.

21.5 SIGNIFICANCE OF MISMATCHES

There is a substantial difference in the *significance* of these two kinds of mismatches. As described earlier, the mismatch in the U.S.A. was rectifiable very easily—simply by providing, on printers or displays, the left bracket and right bracket in place of the cent sign and exclamation point. But the mismatch between alphabetic extenders (in Germany and in the Scandinavian countries) could not be rectified in such a simple way.

To begin to understand the significance of these two different kinds of mismatches, consider the effect on, for example, the IBM 029 keypunch. Let us look first at the mismatch in the U.S.A. To modify the keypunch, two "fixes" would have to be made:

- Provide keytops (and the emphasis here is on the engraved symbols on the tops of the keys) engraved [and] to replace the keytops engraved ¢ and !.
- The mechanism that interprets along the top of a keypunched card is called a "code plate." Provide a new code plate that, for the hole patterns 12-8-2 and 11-8-2, would interpret [and].

Both of these fixes could be made to keypunches in the field and the cost would be moderate.

Before considering the equivalent fixes for an IBM 029 keypunch for Germany, it should be recalled (as was described in Chapter 10) that one of the keytops has no graphic engraved on it, but instead has 0-8-2 engraved. When this key is depressed, the 0-8-2 hole pattern is punched in the card, and no symbol is interpreted on the card; that is to say, for that column of the card, it would appear to a human viewer that, from an interpretation point of view, the Space bar had been depressed. The fixes, then, for the German 029 keypunch would be

- Provide keytops engraved

  Ä Ö Ü # @ $

  to replace, respectively, the keytops engraved

  ö 0-8-2 ü Ä Ö Ü

- Provide a new code plate which would interpret

  Ä Ö Ü # @ $

  for the hole patterns 12-8-2, 0-8-2, 11-8-2, 8-3, 8-4, 11-8-3, respectively.

Both of these fixes could be made in the field, at very close to the same moderate cost of making the fixes in the U.S.A.

If we look at another aspect, output printing from the CPU, and consider the printer to be a chain/train printer, then the fix would be simply a new chain/train, with the necessary substitutions. The cost would be very moderate.

It can be concluded that the cost to modify equipment would be moderate. But the cost to a user to fix existing punched card data files could be horrendous. The user would have to convert *all* card files which contained alphabetic information so that they could be processed correctly by the modified equipment.

In addition, any user programs that are code dependent on the alphabetics (and many users have many such programs) would have to be reprogrammed.

These two fixes, conversion of card files and reprogramming, could prove to be an uneconomical burden.

21.6 THE FRENCH SOLUTION

It might well be asked if ISO/TC97/SC2 did not foresee this economic consequence. In fact, ISO/TC97/SC2 *did* foresee it. One proposal was put forward (by France) which had a "solution."

It was proposed that a footnote be added to the card-code table, pointing at code-table positions 2/3, 2/4, 4/0, 5/11, 5/12, and 5/13:

> Within six positions, presentations of card codes shall be allowed; i.e., the assignment of the six card codes 8-3, 11-8-3, 8-4, 12-8-2, 11-8-2, and 0-8-2 to the six positions 2/3, 2/4, 4/0, 5/11, 5/12, and 5/13 may vary by permutation when required to take into account well-established national usage.

It is to be noted that this footnote would not point at code-table positions 7/11, 7/12, 7/13, The reason was that, at this time, there was little if any usage of small alphabetics, and therefore of small alphabetic extenders, in punched-card applications. In consequence, little if any economic consequence was foreseen for users in respect to small alphabetic extenders.

Figure 21.6 shows an example of how this could be implemented in Country 1, "where no established usage opposes," and in Countries 2 and 3 "to comply with established usage." In Country 2, a single national character is used in both upper and lower case. In Country 3, three national characters are used in both upper and lower case.

This proposal clearly would have met the requirement to comply with established usage *within* a country, but there were two undesirable aspects.

Position in the ISO 7-Bit Table	Card code to be used when no established usage opposes	Character and card code assigned to this character to comply with established usage			
	Country 1	Country 2		Country 3	
2/3	8-3	£	8-3	£	12-8-2
2/4	11-8-3	$	11-8-3	$	0-8-2
4/0	8-4	@	0-8-2	@	11-8-2
5/11	12/8-2	[	12-8-2	Ä	8-3
5/12	0-8-2	Ñ	8-4	Ö	11-8-3
5/13	11-8-2	]	11-8-2	Ü	8-4
7/11	12-0	{	12-0	ä	12-0
7/12	12-11	ñ	12-11	ö	12-11
7/13	11-0	}	11-0	ü	11-0

Fig. 21.6 Permutation of card codes

21.6.1 Undesirable Aspects of the French Solution

A. Consider the graphics @ # and $. Under this proposal each of these graphics would have different hole patterns in different countries, with the following consequences:

- For manufacturers of punched-card equipment, different and incompatible lines of card equipment.
- For the manufacturer, dual or multiple maintenance and distribution of programming decks.
- For the users in different countries, difficult or impossible interchange of card deck.

B. Under this proposal a given ISO 7-bit bit pattern could be associated with different Hollerith hole patterns, and a given Hollerith hole pattern could be associated with different ISO 7-bit bit patterns, in different countries. The consequence of this, for the manufacturer of card-code to/from bit-code equipment, would be different and incompatible lines of equipment between countries.

As reviewed by ISO/TC97/SC2, these two aspects were deemed sufficiently undesirable that the proposal was rejected.

The alphabetic extender problem remains unsolved to this day. Manufacturers of punched-card equipment continue to implement the card codes "of established usage," not the ISO card codes, for the alphabetic extenders. Users continue with the card codes "of established usage." Short of government intervention to force compliance, the ISO Standard card code will evidently not be implemented in Europe with respect to alphabetic extenders.

22 Graphic Subsets for the Government

22.1 A AND H SUBSETS

As described in Chapters 4, 9, and 10, there were various subsets of BCDIC and of EBCDIC. The most popular of these, the A and H sets, manifested themselves on 48-character trains/chains.

BCDIC			EBCDIC		
A to Z		26	A to Z		26
0 to 9		10	0 to 9		10
. , / * − $		6	. , / + * − $ &		8
⌑	)		<	)	
%	(		%	(	
&	+	5	#	=	4
#	=		@	'	
@	'				
A-Set	H-Set		A-Set	H-Set	
	Total	47		Total	48

It is to be noted that the BCDIC subsets shown above are actually 47-graphic sets. The 48th graphic on the chain/train was sometimes ǂ, sometimes − (repeated), sometimes a company logo, and sometimes something else.

In fact, there were quite a number of 48-graphic trains/chains available. Consider, then, the problem of a computing installation that received from some source outside the installation a report to be listed. This report might come to the installation on punched cards or on magnetic tape. If the source installation had prepared the report using an A-train and if the receiver installation used an H-train to list the report, there would be some unintelligibility in the listing, the amount depending on the extent to which the source installation had used special graphics beyond the period, comma, slash, asterisk, minus sign, and dollar sign.

This kind of confusion was compounded where many installations sent data from one to another to be listed and where a variety of 48-graphic trains/chains were used in the installations. An obvious solution to such situations would have been for somebody in authority over all the installations to issue an edict to all installations to always use the same train/chain.

Such a simplistic solution would probably not have worked very well in real life. Installations had particular trains/chains for a particular reason, and users would have resisted the bureaucratic order to change.

22.2 DEPARTMENT OF DEFENSE SOLUTION

During the early 1960s, a different kind of solution was tried in the Department of Defense. Recognizing that 42 graphics—26 alphabetics, 10 numerics, and 6 specials (period, comma, slash, asterisk, minus sign, and dollar sign)—were common to all trains/chains, an edict was issued that only these 42 graphics could be used on reports. (Incidentally, it was exactly this set of 42 characters, together with the Space character, that formed the "hard core 43" described in Chapter 17.

This solution had moderate success. A countervailing factor was that military part numbers used the left parenthesis, the right parenthesis, and the number sign, and part numbers were in much of the report data interchanged between military installations.

22.3 FIPS PUB 15 SOLUTION

In the late 1960s, a Federal Information Processing Standards Publication (FIPS PUB) 15 was approved. FIPS PUB 15 stated that "all applicable equipment ordered on or after the date of this FIPS PUB must be in conformance with this standard . . .".

"Applicable equipment" included printers, display devices, punched-card equipment, and other data processing or communications equipment.

The standard specified three graphic subsets:

- a 16-character graphic subset,
- a 64-character graphic subset,
- a 95-character graphic subset.

The graphic subsets were derived from ASCII, shown in Fig. 22.1.

The 16-graphic subset consisted of the 10 numerics and 6 specials in column 3 of the code table.

	Column	0	1	2	3	4	5	6	7
	Bit Pattern b7	0	0	0	0	1	1	1	1
	b6	0	0	1	1	0	0	1	1
	b5	0	1	0	1	0	1	0	1
Row	b4 b3 b2 b1								
0	0 0 0 0			SP	0	@	P	`	p
1	0 0 0 1			!	1	A	Q	a	q
2	0 0 1 0			"	2	B	R	b	r
3	0 0 1 1			#	3	C	S	c	s
4	0 1 0 0			$	4	D	T	d	t
5	0 1 0 1			%	5	E	U	e	u
6	0 1 1 0			&	6	F	V	f	v
7	0 1 1 1			'	7	G	W	g	w
8	1 0 0 0			(	8	H	X	h	x
9	1 0 0 1			)	9	I	Y	i	y
10	1 0 1 0			*	:	J	Z	j	z
11	1 0 1 1			+	;	K	[	k	{
12	1 1 0 0			,	<	L	\	l	¦
13	1 1 0 1			-	=	M	]	m	}
14	1 1 1 0			.	>	N	^	n	~
15	1 1 1 1			/	?	O	_	o	

Fig. 22.1 ASCII

The 64-graphic subset consisted of the Space character and the 63 graphics in columns 2 through 5 of the code table.

The 95-graphic subset consisted of the Space character and the 94 graphics in columns 2 through 7 of the code table.

The wording of FIPS PUB 15 was such that these *graphic* subsets were to be provided *regardless* of the code of the equipment. Thus, for example, FIPS PUB 15 was applicable not only to ASCII-based equipment, but also to EBCDIC-based equipment, or, indeed, to equipment based on any code whatsoever.

Since this federal standard applied to all equipment entering the federal inventory, manufacturers had to supply these graphic subsets if they wished to market to the federal government. FIPS PUB 15 had, therefore, considerable clout.

22.4 FIPS PUB 15 TRADE-OFF

The federal government had made an interesting trade off here. As described in Chapter 10, the nominal printing speeds (LPM, lines printed per minute) for train/chain printers, depending on the number of graphics in the repeated sets of the train/chain, were as follows:

Number of graphics	Repeated sets	Nominal printing speed
40	6	1250 LPM
48	5	1100 LPM
60	4	950 LPM
120	2	570 LPM
240	1	300 LPM

The number 64 does not divide evenly into 240, so for a 64 (really 63 plus Space) graphic set, the "preferred" graphic approach would have to be used. For the 64-graphic subset, the nominal printing speed would be approximately 940 LPM. Assume, for the purposes of discussion, that the train/chain printer would run without stopping for 8 hours. Then the 48-graphic printer would produce 528,000 lines of print, whereas the 64-graphic printer would produce approximately 451,000 lines of print; that is, approximately 77,000 lines of print less during 8 hours.

The federal government was making a trade-off between productivity of lines printed and intelligibility of interchanged data as printed.

FIPS PUB 15 did recognize that the bulk of printing in an installation would probably not come from data interchanged from another installation, and that productivity of printed lines for such noninterchanged data should not be reduced. FIPS PUB states:

> Printers of the "chain" or "train" or other replaceable symbol technology must be provided with the ability to conform to one of the subsets herein but may also be provided with optional subsets having a different number of characters than those specified herein in order to increase either the printer repertoire of symbols or the printer speed in local use.

That is to say, federal departments and agencies could order 48-character train/chain printers, as long as a 64-character train/chain could be mounted if needed. And that is an easy and simple thing to do with train/chain printers.

23
Which ASCII?

23.1 ASCII–1963

A national standard, even when approved, may nevertheless not remain fixed and unchanging. When ASCII became an approved American standard in 1963, it was not complete. As may be seen from Fig. 23.1, 28 code positions in columns 6 and 7 were not filled. In addition, the control character in position 0/8 was defined very broadly as a "format effector" (in contrast with the other "format effectors" Horizontal Tab, Line Feed, Vertical Tab, Form Feed, and Carriage Return that were defined very specifically), and the control characters in positions 0/8 through 015 were broadly called "separators."

23.2 ASCII–1965

As can be seen from Fig. 23.2, ASCII in 1965 was changed from ASCII–1963. Some of these changes, such as the addition of small alphabetics in columns 6 and 7, the change of name without change of essential meaning of some control characters—Start of Message (SOM) changed to Start of Header (SOH); End of Message (EOM) changed to End of Text (ETX)—can be described as evolutionary, in that they did not change what existed before, but simply added to it.

Other changes, such as moving Escape (ESC) from position 7/14 to position 1/11; "... Are you?" (RU) in position 1/6 being replaced by Acknowledge (ACK); changing \ ↑ and ← to ˜ ˆ and _ in positions 5/12, 5/14, and 5/15, respectively, can be described as revolutionary, in that they *did* change what existed before.

	Column	0	1	2	3	4	5	6	7
	Bit Pattern b7	0	0	0	0	1	1	1	1
	b6	0	0	1	1	0	0	1	1
	b5	0	1	0	1	0	1	0	1
Row	b4 b3 b2 b1								
0	0 0 0 0	NULL	DC0	b̸	0	@	P		
1	0 0 0 1	SOM	DC1	!	1	A	Q		
2	0 0 1 0	EOA	DC2	"	2	B	R		
3	0 0 1 1	EOM	DC3	#	3	C	S		
4	0 1 0 0	EOT	DC4	$	4	D	T		
5	0 1 0 1	WRU	ERR	%	5	E	U		
6	0 1 1 0	RU	SYNC	&	6	F	V		
7	0 1 1 1	BELL	LEM	'	7	G	W		
8	1 0 0 0	FE0	S0	(	8	H	X		
9	1 0 0 1	HT/SK	S1	)	9	I	Y		
10	1 0 1 0	LF	S2	*	:	J	Z		
11	1 0 1 1	VTAB	S3	+	;	K	[		
12	1 1 0 0	FF	S4	,	<	L	\		ACK
13	1 1 0 1	CR	S5	-	=	M	]		(2)
14	1 1 1 0	SO	S6	.	>	N	↑		ESC
15	1 1 1 1	SI	S7	/	?	O	←		DEL

2

Fig. 23.1 ASCII–1963

As described in Chapter 16, a proposed American national standard which specified a card code radically different from the well-established Hollerith Card Code was resolutely voted down at X3 by users who clearly foresaw the substantial economic impact of such a revolutionary change. As described in Chapter 21, a draft ISO standard that specified a change to well-established card hole patterns for alphabetic extenders was approved, but has not been implemented—again because of the economic impact on users.

	Column	0	1	2	3	4	5	6	7
	Bit Pattern b7	0	0	0	0	1	1	1	1
	b6	0	0	1	1	0	0	1	1
	b5	0	1	0	1	0	1	0	1
Row	b4 b3 b2 b1								
0	0 0 0 0	NUL	DLE	SP	0	`	P	@	p
1	0 0 0 1	SOH	DC1	!	1	A	Q	a	q
2	0 0 1 0	STX	DC2	"	2	B	R	b	r
3	0 0 1 1	ETX	DC3	#	3	C	S	c	s
4	0 1 0 0	EOT	DC4	$	4	D	T	d	t
5	0 1 0 1	ENQ	NAK	%	5	E	U	e	u
6	0 1 1 0	ACK	SYN	&	6	F	V	f	v
7	0 1 1 1	BEL	ETB	'	7	G	W	g	w
8	1 0 0 0	BS	CAN	(	8	H	X	h	x
9	1 0 0 1	HT	EM	)	9	I	Y	i	y
10	1 0 1 0	LF	SS	*	:	J	Z	j	z
11	1 0 1 1	VT	ESC	+	;	K	[	k	{
12	1 1 0 0	FF	FS	,	<	L	~	l	¬
13	1 1 0 1	CR	GS	-	=	M	]	m	}
14	1 1 1 0	SO	RS	.	>	N	^	n	|
15	1 1 1 1	SI	US	/	?	O	_	o	DEL

Fig. 23.2 ASCII–1965

23.3 ECONOMIC IMPACTS

It is reasonable to inquire, therefore, whether changes to ASCII, such as those described above, have had similar economic impacts. Two examples, to be described below, will give insight on this question. (Both these examples are drawn from the author's experience.) However, the following explanation must preface the examples.

Reference was made above, and will be made below, to ASCII–1965. Some rather unusual circumstances surround ASCII–1965. ASCII–1963 had been approved and published in June 1963. Since that time, the

American subcommittee X3.2 (now X3L2) had been working to complete the code table. There was considerable interplay between the American subcommittee and the ISO subcommittee ISO/TC97/SC2 (which will be referred to in the remainder of this chapter as SC2). SC2 was working on the ISO Draft Proposal for a 7-Bit Code. The two subcommittees, X3.2 and SC2, were striving to achieve compatible 7-bit codes. Some graphics, and some controls, were under contention and controversy.

By January of 1965, X3.2 had completed its "revision" of ASCII and forwarded it to X3 for further processing. During the X3 balloting, a controversy arose with respect to graphics for Logical OR and Logical NOT (see Chapter 24). This controversy delayed further processing of the revised ASCII at that time.

Meanwhile X3.2 had received information that, at the upcoming April 1966 meeting of SC2, changes would be made to the ISO Draft Proposal. These changes would create incompatibilities between the ISO 7-Bit Code and ASCII.

Therefore, X3.2 requested X3 to request ASA to delay publication of the revised ASCII until after the SC2 meeting. This request was granted and ASCII–1965, although approved by ASA, was, in fact, never published or distributed.

23.4 THE 2260 DISPLAY STATION

IBM's first ASCII transmission product, the 2260 Display Station and 2848 Display Control, announced in 1965, was based on ASCII–1965. As can be seen from Fig. 23.3, not every character of ASCII–1965 was implemented. Eight control characters (sufficient for the data communications protocols of the day), the Space character, and 59 graphic characters were implemented. These 59 graphics, shown below, were the graphic set of the programming language PL/I:

26 capital letters	A to Z
3 alphabetic extenders	# $ @
10 numerics	0 to 9
20 syntactics	( ) < = > . , : ; ?
	¬	% & ' / * + - _

The symbols shown below the code table of Fig. 23.3 had the following meanings:

[1]Displays on the 2260 Display as ▬ (End of Message symbol). Prints on 1053 Printer as ! (exclamation mark).

[2]Displays on the 2260 Display as ■ (Check symbol). Prints on the 1053 Printer as " (quotation marks).

Row	Column	0	1	2	3	4	5	6	7
	Bit Pattern b7	0	0	0	0	1	1	1	1
	b6	0	0	1	1	0	0	1	1
	b5	0	1	0	1	0	1	0	1
Row	b4 b3 b2 b1								
0	0 0 0 0			SP	0		P	@	
1	0 0 0 1	SOH		[1]	1	A	Q		
2	0 0 1 0	STX		[2]	2	B	R		
3	0 0 1 1	ETX		#	3	C	S		
4	0 1 0 0	EOT		$	4	D	T		
5	0 1 0 1		NAK	%	5	E	U		
6	0 1 1 0	ACK		&	6	F	V		
7	0 1 1 1			'	7	G	W		
8	1 0 0 0		CAN	(	8	H	X		
9	1 0 0 1			)	9	I	Y		
10	1 0 1 0	LF [3]		*	:	J	Z		
11	1 0 1 1			+	;	K			
12	1 1 0 0			,	<	L			¬
13	1 1 0 1			-	=	M	[4]		
14	1 1 1 0			.	>	N			|
15	1 1 1 1			/	?	O	_		

	Display	1053 Printer
1	▬	!
2	■	"
3	◢	New Line
4	▶	¢

Fig. 23.3 2260–1965

[3]Displays on the 2260 Display as ◢ (New Line symbol). Causes the 1053 Printer to execute a New Line function.

[4]Displays on the 2260 Display as ▶ (Start Manual Input symbol). Prints on the 1053 Printer as ¢ (cent sign).

23.5 THE 1053 PRINTER

It is to be noted, then, that 62 graphics were printable on the 1053 Printer. The 1053 Printer was a typewriter-based product with a printing capability of 88 graphics. However, the capital letters were duplicated, that is, they printed whether the typewriter was in upper-case or lower-case shift. Accordingly, $88-26=62$.

23.6 ASCII–1967

The standards committee X3.2 continued with its work to arrive at an agreed-upon ASCII, and ASCII–1967 was the result.

23.7 ASCII–1965 VERSUS ASCII–1967

A comparison of ASCII–1965 (Fig. 23.2) and ASCII–1967 (Fig. 23.4) shows changes in 6 code positions:

Code position	ASCII–1965	ASCII–1967
1/10	SS	SUB
4/0	`	@
5/12	~	\
6/0	@	`
7/12	¬	|
7/14	|	~

23.8 THE 2265 DISPLAY STATION

The changes in code positions 1/10 and 5/12 do not affect the discussion in this chapter, since they had not been implemented on the 2260. But a follow-on product to the 2260, the 2265 Display Station and 2845 Display Control, was being developed, and the pertinent question was

> Should the 2265 be compatible with the 2260 and hence in nonconformance to ASCII–1967, or should the 2265 be in conformance to ASCII–1967 and hence incompatible with the 2260?

Since the 2265 was planned to replace installed 2600's, compatibility was a requirement. The question then became "Which was more important, compatibility or conformance?" In the end, it was decided that compatibility was more inportant. The nonconformance to ASCII–1967 (but conformance to ASCII–1965) was carefully explained in the 2265 manuals.

A dilemma of a different kind was posed because of an incorrect guess on the ultimate shape of a draft standard. Before discussing this

	Column	0	1	2	3	4	5	6	7
	Bit Pattern b7	0	0	0	0	1	1	1	1
	b6	0	0	1	1	0	0	1	1
	b5	0	1	0	1	0	1	0	1
Row	b4 b3 b2 b1								
0	0 0 0 0	NUL	DLE	SP	0	@	P	`	p
1	0 0 0 1	SOH	DC1	!	1	A	Q	a	q
2	0 0 1 0	STX	DC2	"	2	B	R	b	r
3	0 0 1 1	ETX	DC3	#	3	C	S	c	s
4	0 1 0 0	EOT	DC4	$	4	D	T	d	t
5	0 1 0 1	ENQ	NAK	%	5	E	U	e	u
6	0 1 1 0	ACK	SYN	&	6	F	V	f	v
7	0 1 1 1	BEL	ETB	'	7	G	W	g	w
8	1 0 0 0	BS	CAN	(	8	H	X	h	x
9	1 0 0 1	HT	EM	)	9	I	Y	i	y
10	1 0 1 0	LF	SUB	*	:	J	Z	j	z
11	1 0 1 1	VT	ESC	+	;	K	[	k	{
12	1 1 0 0	FF	FS	,	<	L	\	l	¦
13	1 1 0 1	CR	GS	-	=	M	]	m	}
14	1 1 1 0	SO	RS	.	>	N	^	n	~
15	1 1 1 1	SI	US	/	?	O	_	o	DEL

Fig. 23.4 ASCII–1967

problem, it is necessary to understand the operation of Pack/Unpack and Decimal Arithmetic instructions on the IBM System/360.

23.9 SYSTEM/360 DECIMAL ARITHMETIC

In the System/360, decimal arithmetic is performed with operands in the packed format. What does this mean?

In what is called the "zoned format" for numerics, each numeric occupies an 8-bit byte, with the high-order 4 bits called the "zone" and the low-order 4 bits being the familiar BCD representation for numerics. Referring to an EBCDIC code chart (Fig. 23.5), we see that the representation of the numerics is as follows:

Numeric	Zone	BCD numeric
0	1111	0000
1	1111	0001
2	1111	0010
3	1111	0011
4	1111	0100
5	1111	0101
6	1111	0110
7	1111	0111
8	1111	1000
9	1111	1001

<table>
<tr><th></th><th>Column</th><th>0</th><th>1</th><th>2</th><th>3</th><th>4</th><th>5</th><th>6</th><th>7</th><th>8</th><th>9</th><th>A</th><th>B</th><th>C</th><th>D</th><th>E</th><th>F</th></tr>
<tr><th></th><th>Bit Pat. →</th><th colspan="4">00</th><th colspan="4">01</th><th colspan="4">10</th><th colspan="4">11</th></tr>
<tr><th>Row</th><th>↓</th><th>00</th><th>01</th><th>10</th><th>11</th><th>00</th><th>01</th><th>10</th><th>11</th><th>00</th><th>01</th><th>10</th><th>11</th><th>00</th><th>01</th><th>10</th><th>11</th></tr>
<tr><td>0</td><td>0000</td><td></td><td></td><td></td><td></td><td>SP</td><td>&</td><td>-</td><td></td><td></td><td></td><td></td><td></td><td></td><td></td><td></td><td>0</td></tr>
<tr><td>1</td><td>0001</td><td></td><td></td><td></td><td></td><td></td><td></td><td>/</td><td></td><td></td><td></td><td></td><td></td><td>A</td><td>J</td><td></td><td>1</td></tr>
<tr><td>2</td><td>0010</td><td></td><td></td><td></td><td></td><td></td><td></td><td></td><td></td><td></td><td></td><td></td><td></td><td>B</td><td>K</td><td>S</td><td>2</td></tr>
<tr><td>3</td><td>0011</td><td></td><td></td><td></td><td></td><td></td><td></td><td></td><td></td><td></td><td></td><td></td><td></td><td>C</td><td>L</td><td>T</td><td>3</td></tr>
<tr><td>4</td><td>0100</td><td></td><td></td><td></td><td></td><td></td><td></td><td></td><td></td><td></td><td></td><td></td><td></td><td>D</td><td>M</td><td>U</td><td>4</td></tr>
<tr><td>5</td><td>0101</td><td></td><td></td><td></td><td></td><td></td><td></td><td></td><td></td><td></td><td></td><td></td><td></td><td>E</td><td>N</td><td>V</td><td>5</td></tr>
<tr><td>6</td><td>0110</td><td></td><td></td><td></td><td></td><td></td><td></td><td></td><td></td><td></td><td></td><td></td><td></td><td>F</td><td>O</td><td>W</td><td>6</td></tr>
<tr><td>7</td><td>0111</td><td></td><td></td><td></td><td></td><td></td><td></td><td></td><td></td><td></td><td></td><td></td><td></td><td>G</td><td>P</td><td>X</td><td>7</td></tr>
<tr><td>8</td><td>1000</td><td></td><td></td><td></td><td></td><td></td><td></td><td></td><td></td><td></td><td></td><td></td><td></td><td>H</td><td>Q</td><td>Y</td><td>8</td></tr>
<tr><td>9</td><td>1001</td><td></td><td></td><td></td><td></td><td></td><td></td><td></td><td>`</td><td></td><td></td><td></td><td></td><td>I</td><td>R</td><td>Z</td><td>9</td></tr>
<tr><td>A</td><td>1010</td><td></td><td></td><td></td><td></td><td>¢</td><td>!</td><td>¦</td><td>:</td><td></td><td></td><td></td><td></td><td></td><td></td><td></td><td></td></tr>
<tr><td>B</td><td>1011</td><td></td><td></td><td></td><td></td><td>.</td><td>$</td><td>,</td><td>#</td><td></td><td></td><td></td><td></td><td></td><td></td><td></td><td></td></tr>
<tr><td>C</td><td>1100</td><td></td><td></td><td></td><td></td><td><</td><td>*</td><td>%</td><td>@</td><td></td><td></td><td></td><td></td><td></td><td></td><td></td><td></td></tr>
<tr><td>D</td><td>1101</td><td></td><td></td><td></td><td></td><td>(</td><td>)</td><td>_</td><td>'</td><td></td><td></td><td></td><td></td><td></td><td></td><td></td><td></td></tr>
<tr><td>E</td><td>1110</td><td></td><td></td><td></td><td></td><td>+</td><td>;</td><td>></td><td>=</td><td></td><td></td><td></td><td></td><td></td><td></td><td></td><td></td></tr>
<tr><td>F</td><td>1111</td><td></td><td></td><td></td><td></td><td>|</td><td>¬</td><td>?</td><td>"</td><td></td><td></td><td></td><td></td><td></td><td></td><td></td><td></td></tr>
</table>

Fig. 23.5 EBCDIC

A string of zoned numerics, then, occupies a string of 8-bit bytes as follows:

Zone	Digit	Zone	Digit	------------	Zone	Digit	Sign	Digit

Note that in the rightmost (low-order) byte, the high-order 4 bits are not a "zone" but are a "sign." More will be said about this later.

When the Pack instruction is executed on an operand in the zoned format above, a "packed decimal number" results as shown below:

Digit	Digit	Digit	--------------------	Digit	Digit	Digit	Sign

The high-order 4 bits of the low-order 8-bit byte from the zoned format now occupy the low-order 4 bits of the packed decimal format, all zones have been removed, and the 4 bits of the decimal numbers are now "packed" together from right to left. If necessary, four zero-bits are filled in to the extreme left 4 bits of the resultant high-order 8-bit byte.

23.10 PACKED DECIMALS

This, then, is a packed decimal number, and on such numbers the System/360 performs decimal arithmetic (which includes comparison) instructions. As described elsewhere in this book (Chapter 19), the concept of signed numerics was incorporated into EBCDIC. A number with a zone of 1100 was considered a positive numeric, a number with a zone of 1101 was considered a negative numeric, and a number with a zone of 1111 was considered an absolute numeric. The zones over the low-order numeric in a string of zoned numerics, when that string was packed, became the sign of the packed numeric string. The arithmetic circuitry of the CPU would recognize these signs during execution of decimal arithmetic instructions, and would also generate the appropriate sign for the result.

23.11 USASCII-8

In order to appreciate the significance of this zone-to-sign relationship, consider another CPU code implemented on the System/360, called USASCII-8 (Fig. 23.6).

USASCII-8 was an 8-bit representation of ASCII–1963 (Fig. 23.1). More will be said later about why this particular 8-bit representation was chosen. For now, however, consider the zone-to-sign relationship. In Fig. 23.6, the numerics are in column hex 5, with high-order zone bits of 0101; therefore, 0101 was chosen as the zone for absolute numerics.

For reasons which will be described a little later, column hex A, with high-order zone bits of 1010, and column hex B, with high-order zone

Row	Column	0	1	2	3	4	5	6	7	8	9	A	B	C	D	E	F
	Bit Pat.	00				01				10				11			
		00	01	10	11	00	01	10	11	00	01	10	11	00	01	10	11
0	0000	NUL	DC0			SP	0					@	P				p
1	0001	SOM	DC1			!	1					A	Q			a	q
2	0010	EOA	DC2			"	2					B	R			b	r
3	0011	EOM	DC3			#	3					C	S			c	s
4	0100	EOT	DC4			$	4					D	T			d	t
5	0101	WRU	ERR			%	5					E	U			e	u
6	0110	RU	SYN			&	6					F	V			f	v
7	0111	BEL	LEM			'	7					G	W			g	w
8	1000	BS	S0			(	8					H	X			h	x
9	1001	HT	S1			)	9					I	Y			i	y
A	1010	LF	S2			*	:					J	Z			j	z
B	1011	VT	S3			+	;					K	[			k	
C	1100	FF	S4			,	<					L	\			l	
D	1101	CR	S5			-	=					M	]			m	
E	1110	SO	S6			.	>					N	↑			n	ESC
F	1111	SI	S7			/	?					O	←			o	DEL

Fig. 23.6 USASCII-8, 1964

bits of 1011, were chosen as zones for positive and negative numerics respectively. We have then, the following:

	EBCDIC	USASCII-8
Absolute	1111	0101
Positive	1100	1010
Numeric	1101	1011

The Pack and Unpack instructions operated independently of whether the numeric data was EBCDIC or USASCII-8. But the decimal arithmetic instructions had to take the differences in signs into account. This was controlled by bit 12 in what was called the Program Status Word (PSW). If bit 12 in the PSW was set to zero, the decimal arithmetic instructions assumed EBCDIC signs on input to the arithmetic circuits, and generated EBCDIC signs on output from the arithmetic circuits. If bit 12 in the PSW was set to one, the decimal arithmetic instructions assumed and generated USASCII-8 signs. (Bit 12 of the PSW was normally set at zero (EBCDIC). Setting it to one (USASCII-8) was under control of the System/360 Operating System, but a discussion of that is beyond the scope of this book.)

In any event, the particular bit patterns chosen for absolute, positive, and negative numerics in USASCII-8 established the code structure of USASCII-8. More precisely speaking, the code structure of USASCII-8 established the bit patterns for absolute, positive, and negative numerics. Why was that code structure chosen? This question is answered in some detail in Chapter 20, and need not be discussed further here.

It is clear why column hex 5, which contained the numerics, was chosen for the zone for absolute numerics. But why columns hex A and B for positive and negative numerics?

23.12 DECIMAL ASCII

As described in Chapter 16, a new card code had been proposed to the standards committees in November 1963 that came to be called Decimal ASCII. This card code, which was initially accepted by the ANSI, ECMA, and ISO standards committees, was planned to be the card code for the System/360 when functioning in the USASCII-8 mode.

In those days, the input/output card reader/punch was considered to be a vital part of a computing system; a keypunch to prepare input card data was considered as equally essential. One assumption was made. The practice, widespread in Hollerith punched card applications, of over-punching the units position of a numeric field with a 12-punch or 11-punch to indicate a positive or negative numeric field would be carried over to Decimal ASCII punched card applications.

Since, in the Decimal ASCII card code, the 12-punch was associated with the alphabetics A through I, and the 11-punch with alphabetics P through Y, the association of these card columns in USASCII-8 with positive and negative numerics was the natural choice.

23.13 COMPILERS

Another part of the computing system which may be, and usually is, dependent on the CPU code is the programming system.

For example, when a FORTRAN compiler is scanning FORTRAN statements, it "looks" for left and right parentheses, which enclose FORTRAN expressions. Actually, of course, it looks for the bit patterns which represent left and right parentheses; they would be 01001101 and 01011101, respectively, for EBCDIC and 01001000 and 01001001, respectively, for USASCII-8.

All compilers are similarly code dependent for all the bit patterns for which they "look."

In the spring of 1964, work was underway in IBM designing the Decimal ASCII keypunch and input/output card reader/punch. Also, programmers had been instructed to identify all code-dependent parts of their programs so that programming systems dependent on USASCII-8 could be developed in due course.

In June of 1964, at the meeting of X3.2.3 (the task group responsible for developing the punched card code standard), strong words against Decimal ASCII and for Hollerith were spoken (by the UNIVAC representative).

A review of the merits of Decimal ASCII versus Hollerith was commenced vis-a-vis customers. It was concluded that, in general, customers would *not* accept Decimal ASCII. (Ultimately, as demonstrated by the rejection of the draft Decimal ASCII standard by users at the X3 level, this conclusion turned out to be correct.) Work on the Decimal ASCII card equipment was halted, and never resumed.

As explained in Chapter 16, the main concept behind the Decimal ASCII card code was a simple translation to ASCII. It is quite feasible to translate from the Hollerith card code to ASCII and hence also to USASCII-8, but the translation is very complex and the translation hardware would, in those days, have been costly. Also, signed numerics would present a problem. For example, in the Hollerith card code, the 11-2 hole pattern represents K, and so it should be translated, for USASCII-8, to hex position AA (see Fig. 23.6). But if the 11-2 hole pattern represents −2 as a signed numeric, it should be translated to hex position B2. And there is no way to tell, just from the hole pattern, whether it represents K or −2.

A more serious problem arose. The 8-bit representation of ASCII known as USASCII-8 had been proposed to X3.2 and, apparently, accepted. But in mid-1965, this representation was opposed in X3.2, and eventually another representation was adopted (see Chapter 20).

Further work to support USASCII-8, therefore, was not done. The CPU hardware described above for decimal arithmetic was provided in every model of the System/360, but without programming systems support. The guess on the 8-bit representation of ASCII had turned out to be incorrect.

24 Logical OR, Logical NOT

24.1 ASCII–1963

When ASCII became an approved American National Standard in 1963, it was not complete. There were 28 code positions in columns 6 and 7 that had no assigned meaning. There was some controversy on whether small alphabetics or additional control characters should be assigned to these positions. This controversy was resolved when, at the 1963 October meeting of ISO/TC97/SC2, it was decided to assign small alphabetics in the ISO 7-Bit Code then being developed.

24.2 ASCII–1965

By January 1965, X3.2 had completed work on the Proposed Revised ASCII, which was compatible with the ISO 7-Bit Code. The code table is shown in Fig. 24.1.

24.3 PL/I

In this code table, a problem was perceived with respect to PL/I, the new programming language which had been announced with the System/360 in April 1964. The graphics of PL/I were of five kinds:

1	space	
10	numerics	0 to 9
26	alphabetics	A to Z
3	alphabetic extenders	# $ @
20	syntactics	/ * + − = &	¬ < > ' , . ; : ( ) % _ ?

	Column	0	1	2	3	4	5	6	7
	Bit Pattern b7	0	0	0	0	1	1	1	1
	b6	0	0	1	1	0	0	1	1
	b5	0	1	0	1	0	1	0	1
Row	b4 b3 b2 b1								
0	0 0 0 0	NUL	DLE	SP	0	`	P	@	p
1	0 0 0 1	SOH	DC1	!	1	A	Q	a	q
2	0 0 1 0	STX	DC2	"	2	B	R	b	r
3	0 0 1 1	ETX	DC3	#	3	C	S	c	s
4	0 1 0 0	EOT	DC4	$	4	D	T	d	t
5	0 1 0 1	ENQ	NAK	%	5	E	U	e	u
6	0 1 1 0	ACK	SYN	&	6	F	V	f	v
7	0 1 1 1	BEL	ETB	'	7	G	W	g	w
8	1 0 0 0	BS	CAN	(	8	H	X	h	x
9	1 0 0 1	HT	EM	)	9	I	Y	i	y
10	1 0 1 0	LF	SS	*	:	J	Z	j	z
11	1 0 1 1	VT	ESC	+	;	K	[	k	{
12	1 1 0 0	FF	FS	,	<	L	~	l	¬
13	1 1 0 1	CR	GS	-	=	M	]	m	}
14	1 1 1 0	SO	RS	.	>	N	^	n	|
15	1 1 1 1	SI	US	/	?	O	_	o	DEL

Fig. 24.1 ASCII–1965

The term *syntactic* meant that the programming language would assign some specific function to a graphic when that graphic was used in a programming language source statement. For example,

* would mean multiplication,
/ would mean division,
& would mean Logical AND,
| would mean Logical OR,
¬ would mean Logical NOT.

24.4 THE PROBLEM

The problem that was perceived was based on the assumption that, some day, PL/I would become a candidate for international standardization. (That assumption turned out to be correct.) At that time, among the many aspects of PL/I that would be reviewed would be the character set. Presumably, the committee would agree on the 29 alphabetic, 10 numeric, and 20 syntactic *functions*, but they might well debate the actual graphics to be associated with the syntactic functions. For example, is ¬ the appropriate graphic to be associated with the syntactic function of "Logical NOT"?

A further assumption was made, which was that the committee would set two ground rules with respect to the graphic character set of PL/I.

24.5 GROUND RULES

Ground rule 1

All graphics for the PL/I character set would be chosen from the set of 63 graphics in the center four columns of the ISO 7-Bit Code.

Ground rule 2

The graphics associated with the syntactic of PL/I should not be associated with a code position reserved in the ISO 7-Bit Code for alphabetic extenders.

The reasoning behind these ground rules was as follows. For Ground rule 1, it was the judgment of many people at that time that most upcoming printing and display devices would have a repertoire of 63 graphics and Space. The 63 graphics, in fact, would be those in columns 2, 3, 4, and 5 (the center four columns) of the 7-Bit Code. For Ground rule 2, although alphabetic extender code positions might have graphics like [ ] and \ in English-speaking countries, in Germany and the four Scandinavian countries, alphabetics would indeed be in those positions. Therefore, no matter how appealing a graphic in an alphabetic extender position of the code for English-speaking countries might be, it could not be used as a syntactic for any programming language.

As can be seen from Fig. 24.1, three of the PL/I graphics, @ | and ¬, were not in the center four columns (Ground rule 1), and also ¬ and |, PL/I syntactics, were in alphabetic extender positions (Ground rule 2).

The reason for @ being in code position 6/0 is interesting. It was forecast that, in the French national variant of the ISO 7-Bit Code, @ would be replaced by à. Since à is an accented small letter, it should be in columns 6 or 7 where the other small alphabetics were positioned. With the U.S.A. requesting that @ be in code position 4/0, and with France requesting that it be in 6/0, it actually moved back and forth at successive meetings of ISO/TC97/SC2. Ultimately, it was agreed to position it in 4/0, and thus this part of the PL/I graphic set problem resolved itself satisfactorily.

The problem with | (Logical OR) and ¬ (Logical NOT) remained. User groups SHARE, GUIDE, and COMMON, becoming aware of this problem, became concerned. Letters were written from various companies in these user groups to the Chairman of X3 requesting that the problem be solved. Representatives from these user groups attended X3.2 meetings and X3 meetings to lobby for their requirement.

	Column	0	1	2	3	4	5	6	7
	Bit Pattern b7 b6 b5	0 0 0	0 0 1	0 1 0	0 1 1	1 0 0	1 0 1	1 1 0	1 1 1
Row	b4 b3 b2 b1								
0	0 0 0 0	NUL	DLE	SP	0	@	P	`	p
1	0 0 0 1	SOH	DC1	!	1	A	Q	a	q
2	0 0 1 0	STX	DC2	"	2	B	R	b	r
3	0 0 1 1	ETX	DC3	#	3	C	S	c	s
4	0 1 0 0	EOT	DC4	$	4	D	T	d	t
5	0 1 0 1	ENQ	NAK	%	5	E	U	e	u
6	0 1 1 0	ACK	SYN	&	6	F	V	f	v
7	0 1 1 1	BEL	ETB	'	7	G	W	g	w
8	1 0 0 0	BS	CAN	(	8	H	X	h	x
9	1 0 0 1	HT	EM	)	9	I	Y	i	y
10	1 0 1 0	LF	SUB	*	:	J	Z	j	z
11	1 0 1 1	VT	ESC	+	;	K	[	k	{
12	1 1 0 0	FF	FS	,	<	L	\	l	|
13	1 1 0 1	CR	GS	-	=	M	]	m	}
14	1 1 1 0	SO	RS	.	>	N	^	n	~
15	1 1 1 1	SI	US	/	?	O	_	o	DEL

Fig. 24.2 ISO 7-Bit Code

Various proposals were made to solve the problem. For example, it was proposed that | and ¬ replace ! and ^, respectively, in the code table. All such proposals to change the set of graphics in the ASCII code table were rejected by a majority of the X3.2 members. X3.2 completed its work on the draft Proposed Revised ASCII and forwarded it to X3.

On the X3 letter ballot, the Joint Users Group (which included SHARE and COMMON among its members) voted no. A sufficient majority of X3 affirmative votes was received, however, and the draft standard was forwarded to ASA. In December 1965, the Information Processing Standards Board of ASA approved the Revised ASCII.

In January 1966, X3.2 received information that changes would be made to the 4th Draft ISO Proposal for (6 and) 7-Bit Codes. These changes would not speak to the Logical OR/Logical NOT problem; they would be with respect to some control characters. The consequence would be that ASCII would then be incompatible to the ISO 7-Bit Code. Therefore, X3.2 requested X3 to request ASA to delay publication of the revised ASCII until after the April 1966 meeting of ISO/TC97/SC2. This request was granted. In fact, ASCII–1965, although approved by ASA, never was published.

Changes were indeed made to the ISO 7-Bit Code at the April 1966 meeting. The SS (Start of Special) character in position 1/10 was replaced by SUB (Substitute); @ and ` flip-flopped again, @ ending up in 4/0 and ` in 6/0; ? replaced ˜ in position 5/12; | was moved from 7/14 to 7/12; and ¯ (overline, or tilde) was placed in 7/14. This was the final version of the ISO 7-Bit Code. It became an approved ISO Recommendation, R646, in 1967 (see Fig. 24.2).

The Logical OR/Logical NOT problem was discussed at this meeting and a solution (of sorts) was set into the document. (This solution will be described later in this chapter.)

24.6 REVISED ASCII

ASCII itself now had to be revised to bring it into line with the changes made to the ISO 7-Bit Code. Once again, a draft Proposed Revised ASCII was under preparation by X3.2. The Logical OR/Logical NOT problem continued to concern X3.2.

24.7 THE SOLUTION FOR ASCII

Ultimately a solution was found. The solution has two parts. The first part is the inclusion of the following text in Section 6.4 of the ASCII standard:

> 6.4
> No specific meaning is prescribed for any of the graphics in the code table except that which is understood by the users. Furthermore, this

> standard does not specify a type style for the printing or display of the various graphic characters. In specific applications, it may be desirable to employ distinctive styling of individual graphics to facilitate their use for specific purposes as, for example, to stylize the graphics in code positions 2/1 and 5/14 into those frequently associated with Logical OR | and Logical NOT ¬, respectively.

This text was taken to mean that manufacturers could, if they wished, substitute the graphics | for !, and ¬ for ^.

24.8 THE SOLUTION FOR THE ISO 7-BIT CODE

It should be pointed out that positioning the Logical NOT graphic in position 5/14 does not strictly satisfy Ground rule 2, since position 5/14 is an alphabetic extender position. However, the 10 positions designated as alphabetic extender positions in the ISO 7-Bit Code are viewed as being divided into "primary" and "secondary" by virtue of the footnotes which speak to them. For the 7 "primary" positions, 4/0, 5/11, 5/12, 7/11, 7/12, 7/13, the footnote reads (in part):

> Reserved for National Use. These positions are primarily intended for alphabetic extensions. If they are not required for that purpose, they may be used for symbols . . .

By contrast, for the 3 "secondary" positions, 5/14, 6/0, and 7/14, the footnote reads (in part):

> Positions 5/14, 6/0, and 7/14 . . . are normally provided for the diacritical signs "circumflex," "grave accent," and "overline." However, these positions may be used for other graphical symbols when it is necessary to have 8, 9, or 10 positions for national use.

It was reckoned that, in Germany and in the four Scandinavian countries where the Logical OR/Logical NOT problems would exist, these three "secondary" positions would not be needed for national use, so Ground rule 2 would not really apply to these code positions.

The second part of the solution had to do with the graphic | (Vertical Line) in code position 7/12 (Fig. 24.2). Suppose a manufacturer implemented | (Logical OR) in code position 2/1, as permitted by Section 6.4 of the ASCII standard. These two graphics, as printed or displayed, would be indistinguishable to the human eye. The solution to this problem was to change the actual graphic in position 7/12 slightly. It became ¦ (still called Vertical Line) but now clearly distinguishable from | (Logical OR).

As an interesting sidelight, IBM had previously called the graphic | Logical OR or Vertical Bar in its internal EBCDIC standard and in reference manuals. From this point in time, IBM called it Logical OR, to avoid confusion in nomenclature.

The "solution" previously mentioned for the ISO 7-Bit Code was the following text in Section 4.3 of ISO R–646:

> 4.3 Interpretation of graphics
> The meaning of the graphics is not defined by this ISO Recommendation. It will be necessary to reach agreement on the meaning and this will depend upon the particular application except in cases where other ISO Recommendations already exist. However no interpretation may be chosen which is contradictory to the customary meaning. A graphical symbol can have more than one meaning, e.g., the graphical symbol – (minus) also can have the meaning of hyphen or separation mark. The font design of the symbol is not part of this ISO Recommendation.

It is to be noted that, in contrast to the explicit solution in ASCII, this is an implicit solution based on the following point. The last sentence of Section 4.3 leaves the question of "font design" open; that is, a manufacturer could design ! to look like | and ˆ to look like ¬.

The Logical OR/Logical NOT problem had finally been solved.

25 A Comparison of Contiguous, Noncontiguous, and Interleaved Alphabets

25.1 THE COMPILER

For programming languages, PL/I, COBOL, FORTRAN, ALGOL, and so on, a compiler is a software product which takes a user's program and turns it into the set of CPU instructions that, when executed, will perform the calculations or operations specified in the user's program. The user's program, written in source language, is mapped into object language by the compiler.

One task performed by the compiler is the analysis of the graphics in the user's source language. While we may, anthropomorphically, view the compiler as "looking for" a left parenthesis, or a right parenthesis, the compiler really "looks for" the bit patterns representing those graphics. Compilers, then, are based on the CPU code of the computing system. One of the requirements of the compilation process is the ability to separate, that is, distinguish between, the bit patterns of alphabetics, numerics, and nonalphamerics.

25.1.2 Separability Requirement

There are operations during the compilation process for programming languages that require the determination of whether a bit pattern is or is not in the set of bit patterns for alphabetics; that is, whether a character is or is not an alphabetic. "Variables" or "names" in programming languages are permitted to be alphameric. Names such as PAYFILE, MAN-NUMBER, CHKPT, SAM, A123, A124, JUMP2, JUMP3 are acceptable. Such names are called symbolic. During the compilation process, they will be converted into absolute numeric memory addresses (and relative numeric addresses, in today's technology).

The programmer may also use absolute addresses, which will be pure numerics, such as 17326 or 4653. The compilation process must, in an early phase, be able to distinguish between absolute and symbolic addresses. Absolute addresses are always pure numerics; symbolic addresses may be pure alphabetics or mixed alphamerics. The rule is that absolute addresses must be pure numeric, and symbolic addresses must have the first character an alphabetic. The rule for distinguishing, then, applies to the first character: if it is numeric, the address is absolute; if it is alphabetic, the address is symbolic. The need to determine whether a character is or is not alphabetic, and whether a character is or is not a numeric, is fundamental to the compilation process.

Note that this rule must be quite rigorous. If the first character of an address is not a numeric, it does not follow that it is necessarily an alphabetic. An error may have introduced an initial character that is neither a numeric nor an alphabetic, and the compilation process must be able to detect such errors.

It is in the context of this compilation requirement that the contiguous alphabet of ASCII and the noncontiguous alphabet of EBCDIC may be compared.

25.2 ASCII AND EBCDIC

The 7-Bit Code, ASCII, and the 8-Bit Code, EBCDIC, are structurally dissimilar. Both codes, generically, may be termed BCD codes; that is, the four low-order bits of the bit patterns for numerics are binary-coded decimal. The structural dissimilarity arises from the bit patterns assigned to alphabetics. For ASCII, the alphabetics are assigned to a contiguous set of bit patterns. For EBCDIC, the alphabetics are noncontiguous.

Computers process bit patterns. Let us examine the bit patterns of the alphabetics in ASCII and in EBCDIC, and apply them to the compilation requirement. The fact that ASCII is a 7-bit code, and EBCDIC an 8-bit code, is immaterial to this discussion.

Consider Fig. 25.1, which shows the bit patterns for the alphabetics of ASCII and of EBCDIC. The alphabetic bit patterns for ASCII run in a contiguous block from 100 0001 for A to 101 1010 for Z. The alphabetics for EBCDIC run in three blocks, contiguous within blocks, but with gaps between blocks:

1100 0001 for A to 1100 1001 for I,
a gap from 1100 1010 to 1101 0000,
1101 0001 for J to 1101 1001 for R,
a gap from 1101 1010 to 1110 0001,
1110 0010 for S to 1110 1001 for Z.

ASCII		EBCDIC	
Bit pattern	Meaning	Bit pattern	Meaning
000 0000	Null	0000 0000	Null
.	.	.	.
.	.	.	.
.	.	.	.
100 0000	@	1100 0000	{
100 0001	A	1100 0001	A
.	.	.	.
.	.	.	.
.	.	.	.
.	.	1100 1001	I
.	.		
.	.	1100 1010	Unassigned
101 1010	Z	.	.
		.	.
101 1011	[	.	.
.	.	1101 0000	}
.	.		
.	.	1101 0001	J
111 1111	Delete	.	.
		.	.
		.	.
		1101 1001	R
		1101 1010	Unassigned
		.	.
		.	.
		.	.
		1110 0000	\
		1110 0001	Unassigned
		1110 0010	S
		.	.
		.	.
		.	.
		1110 1001	Z
		1110 1010	Unassigned
		.	.
		.	.
		.	.
		1111 1111	Eight ones

Fig. 25.1 ASCII and EBCDIC alphabetics

25.2.1 Tests for Alphabetics

There are three methods by which the determination of alphabetics may be made:

Method 1. High-order bit test.
Method 2. Bracket test.
Method 3. Translate and test.

High-order bit test

It is not uncommon to find the statement: "The contiguous alphabet of ASCII may be determined by a high-order bit test, whereas the noncontiguous alphabet of EBCDIC cannot." This is an imprecise, and indeed, incorrect statement.

It was certainly an objective, in the design of ASCII, that the alphabetics be determined by a "single high-order bit test." However, this objective was not accomplished.

In order to examine this aspect of ASCII and EBCDIC, it will be necessary to introduce some elementary Boolean notation and concepts. Three Boolean operators will be used, NOT, Exclusive OR, and AND, as defined in Chapter 2. (*Note:* The derivation is not given for the Boolean equations that follow.)

If we interpret the meaning of "alphabetic" to mean "columns in the code table which contain alphabetics, upper or lower case," then it can be shown that

for ASCII $\text{alphabetic} = \text{b7}$ (Fig. 2.26),
for EBCDIC $\text{alphabetic} = \overline{\text{e0}} \wedge (\overline{\text{e2}} \veebar \overline{\text{e3}})$ (Fig. 2.28).

If we interpret the meaning of "alphabetic" to mean "columns in the code table which contain upper-case alphabetics" (this is closer to the requirement for the compilation process), it can be shown that

for ASCII $\text{alphabetic} = \text{b7} \wedge \overline{\text{b6}}$,
for EBCDIC $\text{alphabetic} = \text{e0} \wedge [\text{e1} \wedge (\overline{\text{e2}} \veebar \overline{\text{e3}})]$.

In the first interpretation, it could be said that ASCII meets the objective of a "high-order bit test." Perhaps this is what is meant when it is stated that "ASCII had a high-order bit test for alphabetics."

Both of the interpretations above, however, are inadequate for the requirement of the compilation process. There is a nonalphabetic in column 4, and there are five nonalphabetics in column 5 of ASCII. And there are nonalphabetics in each of hex columns C, D, and E of EBCDIC. There are six nonnumerics in column 3 of ASCII and six in hex column F of EBCDIC. It is necessary, in the compilation process, to determine precisely and rigorously whether a particular bit pattern is, or is not, in the sets of alphabetic or numeric bit patterns.

If the high-order bit test were actually used by compilers (it isn't—because there are more efficient and less cumbersome methods available), it would be necessary to evaluate Boolean equations. As a matter of interest, the Boolean equations for alphabetics, numerics, and alphamerics for ASCII and EBCDIC are shown (with some simplification, taking common expressions into account).

EBCDIC

Common Expressions

$$A = e0 \wedge e1$$
$$B = e2 \wedge e3$$
$$C = \overline{e4} \wedge (\overline{e5} \wedge \overline{e6})$$
$$D = [\overline{e4} \wedge e5] \vee [\overline{e5} \wedge (e4 \wedge e6)]$$
$$\text{alphabetic} = A \wedge \{[B \wedge D] \vee [\overline{e2} \wedge (C \wedge e7)]\}$$
$$\text{alphameric} = A \wedge \{D \vee \{C \wedge [(\overline{e2} \wedge e7) \vee B]\}\}$$

ASCII

Common Expressions

$$E = \overline{b3} \wedge \overline{b2}$$
$$F = [\overline{b7} \wedge (b6 \wedge b5)] \wedge [\overline{b4} \vee (b4 \wedge E)]$$
$$G = b7 \wedge b6 \wedge (\{\{\overline{b5} \wedge \overline{b4}\} \wedge \{E \wedge \overline{b1}\}\} \vee \{\{b5 \wedge b4\} \wedge \{[\overline{b3} \wedge (b2 \wedge b1)] \vee b3\}\}])$$
$$\text{alphabetic} = G$$
$$\text{alphameric} = F \vee G$$

For the compilation process described above, the programming flow chart* would be as follows:

Is it an alphameric? $\xrightarrow{\text{no}}$ Error routine
↓ yes
Is it an alphabetic? $\xrightarrow{\text{no}}$ Numeric routine
↓ yes
Alphabetic routine

The relative complexity of the programs for EBCDIC and ASCII can be estimated by counting the number of Boolean operators $\wedge$ for AND, $\vee$ for OR, and — for NOT, in the equations.

*The determination of "numeric" in the above flow chart was achieved not by evaluating a specific Boolean equation but as a consequence of determining first "alphameric," then "alphabetic" or "nonalphabetic."

	EBCDIC			ASCII		
	AND	OR	NOT	AND	OR	NOT
Common expressions	7	1	5	14	3	10
Alphameric	3	2	1	0	1	0
Alphabetic	4	1	2	0	0	0
Totals	14	4	8	14	4	10

It may be seen, therefore, that the relative complexity of determining alphamerics, alphabetics, and numerics for ASCII and EBCDIC by high-order bit test is approximately the same.

Bracket test

In this method, the question is whether a bit pattern under examination is within the outermost bit patterns of a contiguous block of bit patterns. Suppose the minimum bit pattern of the block is Emin, the maximum bit pattern of the block is Emax, and the bit pattern of the character under examination is X. Then four computer-comparison instructions—two comparison and two branch instructions—will determine if X is in the set bracketed by Emin and Emax.

Step 1.	Is X less than,	equal to,	or greater than Emin?
	↓ yes	↓ yes	↓ yes
	X is not in set	X is in set	go to step 2.
Step 2.	Is X less than,	equal to,	or greater than Emax?
	↓ yes	↓ yes	↓ yes
	X is in set	X is in set	X is not in set.

A bracket test consisting of four instructions determines if a bit pattern is within a set of contiguous bit patterns. Note then that to make the rigorous determination, two bracket tests (eight instructions) are required for ASCII—one for the alphabetic block and one for the numeric block. For EBCDIC, four bracket tests (sixteen instructions) are required—one for each of the three alphabetic blocks and one for the numeric block.

Translate and test

In some modern computers, there is an instruction (in the System/360 or System/370, this instruction is called "Translate and Test") that, by reference to a table of bit patterns stored in memory, can determine whether a bit pattern is or is not in a set of bit patterns, regardless of whether the set is contiguous or noncontiguous. That is to say, to make the rigorous determination, we have one instruction for ASCII or one instruction for EBCDIC.

Is it an alphabetic,	a numeric,	or a nonalphameric?
↓yes	↓yes	↓yes
alphabetic routine	numeric routine	error routine

In summary, the high-order bit test would be approximately the same for ASCII and EBCDIC; the bracket test requires eight instructions for ASCII and sixteen instructions for EBCDIC; translate and test requires one instruction for either ASCII or EBCDIC. Note that while the bracket test was the method that would be used with older 6-bit computers, it is doubtful if any of those 6-bit computers now process 7-bit ASCII data.

It may be concluded that while contiguity and noncontiguity of alphabetics certainly characterize a difference between ASCII and EBCDIC, this characteristic cannot be used to determine superiority or inferiority in any determinative way.

25.2.2 Translation to Hollerith Card Code

There is another aspect of the ASCII and EBCDIC alphabets that serves as a basis for comparison. This aspect is the translation relationship to the Hollerith card code. (See Figs. 2.26, 2.28, and 17.23.)

The well-known BCD relationship holds for numerics 1 through 9 in *both* ASCII and EBCDIC.

Digit punch	Low-order four bits
1	0001
2	0010
3	0011
4	0100
5	0101
6	0110
7	0111
8	1000
9	1001

For EBCDIC, this BCD relationship holds for alphabetics A through I, J through R, and S through Z. For ASCII, it holds only for alphabetics A through I. In building a hardware translator (card code to/from bit code) the BCD relationship must be provided for numerics 1 through 9 for both ASCII and EBCDIC. Then, for EBCDIC, this same part of the translator can be used for *all* 26 alphabetics, while for ASCII it can be used only for alphabetics A through I, and the alphabetics J through Z require additional hardware circuitry.

As long as the punch card persists both as a data entry and as a data storage medium (volume of punched card sales continues to increase each year, at least at the time of writing this book, despite the competition of other media), EBCDIC will be a less complex code than ASCII to implement in hardware. And the lesser complexity is attributable specifically to the particular noncontiguous alphabet for EBCDIC.

25.2.3 Collating Sequence of Alphabetics

In both ASCII and EBCDIC, the collating sequence for alphabetics corresponds to the bit sequence either directly or relatively. American manufacturers who market computing systems in Europe must consider the ISO 7-Bit Code, and the 29-letter alphabets of Germany, Norway, Sweden, Denmark, and Finland. The three additional letters for these countries are as shown in Fig. 25.2.

Germany	Ä	Ö	Ü
Denmark	Æ	Ø	Å
Norway	Æ	Ø	Å
Sweden	Ä	Ö	Å
Finland	Ä	Ö	Å

Fig. 25.2 Diacritical letters

In developing the ISO 7-Bit Code, provision was made for these additional alphabetics by assigning three code positions contiguously following the letter Z in the code table. Provision for these diacritical letters is also made in EBCDIC, but not in code positions contiguous to the other alphabetics.

The fact that the bit patterns for these 29 letters are not in relative binary sequence in EBCDIC means that additional steps must be taken in sorting passes.

Interestingly, the fact that the 29 bit patterns are in relative binary sequence in the ISO 7-Bit Code does not improve the sorting situations by an iota. For Germany, the three umlaut letters Ä, Ö, Ü collate adjacently to the non-umlaut letters A, O, U and do not follow the letter Z as shown in the code table.

In Sweden, by contrast, the diacritical letters *do* collate immediately after the letter Z. But, by a strange quirk (the discussion of which is beyond the scope of this book), the Swedish national standard for the 7-Bit Code positions the three diacritical letters following the letter Z, but in a sequence *different* from their official collating sequence. This anomalous situation also exists in Finland.

Exactly the same extra steps must be taken in sorting passes to sort ISO 7-Bit Code data in Germany, Sweden, and Finland as are taken to sort EBCDIC data.

25.2.4 Signed Numerics

Another factor for comparison is the inherent inability of ASCII, and the inherent ability of EBCDIC, to provide for signed numerics. This has been discussed in Chapter 19. The inability of ASCII and the ability of EBCDIC are direct consequences of their contiguous and noncontiguous alphabets. The noncontiguous alphabet of EBCDIC is based on the Hollerith Card Code, and it was from punched card applications that the practice of overpunching numerics to represent signed numerics arose.

25.3 INTERLEAVED ALPHABETS

During the early days of code development and standardization, the concept of interleaving small and capital letters in the code was frequently proposed. Indeed, as described in Chapter 3, the Stretch Code (Fig. 25.3) and the Information Processing Code (Fig. 25.4) did provide interleaved alphabets.

The first question for an interleaved alphabet was whether the small or capital letter should precede within the alphabetic pair. The primary reason cited for interleaving the alphabetics was to make sorting and collating more efficient. It is engaging that the designers of the Stretch Code decided that the small letter should precede the capital letter in the pair, whereas the designers of the Information Processing Code chose that the capital letter should precede the small letter.

During the development of ASCII, it was proposed that the alphabets should be interleaved. Various factors spoke against such a decision (not the least of which was the fact that the ASCII designers had not, at that time, decided whether or not to include small letters in the code), but the example which spoke most strongly against interleaving was the "Telephone Directory Problem."

25.3.1 The Telephone Directory Problem

When the telephone directories of different cities are studied, it will be observed that there is no common rule for sequencing names. Different

Row	Column	0	1	2	3	4	5	6	7	8	9	A	B	C	D	E	F
	Bit Pat. →	00				01				10				11			
	↓	00	01	10	11	00	01	10	11	00	01	10	11	00	01	10	11
0	0000	SP	[	&	c	k	s	0	8								
1	0001	±	⊃	+	C	K	S	0	8								
2	0010	→	]	$	d	1	t	1	9								
3	0011	≠	∘	=	D	L	T	1	9								
4	0100	∧	←	*	e	m	u	2	.								
5	0101	{	≡	(	E	M	U	2	:								
6	0110	↑	¬	/	f	n	v	3	–								
7	0111	}	√	)	F	N	V	3	?								
8	1000	∨	%	,	g	o	w	4									
9	1001	⍱	\	;	G	O	W	4									
A	1010	↓	◇	'	h	p	x	5									
B	1011	||	|	"	H	P	X	5									
C	1100	>	#	a	i	q	y	6									
D	1101	≥	!	A	I	Q	Y	6									
E	1110	<	@	b	j	r	z	7									
F	1111	≤	~	B	J	R	Z	7									

Fig. 25.3 Stretch, 120-character set

cities have different rules. For the purposes of illustration, consider the following rules:

A. A name shall be given in the following sequence:

   First, the last name;
   then, the first name or initial;
   then, the second name or initial;
   and so on.

B. Names and initials shall be separated by a space, but with no periods or commas.

C. "Space" shall collate low to *all* alphabetics.

D. The alphabetics shall collate in their natural sequence. Thus, A, B, C, D, . . . , X, Y, Z, or a, b, c, d, . . . , x, y, z.

E. A capital letter shall collate low to its corresponding small letter. Thus, "MacDonald" shall collate low to "Macdonald." And "Macdonald Peter" shall collate low to "Macdonald Robert."

	Column	0	1	2	3	4	5	6	7
	Bit Pattern b7	0	0	0	0	1	1	1	1
	b6	0	0	1	1	0	0	1	1
	b5	0	1	0	1	0	1	0	1
Row	b4 b3 b2 b1								
0	0 0 0 0	0	C	K	S	(	α	Σ	[3]
1	0 0 0 1	1	c	k	s	!	×	¼	[2]
2	0 0 1 0	2	D	L	T	?	β	≤	Ⓔ
3	0 0 1 1	3	d	l	t	#	÷	½	ⓔ
4	0 1 0 0	4	E	M	U	°	=	≥	Bk_1
5	0 1 0 1	5	e	m	u	/	–	¾	Bk_2
6	0 1 1 0	6	F	N	V		√	∞	Bk_3
7	0 1 1 1	7	f	n	v		∫	↓	Bk_4
8	1 0 0 0	8	G	O	W	*	:	θ	.
9	1 0 0 1	9	g	o	w	)	;	↑	C_1
10	1 0 1 0	SP	H	P	X	•	@	ϕ	C_2
11	1 0 1 1	RES	h	p	x	,	χ	→	C_3
12	1 1 0 0	A	I	Q	Y	π	"	κ	C_4
13	1 1 0 1	a	i	q	y	_	'	←	C_5
14	1 1 1 0	B	J	R	Z	ω	$	[	C_6
15	1 1 1 1	b	j	r	z	+	ç	]	C_7

[2]
[3]

Fig. 25.4 IPC, 7-bit subset

Then, under these rules, consider a name with five spellings:

Van De Water
Van de Water
van De Water
van de Water
Vandewater

And consider another name also with five spellings:

Van De Wenter
Van de Wenter
van De Wenter
van de Wenter
Vandewenter

And, finally, suppose that for *each* of these ten names, there exists a "John" and a "Peter."

What will be the sequence for these twenty names? (We must also assume, of course, that there are other names, and these are indicated below by dots.)

* Van De Water John
Van De water Peter
.
.
.
Van De Wenter John
Van De Wenter Peter
.
.
.
* Van de Water John
Van de Water Peter
.
.
.
Van de Wenter John
Van de Wenter Peter
.
.
.
* Vandewater John
Vandewater Peter
.
.
.
Vandewenter John
Vandewenter Peter
.
.
.

* van De Water John
van De Water Peter
.
.
.
van De Wenter John
van De Wenter Peter
.
.
.
* van de Water John
van de Water Peter
.
.
.
van de Wenter John
van de Wenter Peter
.
.
.

Consider now the problem of a stranger to the city who wants to look up a name in the telephone directory. He knows he wants to find, let us say, John Van De Water, but is not sure which of the five possible spellings is the correct one. Then he must search the telephone directory in *five* separate sections, as indicated by the asterisk above. The separation of these sections depends on how many intervening names there are. If he checks all five sections, he will locate five names, one of which is the one he wants. By a process of phoning and elimination, he can thus locate the one he wants.

There are two adverse attributes of this particular sequence: (1) the five possible names may be widely separated, and thus not easy to find. (2) Unless the stranger *knows* that there are five possible spellings, he may possibly not locate all five names, and may, in fact, not find the actual name that would turn out to be the correct one.

Is it possible to construct different rules for sequencing names so that the particular problem above is simplified? Consider the following:

A. A name shall be given in the following sequence:
First, the last name;
then, the first name or initial;
then, the second name or initial;
and so on.

B. Names and initials, *as printed*, shall be separated by a space, but with no periods or commas. However, on sorting or collating operations, the spaces shall be ignored; that is, the names and initials will be treated as if concatenated with "space" characters removed.

C. The alphabetics shall collate in their natural sequence. Thus, A, B, C, D, . . . , X, Y, Z, or a, b, c, d, . . . , x, y, z.

D Capitalization will be ignored in sorting and collating operations, unless two names are otherwise identical, in which case a capital letter shall collate low to its corresponding small letter. Thus "MacDonald John" and "Macdonald John" will both collate low to "Macdonald Peter," but "MacDonald John" will collate low to "Macdonald John."

Under these rules, the twenty names of the example will collate as follows:

* Van De Water John
Van de Water John
Vandewater John
van De Water John
van de Water John

* Van De Water Peter
Van de Water Peter
Vandewater Peter
van De Water Peter
van de Water Peter

* Van De Wenter John
Van de Wenter John
Vandewenter John
van De Wenter John
van de Wenter John

* Van De Wenter Peter
Van de Wenter Peter
Vandewenter Peter
van De Wenter Peter
van de Wenter Peter

The stranger's problem is clearly much simplified by such a sequence, for all names that "sound" the same (which is what he knows) appear in a block, regardless of idiosyncratic spelling with capital/small letters or spaces.

This example has been given, not in order to champion any particular set of rules for sequencing names, but rather to point out that there *can* be different rules, depending on particular requirements. It may be pointed out that the first set of presented rules above makes for simple sorting and collation algorithms but a complex look-up algorithm, while the second set of rules makes for a simple look-up algorithm but complex sorting and collation algorithms. Indeed, if the second set of rules was implemented in a data processing application, it would be helpful to carry the "name" twice, first *without* spaces for collating purposes, and again *with* spaces for printing purposes. Then the ignoring of capitals except when two names are otherwise identical can be programmed more simply.

It was in the light of examples such as the above that it was recognized that whether capital letters should collate low or high to the corresponding small letters was a matter of taste (that is, depended on the particular application), and that the interleaving of capital and small letters in a coded character set might be useful in some applications but, in general, would not serve a useful purpose.

In fact, a more general realization emerged. With respect to alphabetics, numerics, and specials, the collating sequence depends on the application and will be different for different applications. It was this realization that said the interleaving of alphabetics was much less significant than other code criteria that spoke against interleaving.

26 Code Extension and Examples

A problem that eventually arises with almost any code is that the code positions become full, but new and additional requirements are put on the code. New equipment designed to operate with the code may need new control functions, or applications for new or old equipment may require additional graphic characters. How may these additional characters be provided, if there are no unused bit patterns in the code? This problem and its solution are the subject of what is called code extension. The solutions generally fall under the headings of substitution, precedence codes, and Escape sequences.

26.1 SUBSTITUTION

Examples of substitution, particularly to provide graphic characters, are common. Some have been described in this book; the "scientific" and "commerical" duals of BCDIC (Chapter 4); the alphabetic extenders for certain European alphabets (Chapter 4); Katakana and other non-Latin alphabets (Chapter 18).

Substitution to provide additional control functions is less frequent than graphic substitution, but not unknown. In fact, the designers of coded character sets, realizing that such a requirement will eventually, if not initially, be put on the code, have placed what are called general purpose control characters in the code.

For example, in the 7-Bit Code and in EBCDIC, there are four control characters called Device Control 1, Device Control 2, Device Control 3, and Device Control 4 (DC1, DC2, DC3, and DC4, respectively). The definition of these control characters is intentionally broad and unspecific—"A device Control character is used for the control of a device." The nature of the control is not specified. When a particular device needed one or more of the Device Control characters, the code designers would define them specifically and for functions peculiar to that device. Some other kind of device would use one or more of the DC's for some specific control functions, but the functions for the DC's of one device need not, and probably would not, match the functions for the DC's of the other devices.

Under this philosophy, the code designers realized that interchange of data between these different kinds of devices would then be expected to be difficult or impossible without human intervention, but they presumed that interchange of data between unlike devices would seldom if ever be required. When rare instances arose where such interchange was required, the humans operating the different devices would have to understand the difference in the DC's, and accommodate it in some fashion.

Four additional general-purpose characters, the so-called information separators, were designed into the 7-Bit Code and into EBCDIC. File Separator, Group Separator, Record Separator, and Unit Separator were defined broadly to be used to separate blocks of information. But how they were to be used to separate blocks, what philosophy of file and record structuring was to be used, was intentionally not specified. Such detailed specification would be left to the particular data processing application in which the separators would be used. Initially, a hierarchial philosophy of structuring information blocks was defined. A "file" was larger than, and would enclose, "groups." A "group" was larger than, and would enclose, "records." And a "record" was larger than, and would enclose, "units." Eventually, the standards committees made this hierarchial specification optional; that is, the separators need not be used hierarchially, but *if* they were, then the hierarchy would be as described above. The standards committees realized that, as with the Device Controls, the unspecificity of the information separators could lead to difficulty of information interchange, but such difficulties could be worked out in the rare instances when they arose.

26.2 PRECEDENCE CODES

Another general technique for extending the repertoires of codes was the technique of precedence or shift characters. This technique has been described with respect to CCITT #2 (Chapter 3), and with respect to PTTC (Chapter 6). Under this technique, the meanings of a bit pattern in a specific subset of bit patterns (of the total set of different bit patterns) depends not only on the bit patterns itself, but also on which precedence character preceded it. By this technique, although the total number of different *bit patterns* of a code is mathematically prescribed, the total number of *meanings* associated with the bit patterns can be extended.

In CCITT #2, a 5-bit code, there are 32 different bit patterns, but 52 graphic meanings and 6 control meanings—a total of 58 different meanings. In PTTC, a 6-bit code with 64 different bit patterns, there are 94 graphic meanings and 17 control meanings—a total of 111 different meanings.

The designers of the 7-Bit Code, realizing that the future might well see the requirement for more than 128 different code meanings, placed two precedence characters in the code, Shift In and Shift Out. These two characters are also included in EBCDIC. The standards committees are (as this book is written) studying the ways in which these two precedence characters may be used for extension of the 7-Bit Code.

26.3 ESCAPE SEQUENCES

Another means of extending the repertoire of meanings of a code is by use of the Escape character. Under this technique, the Escape character and the succeeding character are to be regarded as an entity, defining some control function. The character directly following the Escape character is to be regarded as not having its normal meaning. The two characters to be regarded as an entity are called, in the literature, an Escape sequence. Meaning is associated with the Escape sequence. Escape sequences may consist of more than two characters and may be variable in length. (The philosophy of variable length Escape sequences is not described in this book.) It is to be noted that this technique is a form of a precedence code. It is interesting to observe the difference between this form of precedence code and the one described above.

Under the techniques described for CCITT #2 and for PTTC, the precedence or shift character establishes a mode, which remains in effect until another shift character appears, which establishes, in its turn, its mode. Such characters are described in the literature as "locking shift characters"; that is, a shift character "locks" a mode, which remains "locked" until the other shift character "unlocks" that mode and "locks" its mode. By constrast the Escape character affects the meaning of only the following character. It has been described in the literature as a nonlocking shift character.

Another character in the code is called Data Link Escape (DLE). The DLE character is to function in a manner similar to the Escape character, but its use, and the meanings assigned to DLE sequences, is for use on data communication products only (hardware or software). DLE sequences will not be discussed in this book.

As an example of the use of Escape sequences, let us look at PTTC. PTTC was designed before the 7-Bit Code was designed, and its nonlocking shift character was called Prefix, instead of Escape. There are 20 Prefix sequences assigned in PTTC. The meanings assigned to them are control meanings, not graphic meanings. The second character of a Prefix sequence is a graphic character (with two exceptions). Since PTTC is a shifted 6-bit code, the graphic bit pattern of a Prefix sequence will have

Prefix sequence			Control meaning
PRE 1	or	PRE =	Printer 1 on
PRE 2		PRE ¤	Printer 2 on
PRE 3		PRE ;	Punch 1 on
PRE 4		PRE :	Punch 2 on
PRE 5		PRE %	Printer 1 off
PRE 6		PRE ′	Printer 2 off
PRE 7		PRE ″	Punch 1 off
PRE 8		PRE *	Punch 2 off
PRE 9		PRE (	Reader 1 on, Reader 2 off
PRE 0		PRE)	Reader 2 on, Reader 1 off
PRE a		PRE A	Ribbon shift up
PRE b		PRE B	Ribbon shift down
PRE c		PRE C	Select single line feed
PRE d		PRE D	Select double line feed
PRE e		PRE E	Card punch duplicate
PRE g		PRE G	Card punch alternate program
PRE h		PRE H	Card punch release
PRE j		PRE J	Reader skip stop
PRE LF		PRE LF	Form feed
PRE SP		PRE SP	Vertical tab

Figure 26.1

two graphic meanings, but this difference in graphic meaning does not affect the meaning assigned to the Prefix sequence itself. That is to say, the control meanings of Prefix sequences (like the control meanings of single character controls) are independent of the preceding shift character (Upper Case or Lower Case). The 20 Prefix sequences of PTTC and their meanings are shown in Fig. 26.1.

Products using either Escape or Prefix sequences not only have to build in the hardware to execute the control meaning assigned to the sequence, but also have to suspend the normal reaction to the second character of the sequence. That is, once either an Escape or Prefix bit pattern has been detected in the data stream by the product, the product must then be set not to react normally to the bit pattern(s) immediately following. For example, when a "PRE A" sequence appears in the data stream on the IBM 1050 (a product implementing PTTC) not only is the typewriter ribbon shifted up, but the letter A is not printed, nor is the typewriter carriage spaced.

Escape sequences of two, and even three, characters have been implemented on modern printers, display devices, and terminals, providing many and varied control functions.

An interesting current development on the standards committees is that an Escape sequence itself (for certain Escape sequences) is being regarded as having locking-shift meaning. That is, a particular Escape sequence is to be regarded as establishing a particular mode of meanings to be associated with subsequent bit patterns in the data stream, which is to remain in effect until some other particular Escape sequence disestablishes that mode and establishes its own mode of meanings to be associated with subsequent bit patterns in the data stream. Further details of this philosophy of code extension are not given in this book. It would take a book itself to explain and describe fully the intricacies of code extension envisaged by the standards committees under the current philosophy of Escape sequences.

26.4 TEXT/360

A particular example of code extension, text processing under Text/360 (an IBM software product), will now be described. It is interesting to appreciate the design criteria placed on this system for text processing and to see how the design criteria affected the system design.

26.4.1 Text Processing Defined

First, it is necessary to understand what is meant by "text processing." Most data processing applications are satisfied from a printing or display point of view with 26 alphabetics, 10 numerics, and a varied number of specials. The vast majority of fast, parallel printers provides character sets up to 64 characters. These various character sets, from 48 to 64, have one aspect in common. They have one set of alphabetics—26 letters in English-speaking countries, up to 29 letters in some European countries. And, in general, the alphabetics are block capitals (sometimes referred to as upper-case letters, but this is a misnomer in the context of only one set of alphabetics).

In everyday life, books, magazines, newspapers, etc., are printed in *two* cases of alphabet, capital letters and small letters. The printing of a particular document may involve many different fonts of letters but only two cases. The question arose, Could such documents be printed by data processing equipment, and would it be economical to do so?

The answer to the first part of the question is affirmative. What is required is a printing element with two alphabetic cases. Such printing elements are entirely feasible. In general, parallel printers with such elements are either more costly or they print more slowly (which increases printing costs), or both. This realization leads naturally to the second part of the question. Since most data processing applications are satisfied with one case of alphabetics and since the provision or use of parallel printing with two alphabetic cases leads to higher costs, it is, in general, uneconomical to use two alphabetic cases.

However, there is a certain class of data processing where it is either economical or necessary to use two alphabetic cases. This class of applications is grouped under the name of text processing. Let us look at the characteristics of text processing.

There are four identifiable requirements. A particular text-processing application may not have all of these requirements:

1. Two cases of letters, for ease of human reading. Humans find that a page of text with capital and small letters is easy to read; that text with small letters only is less but not much less easily readable; and that text with capital letters only is much less easily readable. For example,

   John and Peter went to Poughkeepsie.
   john and peter went to poughkeepsie.
   JOHN AND PETER WENT TO POUGHKEEPSIE.

2. Two cases of letters for unambiguity. In chemical abstracting, for example, carbon monoxide (CO) and cobalt (Co) can be distinguished only if upper and lower case letters are used.
3. A large body of text that is expected to require numerous changes of greater or lesser degree.

4. Symbols not normally found on data processing printers. For example,
   - corners, intersections, vertical and horizontal lines for drawing charts, tables, boxes in flow diagrams;
   - arrows for drawing flow charts and electronic circuits;
   - mathematical symbols, for use in some of the more exotic programming languages;
   - accents and diacritical marks (for European, Russian, etc., names and book titles) in library bibliographic work.

Some of these applications, where duocase alphabetic capability is the only requirement, can be processed without the use of a computer. A skilled operator of a typewriter with some means of storage—paper tape, magnetic tape, tape cassette/cartridge, magnetic cards, punched cards—can, and frequently does, perform text processing, including the frequent changes. (The initial and subsequent drafts of this book were prepared on an IBM Magnetic Tape Selectric Typewriter.)

An available computer will enter the application when the skilled operator and typewriter with storage are not available. With a computer, of course, a program is necessary, as well as an appropriate printing capability.

26.4.2 Development

Text/360 is the name of a program that is the extension of an earlier program, Text 90. Text 90 was a program written for the IBM 7090 computing system. It was developed to meet a requirement which had arisen in internal operations. As a computing system is designed, and goes through succeeding stages of development to completion, the functional specifications change, often from day to day. The document containing the functional specifications changes in consequence.

It was found that the process of typing and retyping these documents was too slow to keep up with the design and development schedules. It was suggested that a computer program, with appropriate input and output equipment, could produce these documents, and could produce them with sufficient rapidity to meet the schedule demands. The program was written, debugged, and used. It was called Text 90. The functional specifications for the System/360 were documented by Text 90. Some System/360 reference manuals for customers were printed by Text 90 (and then reproduced by other printing and publishing methods).

Text/360 was an extension of Text 90 which was written to operate on the System/360. It also was developed for internal operations. But it was judged it would be useful externally and it was therefore announced and made available to customers. The customer reference manual itself for Text/360 was printed by Text/360.

Let us look at the criteria that were set for Text 90, and subsequently for Text/360. These criteria came from constraints on the input character set and requirements on the output (printed) character set.

1. The output character set should contain
   - the Space character,
   - numerics,
   - small alphabetics,
   - capital alphabetics,
   - specials normally found on the date processing printers of the time,
   - symbols for drawing charts, tables, programming flow charts, etc.,
   - symbols for plotting graphs,
   - mathematical symbols beyond those normally found on printers of the time.
2. The input character set should
   - be keypunchable without multipunching; that is, the set should not exceed 48 characters numerically, for the 48-character keypunches of the day. This criterion, was extended for Text/360 to 64 characters, but the 48-character set was retained as an option;
   - be optimum, from a keypunching productivity point of view;
   - contain graphic characters and characters that have printable graphic representation for controlling the various processes of text processing—capitalization (both initial letters of words and complete words), editing, altering, underscoring, etc.—so that the input data could be listed completely.

Two decisions were made before the criteria could be applied: (1) there would be 120 graphic characters and 6 control characters in the output set; and (2) the characters beyond 48 would be represented by either two- or three-character sequences.

26.4.3 System/360 and EBCDIC

The rest of the discussion will be in the context of Text/360, the System/360, and EBCDIC.

There were some considerations that went into the final design.

1. The frequency of use of small letters in text far exceeds the frequency of use of capital letters. In normal text, capital letters are used only as initial letters for sentences, names of people, towns, cities, countries, streets, etc. In text, titles might appear, which might be capitalized in their entirety, but titles are few, relative to lines of text. In a 48-character set, only one of the small or capital alphabetics can be represented as a single character; the other will have to be represented by at least a 2-character sequence. So small letters should be represented by single characters, the alphabetics found on the keypunch.

2. Numerics will appear more frequently than special symbols in text. So numerics should be represented by single characters, the numerics found on the keypunch.

3. Specials such as period and comma, which appear more frequently than other specials, should have single character representation.

4. From keypunching statistics, it was known that numerics and alphabetics are keypunched with a better production speed than specials. Therefore, the final character, or characters, of a two- or three-character sequence should be a numeric or an alphabetic, not a special.

5. From the preceding considerations, there was a conclusion that itself became a consideration. If the numerics and alphabetics were used as single characters (giving rise to 36 characters), and if each of two different precedence characters were used with the alphabetics and numerics (giving rise to an additional 2×36 characters), a maximum of $3 \times 36 = 108$ characters would be required. If more than 108 characters were required, either a third precedence character would be needed, or a double precedence character in a three-character sequence would be needed.

Now some further design decisions were made.

1. The output character set would consist of 120 graphics (to be described later).

2. Graphic representation for the six control operations would be as follows:

   * for single capitalization;

   @ for continued capitalization (beginning and end of capitalization to be represented by the same graphic);

$ for underscoring (beginning and end of underscoring to be represented by the same graphic);

− for editing;

+ for altering; and

/ for graphic set extension.

It is to be noted that the operation of single capitalization is also really graphic set extension, since it will be used to generate the upper-case alphabetics.

3. The graphic set would contain the following 94 graphics of EBCDIC.*

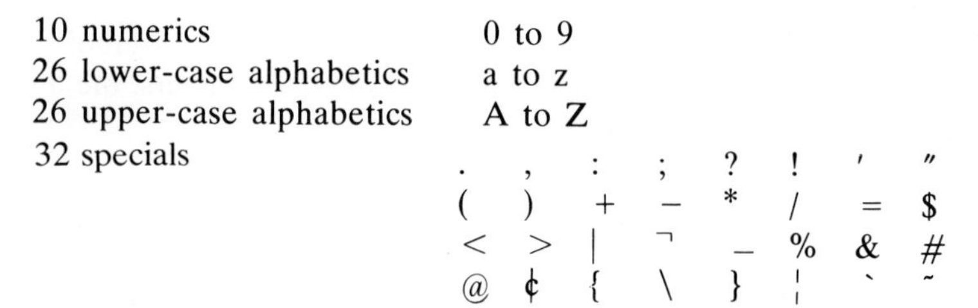

10 numerics	0 to 9
26 lower-case alphabetics	a to z
26 upper-case alphabetics	A to Z
32 specials	. , : ; ? ! ' "
	( ) + − * / = $
	< > | ¬ _ % & #
	@ ¢ { \ } ¦ ` ~

Since it had been decided to provide 120 graphics and since the EBCDIC graphic set had been reduced to 90 characters, 30 additional graphics could be provided. These were as follows:

6 mathematical symbols	± ≤ ≠ [ ] ≥
9 plotting/charting symbols	⌊ ⌋ ⌈ ⌉ − + · ↑ ←
10 superscript numerics	^{0} ^{1} ^{2} ^{3} ^{4} ^{5} ^{6} ^{7} ^{8} ^{9}
1 superscript minus symbol	$^{-}$
3 subscript numerics	$_{1}$ $_{2}$ $_{3}$
1 subscript "n"	$_{n}$

Recall that + − / * @ $ were to be provided as input graphics representing the six *control* characters. However, it was desired that they also be in the output *graphic* set. Therefore, a 2- or 3-character input representation for them as output graphics must be found, even though they also appeared as single-character inputs for control characters.

Some 13 specials had been provided as 2-character input representations in Text 90. These 13 specials were available on the 64 character

---

* This design decision was later slightly aborted (for reasons not known to the author). The later design decision was not to provide ¢ ¦ ` and ~. Actually, ¢ was replaced in its EBCDIC code position (but just for the Text/360 applications) by ←.

keypunches of the time frame of Text/360, and so are represented as single character inputs in Text/360. However, the Text 90 2-character representation for these 13 specials was permitted in Text/360 as an option. Thus there is a single character and also a 2-character input representation for these 13 specials:

: ; ? ! ′ ″ < > = – % & #

The 120-graphic set was provided as shown in Fig. 26.2.

<table>
<tr><th>Output graphics</th><th colspan="2">Input representation</th><th>Total</th></tr>
<tr><td>0 1 2 ··· 9
a b c ··· z
A B C ··· Z</td><td colspan="2">0 1 2 ··· 9
A B C ··· Z
*A *B *C ··· *Z</td><td>10
26
26
4</td></tr>
<tr><td>Superscript $^{0\ 1\ 2 \cdots 9}$</td><td colspan="2">/0 /1 /2 ··· /9</td><td>10</td></tr>
<tr><td>+ * @ $ / –
(See Note 1 following)
± ↑ [ ] ·</td><td colspan="2">/A /X /Q /D /Z /S
/H /C /V /B /P</td><td>11</td></tr>
<tr><td>{ \ } ≤ ≠ ≥
⌊ ⌋ ⌈ ⌉
− +</td><td colspan="2">//V //Z //B //G //O //F
//H //J //Y //U
//L //A</td><td>12</td></tr>
<tr><td>Subscript $_{1\ 2\ 3}$
Subscript $_{n}$
Superscript $^{-}$</td><td colspan="2">//1 //2 //3
//N
//S</td><td>5</td></tr>
<tr><td></td><td>64</td><td>48</td><td></td></tr>
<tr><td>: ; ? !
′ ″ < >
= – % &
←</td><td>: ; ? !
′ ″ < >
= – % &
¢</td><td>/T /E /W /N
/R /Y /G /F
/O /J /K /M
/U /L</td><td>14</td></tr>
<tr><td>| ¬</td><td>| ¬</td><td>//K //C</td><td>2</td></tr>
</table>

Total = 120

Fig. 26.2 Text/360 120-graphic set

Note 1. When + * @ $ / – are required as input graphics to represent controls, they are represented by themselves.

Note 2. The Text/360 chain provides the 120 graphics above. If the Text/360 chain is replaced by another 120 character chain called in IBM literature the IBM TN Chain, seven graphics are replaced as follows:

Text/360 Graphic	TN Graphic
←	¢
\	¤
↑	■
Subscript $_1$	Superscript $^+$
Subscript $_2$	Superscript $^($
Subscript $_3$	Superscript $^)$
Subscript $_n$	° (degree)

The objectives of Text/360 were as follows:

1. Small input set.
2. Large output graphic set, sufficient for most text processing applications.
3. Control characters for text processing.
4. All input representations, graphic and control, printable for debugging purposes.
5. Input set (single, double, and triple character representations) optimized for keypunching productivity.
6. Code compatibility with EBCDIC.

These objectives were achieved.

26.5 SUMMARY

Two general techniques of code extension, substitution and precedence characters, have been discussed. A particular example of code extension, Text/360, which uses both substitution and precedence characters, has been described. It might be observed that the chief difference of this method of code extension from those previously described is that all input representations, control and graphic, are keypunchable with single key depressions (that is, they do not require multipunching), and are printable for debugging purposes.

27 The 96-Column Card Code

27.1 THE SMALL CARD

During 1966, a new medium was being developed for the storage of data—a punched card, but a much smaller punched card than the traditional $3\frac{1}{4} \times 7\frac{3}{8}$ inch card in broad use in the data processing industry. The basic design objective was the smallness of the card. If the card could be made small, the associated card-handling equipment would be correspondingly compact, and costs would be low. The objective was a card approximately one third the size of the normal punched card.

27.2 CRITERIA

All the normal punched card operations were envisaged for the small card; key punching, verifying, sorting, collating, and computer input/output. Design criteria were set for the small card:

Criterion 1

The small card should be capable of receiving as many characters as the regular punched card; that is, at least 80.

Criterion 2

All punching character positions of the card should be capable of being interpreted on the card.

Criterion 3

The "primary" graphic set should be a 6-bit, 64-character set.

Criterion 4

It should be possible to punch 256 different hole patterns on the card; that is, the card should be capable of being an input/output medium for the System/360.

Criterion 5

The numerics should be BCD coded with the capability for positive and negative numerics. Negative numerics should be derivable by overpunching absolute numerics, analagously to the technique for the Hollerith Card Code.

Criterion 6

The Space character should be a no-holes hole pattern, as it was in the Hollerith Card Code. This would provide the capability to leave card columns or fields blank during the keypunching operation so that they could be punched with processed data during subsequent card operation.

Criterion 7

The numerics should have no zone punches.

Criterion 8

The translation relationship, bit code to/from card code, should be as simple as possible.

During the discussions on a code for the small card, it was decided that the code should be an eight-row code, as contrasted with the twelve-row Hollerith code for the regular punched card. If possible, it should be a direct representation of EBCDIC—a hole on the card corresponding to a one bit in EBCDIC and the absence of a hole corresponding to a zero bit. There seemed to be a possibility that the small card code would not be an exact direct representation of EBCDIC. In order to avoid confusion in such an eventuality, the holes or bits of the code were named DCBA8421, from high to low order:

Small card code	D	C	B	A	8	4	2	1
EBCDIC byte	0	1	2	3	4	5	6	7

27.3 THREE TIERS

The necessary size of the holes, and their necessary vertical and horizontal separation, posed a problem with respect to interpretation on the card. The geography of the card would allow for three tiers, each tier designed to contain the eight rows of a character, as shown in Fig. 27.1.

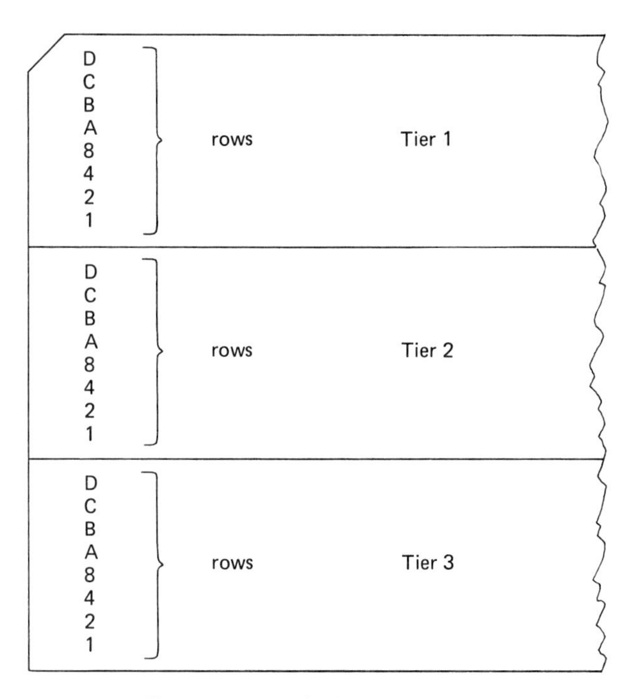

Fig. 27.1 Three-tier card, Version 1

27.4 INTERPRETATION ON THE CARD

Corresponding to the three tiers of punched characters, there would have to be at least three rows of unpunched card space at the top of the card to receive interpreting. The first step to solve this problem was to separate the D and C rows of each tier, as shown in Fig. 27.2.

Criterion 3 called for a 64-character, 6-bit "primary" graphic set. If, for these 64 characters, the D and C bits were zero, no holes would be punched, and the top of the card would be left unpunched to receive interpreting. Out of this realization grew the decision to *interpret* only the 64-character "primary" subset, even though a full 256-character, 8-bit set could be *punched* if necessary. Keypunched input data, which posed the main requirement for interpreting, would consist of 64 graphics only, although output data could consist of 256 characters.

The geography of the card had begun to dictate aspects of the coded character set. As it turned out, four rows were available to receive interpreting at the top of the card. Three tiers of 32 characters (96 characters total) could be punched, but four rows of 32 spaces (128

D, C	Tier 1, high-order rows
D, C	Tier 2, high-order rows
D, C	Tier 3, high-order rows
B, A, 8, 4, 2, 1	Tier 1, low-order rows
B, A, 8, 4, 2, 1	Tier 2, low-order rows
B, A, 8, 4, 2, 1	Tier 3, low-order rows

Fig. 27.2 Three-tier card, Version 2

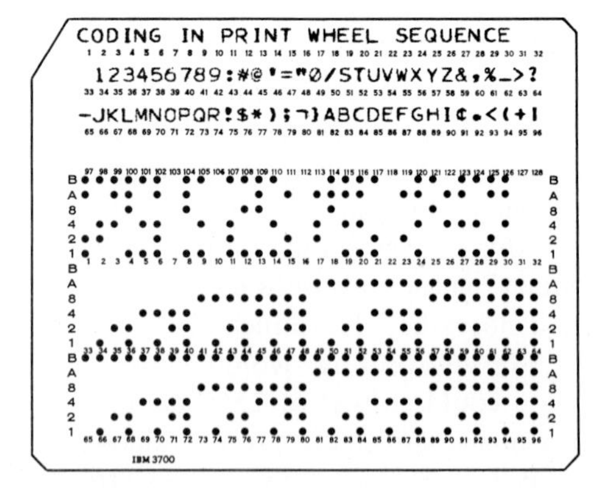

Fig. 27.3 Three-tier card, final version

spaces) were available for interpreting (see Fig. 27.3). Criteria 1, 2, and 4 had also been met by these decisions.

27.5 THE CHARACTER SET

Attention was now turned to the actual character set to be specified. As EBCDIC had been originally specified, it contained 88 graphics and the Space character. The lower-case alphabet comprised 26 of these graphics, leaving a 62 graphic set. These 62 EBCDIC graphics and the Space character would constitute 63 of the 64 small card character set. These graphics, in their code positions, are shown in a partial EBCDIC code chart (Fig. 27.4).

	Column	0	1	2	3	4	5	6	7	8	9	A	B	C	D	E	F
	Bit Pat. →	00				01				10				11			
Row ↓		00	01	10	11	00	01	10	11	00	01	10	11	00	01	10	11
0	0000					SP	-	&									0
1	0001						/							A	J		1
2	0010													B	K	S	2
3	0011													C	L	T	3
4	0100													D	M	U	4
5	0101													E	N	V	5
6	0110													F	O	W	6
7	0111													G	P	X	7
8	1000													H	Q	Y	8
9	1001													I	R	Z	9
A	1010					¢	!		:								
B	1011					.	$	,	#								
C	1100					<	*	%	@								
D	1101					(	)	_	'								
E	1110					+	;	>	=								
F	1111					|	¬	?	"								

Fig. 27.4 EBCDIC graphics

It was observed that if EBCDIC bits 0 and 1 were dropped, this set collapsed neatly into a 6-bit code, as shown in Fig. 27.5, which was virtually the required card code for the small card.

Bit Pattern		0 0	0 1	1 0	1 1
0 0 0 0		SP	&	-	0
0 0 0 1		A	J	/	1
0 0 1 0		B	K	S	2
0 0 1 1		C	L	T	3
0 1 0 0		D	M	U	4
0 1 0 1		E	N	V	5
0 1 1 0		F	O	W	6
0 1 1 1		G	P	X	7
1 0 0 0		H	Q	Y	8
1 0 0 1		I	R	Z	9
1 0 1 0		¢	!		:
1 0 1 1		.	$	,	#
1 1 0 0		<	*	%	@
1 1 0 1		(	)	_	'
1 1 1 0		+	;	>	=
1 1 1 1		|	¬	?	"

Fig. 27.5 6-bit code, Version 1

27.6 APPLICATION OF CRITERION 7

Criterion 7 had specified no zone punches for the numerics. Analyses had shown that numeric data constituted about 75 percent of the data punched on regular cards. It was assumed that the same would hold true for small card applications. In order to have as few holes as possible punched on a small card, it was clear that numerics should have no zone punches. This suggested that the two high-order bits (Fig. 27.5) should be reversed. The result, using the BA8421 bit-naming notation, is shown in Fig. 27.6.

Bit Pattern		BA	B	A	No Zone Bits
No Bits		SP	&	–	0
1		A	J	/	1
2		B	K	S	2
2 1		C	L	T	3
4		D	M	U	4
4 1		E	N	V	5
4 2		F	O	W	6
4 2 1		G	P	X	7
8		H	Q	Y	8
8 1		I	R	Z	9
8 2		¢	!		:
8 2 1		.	$	,	#
8 4		<	*	%	@
8 4 1		(	)	_	'
8 4 2		+	;	>	=
8 4 2 1		[	¬	?	"

Fig. 27.6 6-bit code, Version 2

27.7 APPLICATION OF CRITERIA 5 AND 6

Attention now turned to Criteria 5 and 6. Space should be No Punches, and negative numerics should be accommodated by overpunching positive numerics.

An obvious possibility was to prescribe the code column containing J, K, L, . . . , R as being equivalent to negative numerics −1, −2, −3, . . . , −9, as they are in EBCDIC. Then the – in the top row of this column could stand for −0. But, since negative numerics must be derivable by overpunching absolute numerics, this would require 0 to be

No Punches. Criterion 6 had specified the assignment of No Punches to the Space character. Here was a problem.

The solution was seen when the graphics were rearranged back into their EBCDIC code positions, with the DCBA8421 code superimposed, as in Fig. 27.7 (The code positions with entries of the form [x] and the small numbered squares below the code table will be explained later.)

<table>
<tr><td></td><td></td><td>Column</td><td>0</td><td>1</td><td>2</td><td>3</td><td>4</td><td>5</td><td>6</td><td>7</td><td>8</td><td>9</td><td>A</td><td>B</td><td>C</td><td>D</td><td>E</td><td>F</td><td></td></tr>
<tr><td></td><td>Bit Pat.</td><td></td><td colspan="4">00</td><td colspan="4">01</td><td colspan="4">10</td><td colspan="4">11</td><td></td></tr>
<tr><td></td><td></td><td></td><td>00</td><td>01</td><td>10</td><td>11</td><td>00</td><td>01</td><td>10</td><td>11</td><td>00</td><td>01</td><td>10</td><td>11</td><td>00</td><td>01</td><td>10</td><td>11</td><td></td></tr>
<tr><td></td><td></td><td>Hole Pat.</td><td></td><td></td><td></td><td></td><td></td><td></td><td></td><td></td><td></td><td></td><td></td><td></td><td></td><td></td><td></td><td></td><td>Hole Pat.</td></tr>
<tr><td></td><td></td><td></td><td></td><td></td><td></td><td></td><td>B</td><td>B</td><td></td><td></td><td></td><td></td><td></td><td></td><td>B</td><td>B</td><td></td><td></td><td></td></tr>
<tr><td>Row</td><td></td><td></td><td></td><td></td><td></td><td></td><td>A</td><td></td><td>A</td><td></td><td></td><td></td><td></td><td></td><td>A</td><td></td><td>A</td><td></td><td></td></tr>
<tr><td>0</td><td>0000</td><td></td><td>[1]</td><td>[2]</td><td>[3]</td><td>[4]</td><td>[5] SP</td><td>[6] &</td><td>[7] -</td><td>[8]</td><td>[9]</td><td>[10]</td><td>[11]</td><td>[12]</td><td>[13]</td><td>[14]</td><td>[15]</td><td>[16] 0</td><td></td></tr>
<tr><td>1</td><td>0001</td><td>1</td><td></td><td></td><td>[17]</td><td></td><td></td><td></td><td>[18] /</td><td></td><td></td><td></td><td>[19]</td><td></td><td>A</td><td>J</td><td>[20]</td><td>1</td><td></td></tr>
<tr><td>2</td><td>0010</td><td>2</td><td></td><td></td><td></td><td></td><td></td><td></td><td></td><td></td><td></td><td></td><td></td><td></td><td>B</td><td>K</td><td>S</td><td>2</td><td></td></tr>
<tr><td>3</td><td>0011</td><td>21</td><td></td><td></td><td></td><td></td><td></td><td></td><td></td><td></td><td></td><td></td><td></td><td></td><td>C</td><td>L</td><td>T</td><td>3</td><td></td></tr>
<tr><td>4</td><td>0100</td><td>4</td><td></td><td></td><td></td><td></td><td></td><td></td><td></td><td></td><td></td><td></td><td></td><td></td><td>D</td><td>M</td><td>U</td><td>4</td><td></td></tr>
<tr><td>5</td><td>0101</td><td>4 1</td><td></td><td></td><td></td><td></td><td></td><td></td><td></td><td></td><td></td><td></td><td></td><td></td><td>E</td><td>N</td><td>V</td><td>5</td><td></td></tr>
<tr><td>6</td><td>0110</td><td>42</td><td></td><td></td><td></td><td></td><td></td><td></td><td></td><td></td><td></td><td></td><td></td><td></td><td>F</td><td>O</td><td>W</td><td>6</td><td></td></tr>
<tr><td>7</td><td>0111</td><td>421</td><td></td><td></td><td></td><td></td><td></td><td></td><td></td><td></td><td></td><td></td><td></td><td></td><td>G</td><td>P</td><td>X</td><td>7</td><td></td></tr>
<tr><td>8</td><td>1000</td><td>8</td><td></td><td></td><td></td><td></td><td></td><td></td><td></td><td></td><td></td><td></td><td></td><td></td><td>H</td><td>Q</td><td>Y</td><td>8</td><td></td></tr>
<tr><td>9</td><td>1001</td><td>8 1</td><td></td><td></td><td></td><td></td><td></td><td></td><td></td><td></td><td></td><td></td><td></td><td></td><td>I</td><td>R</td><td>Z</td><td>9</td><td></td></tr>
<tr><td>A</td><td>1010</td><td>8 2</td><td></td><td></td><td>[21]</td><td></td><td>¢</td><td>!</td><td>[22]</td><td>:</td><td></td><td></td><td></td><td></td><td></td><td></td><td></td><td></td><td></td></tr>
<tr><td>B</td><td>1011</td><td>8 21</td><td></td><td></td><td></td><td></td><td>.</td><td>$</td><td>,</td><td>#</td><td></td><td></td><td></td><td></td><td></td><td></td><td></td><td></td><td></td></tr>
<tr><td>C</td><td>1100</td><td>84</td><td></td><td></td><td></td><td></td><td><</td><td>*</td><td>%</td><td>@</td><td></td><td></td><td></td><td></td><td></td><td></td><td></td><td></td><td></td></tr>
<tr><td>D</td><td>1101</td><td>84 1</td><td></td><td></td><td></td><td></td><td>(</td><td>)</td><td>_</td><td>'</td><td></td><td></td><td></td><td></td><td></td><td></td><td></td><td></td><td></td></tr>
<tr><td>E</td><td>1110</td><td>842</td><td></td><td></td><td></td><td></td><td>+</td><td>;</td><td>></td><td>=</td><td></td><td></td><td></td><td></td><td></td><td></td><td></td><td></td><td></td></tr>
<tr><td>F</td><td>1111</td><td>8421</td><td></td><td></td><td></td><td></td><td>|</td><td>¬</td><td>?</td><td>"</td><td></td><td></td><td></td><td></td><td></td><td></td><td></td><td></td><td></td></tr>
<tr><td></td><td></td><td>Hole Pat.</td><td></td><td></td><td></td><td></td><td></td><td></td><td></td><td></td><td></td><td></td><td></td><td></td><td></td><td></td><td></td><td></td><td></td></tr>
</table>

Hole Patterns:

[1] [2] [3] [4] [5] [6] [7] [8] [9] [10] [11] [12] [13] [14] [15] [16] [17] [18] [19] [20] [21] [22]

Block	Hole Patterns at:
	Top and Left
	Bottom and Left
	Top and Right
	Bottom and Right

Fig. 27.7 EBCDIC, Version 1

27.8 HOLE PATTERNS FOR MINUS, ZERO, AND MINUS ZERO

The four code positions of the top row of Fig. 27.6 will have the hole patterns of BA, B, A, and No Punches. The problem is to distribute these four hole patterns along the top row of Fig. 26.7 so that Criteria 5, 6, and 7 are met. Criterion 6 says that the Space character shall have the hole pattern of No Punches. This then leaves hole patterns BA, B, and A to accommodate the characters −, −0, and 0. Criterion 7 specified that all numerics should have no zone punches. This is clearly not possible, if Space is to be the No-Punches hole pattern; that is to say, although numerics 1 to 9 can have no zone punches, 0 *must* have a zone punch, since only hole patterns with zone punches remain, BA, B, and A. Since 0 cannot have a hole pattern with no zone punches, the next best situation is to have a hole pattern with one zone punch only; that is, either B or A.

The objective behind Criterion 7 was to minimize the number of holes in the hole patterns for numerics. A single hole as the hole pattern for 0, either A or B, really meets the spirit of this objective. The fact that the single hole of the hole pattern is a zone punch rather than a digit punch is not as important as the fact that 0 has a minimum number of holes (namely, one hole) in its hole pattern.

So the problem now was to choose between B and A as the hole pattern for 0. Hole patterns for −0 and − also had to be determined. There were two possibilities:

Possibility 1.				**Possibility 2.**			
	A	for	0		B	for	0
	B	for	−		A	for	−
	BA	for	−0		BA	for	−0

As shown in Fig. 27.6, the column containing J, K, L, . . . , R had been decided to have the zone-punch B. But as signed numerics, J, K, L, . . . , R will correspond to −1, −2, −3, . . . , −9. The character in Fig. 27.7 designated by [14] must represent −0. The zone-punch B, then, must clearly represent the overpunched sign for negative numerics; that is, BA will represent −0. If B is to be the overpunch turning 0 into −0, then 0 must start out as A. So of the two, Possibility 1 was preferable. Therefore, in Fig. 27.7, code positions [6], [14], and [16] will have hole patterns of B, BA, and A, respectively.

We now look again at the collapsed 6-bit set of Fig. 27.6. With the assignments for −, −0, and 0 as in the paragraph above, the graphics in the top row of the table will change, as shown in Fig. 27.8.

We observe that & has not yet an assigned hole pattern, and that there is a hole pattern, A82, in the table (shaded) that has no assigned graphic. It has to be concluded that & will be assigned the hole pattern of A82.

Bit Pattern		BA	B	A	
		-0	-	0	SP
1		A	J	/	1
2		B	K	S	2
2 1		C	L	T	3
4		D	M	U	4
4 1		E	N	V	5
4 2		F	O	W	6
4 2 1		G	P	X	7
8		H	Q	Y	8
8 1		I	R	Z	9
8 2		¢	!		:
8 2 1		.	$	,	#
8 4		<	*	%	@
8 4 1		(	)	_	'
8 4 2		+	;	>	=
8 4 2 1		|	¬	?	"

Fig. 27.8 6-bit code, Version 3

27.9 MINUS ZERO

Before returning to consideration of the 8-bit EBCDIC code table, there is another small problem to solve. When data is entered into a computer, and then listed, unaltered, for debugging purposes, those card fields which had overpunched numerics will list as J, K, L, . . . , R for -1, -2, $-3, \ldots, -9$, respectively. The fact that alphabetics list for signed numerics in a debug listing is quite satisfactory to users. The important fact is that a graphic for -1 is distinguishable from a graphic for 1. In final output listings, of course, it is customary to separate out the minus sign, and list it adjacent to the numeric. But for debug purposes, alphabetics are quite acceptable. The problem is, what is to be listed for -0? What graphic should be assigned to EBCDIC code-position hex D0?

The engineers designing the small card system recommended a graphic θ, since this was clearly representative as a graphic for -0. But EBCDIC had a graphic already assigned to code-position Hex D0; namely, the graphic } (closing brace). It was therefore insisted that the graphic to represent -0 be }. This was not attractive to the engineers for the small card system, for two reasons:

1. The graphic } is not representative of the concept -0.
2. To provide the graphic } without providing its companion graphic { seemed bizarre.

Reason 1 was disposed of quickly. After all, J, K, L, . . . , R are not representative of the concepts -1, -2, $-3, \ldots, -9$.

Reason 2 was not disposed of so easily. To begin, the companion graphic { *is* assigned to EBCDIC, in code-position hex C0. So why not include it in the small card system's graphic set? But all 64 graphic positions in the collapsed 6-bit set were assigned. If { were to be assigned, then one of the previously assigned graphics must be left out of the set. Which one? As it turned out, serious consideration was not given to this question, because a more subtle but more important aspect arose—the translation–simplicity aspect of Criterion 8.

27.10 CRITERION 8, TRANSLATION SIMPLICITY

The simplest possible translation relationship would be where the bit code would be on a one-to-one relationship with the card code—a bit in a bit pattern would become a hole in a hole pattern. This relationship had already been aborted by previous design decisions:

- To reverse bit-code zone-bits 2 and 3 for card-code zone-punches B and A. Actually, an inversion of a bit, if applied uniformly to *all* bit patterns, does not make the translation circuitry any more complex.
- Assignments of hole-patterns No Punches, B, BA, and A to EBCDIC code-positions hex 40, 60, D0, and F0, respectively, certainly are exceptions to, and therefore complicate, the bit-code–to–card-code relationships.

- Assignment of hole pattern A82 to EBCDIC code-position hex 50 is an even more complicating exception than the exceptions stated just above.
- Assignment of A1 to / in the 6-bit code table (Fig. 27.8) results in A1 being the assignment for hex 61 in the 8-bit code table (Fig. 27.7).

27.11 THE MUSICAL-CHAIRS EFFECT

It should be realized that exceptions to bit-code–to–card-code translation relationships have a musical-chairs effect. For each exception, there automatically results another one.

For example, hole-pattern A82 had been decided to correspond to hex-position 50. But hole-pattern A82 would "naturally" correspond to hex position A6. Therefore, some other hole patterns must be assigned to hex-position A6, and that assignment will necessarily be an exception also.

The suggestion that { be assigned in the graphic set of the small card system meant that it would have to be included in the collapsed 6-bit set. But, even as the & received the exception hole pattern of A82, the hole pattern for { would have to be an exception also. If, for example, it had been decided to leave out the ¢ so that the { could be included, then the hole pattern for { would have to be the hole pattern previously assigned to ¢; namely, BA82. This would mean that hole-pattern BA82 would be assigned to the EBCDIC hex position for {; namely, hex C0. And this is clearly a translation exception. And, by the musical-chairs effect, a translation exception would automatically be created somewhere else in the EBCDIC code table.

Translation exceptions lead to an increase in translation complexity. An increase in translation complexity leads to an increase in translation circuitry and, hence, to an increase in cost.

So the trade-off situation was

provide } but not {

which might seem bizarre, or

provide { as well as }

and increase the cost of the system. Since a major objective for the small card system was low cost, the cost argument was decisive. Therefore, } *was provided* (for −0) and { *was not provided.*

27.12 THE FINAL 6-BIT SET

All design decisions had now been reached for the collapsed 6-bit set for the small card system. The result is shown in Fig. 27.9. This code table, then, specifies the card hole patterns for the 63 graphics and the Space character of the small card system. The partially completed 256-character, 8-hole card code for the small card is shown in Fig. 27.10. Exception hole patterns are indicated by the small numbers in squares.

Bit Pattern → ↓		BA	B	A	
		}	–	0	SP
1		A	J	/	1
2		B	K	S	2
2 1		C	L	T	3
4		D	M	U	4
4 1		E	N	V	5
4 2		F	O	W	6
4 2 1		G	P	X	7
8		H	Q	Y	8
8 1		I	R	Z	9
8 2		¢	!	&	:
8 2 1		.	$	,	#
8 4		<	*	%	@
8 4 1		(	)	–	'
8 4 2		+	;	>	=
8 4 2 1		|	¬	?	"

Fig. 27.9 6-bit code, Version 4

27.13 COMPLETION OF THE CARD CODE

Attention was now focussed on completing the card code for the small card system. In Fig. 27.10, the specials in hex-columns 4, 5, 6, and 7 had been previously decided to have zone punches of BA, B, A, and No

Row	Bit Pat.	Hole Pat.	0	1	2	3	4	5	6	7	8	9	A	B	C	D	E	F
	Bit Pat. (zone)		00				01				10				11			
			00	01	10	11	00	01	10	11	00	01	10	11	00	01	10	11
		Hole Pat.					B	B							B	B		
							A		A						A		A	
0	0000		[1]	[2]	[3]	[4]	[5] SP	[6] &	[7] -	[8]	[9]	[10]	[11]	[12]	[13]	[14] }	[15]	[16] 0
1	0001	1			[17]				[18] /				[19]		A	J	[20]	1
2	0010	2													B	K	S	2
3	0011	21													C	L	T	3
4	0100	4													D	M	U	4
5	0101	4 1													E	N	V	5
6	0110	42													F	O	W	6
7	0111	421													G	P	X	7
8	1000	8													H	Q	Y	8
9	1001	8 1													I	R	Z	9
A	1010	8 2			[21]		¢	!	[22]	:								
B	1011	8 21					.	$	,	#								
C	1100	84					<	*	%	@								
D	1101	84 1					(	)	_	'								
E	1110	842					+	;	>	=								
F	1111	8421					|	¬	?	"								

Hole Patterns:

[1]	[7] B	[13]	[19]
[2]	[8]	[14] BA	[20]
[3]	[9]	[15]	[21]
[4]	[10]	[16] A	[22]
[5] No Pch	[11]	[17]	
[6] A82	[12]	[18] A1	

Block	Hole Patterns at:
	Top and Left
	Bottom and Left
	Top and Right
	Bottom and Right

Fig. 27.10 96-column card code, Version 1

Zone, respectively (see Fig. 27.9). The same set of zone punches had been assigned to the alphabetics and numerics in hex-columns C, D, E, and F. It was clear, then, that the code positions for the top rows of hex-columns 4, 5, 6, and 7 and the code positions for the bottom rows of hex-columns C, D, E, and F could not have zone-punches BA, B, A, and No-Zone. That is, for these eight code columns, zones for the block of bottom rows would be different than zones for the block of top rows.

It is a fact of the theory of translation relationships that if a zone difference (bottom and top blocks) applies for 8 of the 16 code columns, translation simplicity will be enhanced if zone difference (bottom and top blocks) is also applied to the other 8 code columns.

This then became a further criterion. It would also enhance translation simplicity if the sequence of zone-assignments BA, B, A, and No Zone was applied both to hex-columns 0, 1, 2, and 3 and hex-columns 8, 9, A, and B, respectively.

27.14 FURTHER CRITERIA

The two new criteria are now enunciated:

Criterion 9

The zone difference between hex-rows 0 through 9 and between hex-rows A through F, already decided for hex-columns 4, 5, 6, 7, and C, D, E, F, should also be applied to hex-columns 0, 1, 2, 3 and 8, 9, A, B.

Criterion 10

The sequence of zone-patterns BA, B, A, and No Zone should be applied to hex-columns 0, 1, 2, 3; to hex-columns 4, 5, 6, 7; to hex columns 8, 9, A, B; and to hex-columns C, D, E, F.

Available zone patterns are DCBA, DCB, DCA, DC, CBA, CB, CA, C. A further fact of the theory of translation relationships is that translation simplicity would be enhanced if zone-punch D was applied to hex-columns 0 through 7, and not to hex-columns 8 through F; and if zone-punch C was applied both to hex-columns 0, 1, 2, 3 and hex-columns 8, 9, A, B, and not to hex-columns 4, 5, 6, 7 and hex-columns C, D, E, F. This application would be to hex-rows 0 through 9. By consequence of Criterion 9, the opposite assignment of D and C zones should be applied to hex-rows A through F. With the translation exceptions already noted in Fig. 27.9, this leads to an assignment of zone patterns as shown in Fig. 27.11.

Row	Bit Pat.	Hole Pat.	0	1	2	3	4	5	6	7	8	9	A	B	C	D	E	F
		Column	0	1	2	3	4	5	6	7	8	9	A	B	C	D	E	F
	Bit Pat.		00				01				10				11			
			00	01	10	11	00	01	10	11	00	01	10	11	00	01	10	11
		Hole Pat.	D	D	D	D	D	D	D	D								
			C	C	C	C					C	C	C	C				
			B	B			B	B			B	B			B	B		
			A		A		A		A		A		A		A		A	
0	0000		1	2	3	4	5	6	7	8	9	10	11	12	13	14	15	16
1	0001	1			17				18				19				20	
2	0010	2																
3	0011	21																
4	0100	4																
5	0101	4 1																
6	0110	42																
7	0111	421																
8	1000	8																
9	1001	8 1																
A	1010	8 2			21				22									
B	1011	8 21																
C	1100	84																
D	1101	84 1																
E	1110	842																
F	1111	8421																
		Hole Pat.									D	D	D	D	D	D	D	D
			C	C	C	C					C	C	C	C				
			B	B			B	B			B	B			B	B		
			A		A		A		A		A		A		A		A	

Hole Patterns:

1	7 B	13	19
2	8	14	20
3	9	15	21
4	10	16	22
5 No Pch	11	17	
6 A82	12	18	

1
2

Block	Hole Patterns at:
1	Top and Left
2	Bottom and Left

Fig. 27.11 96-column card code, Version 2

27.15 EXCEPTION TRANSLATIONS

As shown in Fig. 27.11, there were six exception translations. From the musical-chairs effect previously cited, we would expect there to be more translation exceptions than the number shown in Fig. 27.10. And this turns out to be so.

The six hexadecimal positions to which exception hole patterns have been assigned are listed below in Table 27.1. Also shown are the hexadecimal positions from which the exception hole patterns originated, and the hole patterns which would have been expected in the exception positions.

TABLE 27.1 Exception hole patterns

Exception positions (hexadecimal)	Exception hole patterns	Origin of exception hole patterns	Expected hole patterns
40	No Punches	F0	DBA
50	A82	6A	DB
60	B	D0	DA
D0	BA	C0	B
F0	A	E0	No Punches
61	A1	E1	DA1

Examination of the columns "Exception positions" and "Origin of exception hole patterns" in Table 27.1 reveals that hex-positions D0 and F0 are in both columns, but hex-positions 6A, C0, E0, and E1 are four further hex-positions in which the musical-chairs effect will be manifested. Also, examination of columns "Exception hole patterns" and "Expected hole patterns" reveals hole patterns No Punches and B to be in both columns, but hole-patterns DBA, DB, DA, and DA1 are four further hole patterns to be distributed to the four hex positions noted in the previous sentence, to complete the musical-chairs effect.

The musical-chairs effect, then, extends Table 27.1 into Table 27.2 shown below. The ten exception hole patterns, shown in Fig. 27.12, make up the 96-column card code, Version 3.

TABLE 27.2 Musical-chairs effect

Exception positions (hexadecimal)	Exception hole patterns	Origin of exception hole patterns	Expected hole patterns
6A	DBA	40	A82
C0	DA	60	BA
E0	DB	50	A
E1	DA1	61	A1

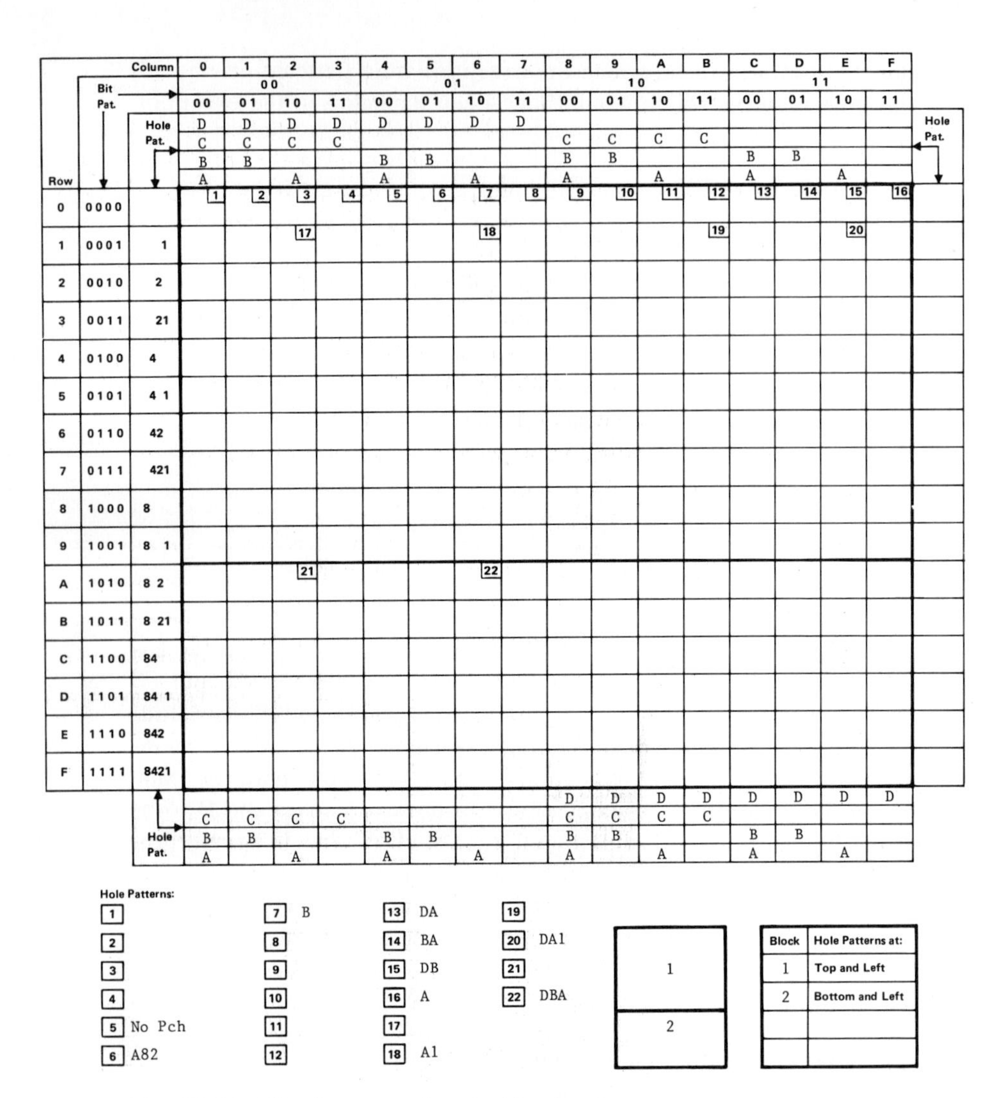

Fig. 27.12 96-column card code, Version 3

Examination of Tables 27.1 and 27.2 combined shows that each hex position in the "Exception positions" columns matches a hex position in the "Origin of exception hole patterns" column, and each hole pattern in the "Exception hole pattern" column matches a hole pattern in the "Expected hole pattern" column.

27.16 REDUCTION OF TRANSLATION COMPLEXITY

It is an interesting anomaly of translation relationships that if one exception is introduced, the consequent translation complexity will be reduced if some additional compensating exceptions are forced in. This is not the musical-chairs aspect previously noted. This anomalous fact is best illustrated with examples. Consider Fig. 27.13, a code chart which would exhibit the optimally simple translation relationship between the eight holes of the 96-column card, D, C, B, A, 8, 4, 2, 1, and the eight bits of an EBCDIC byte, 0, 1, 2, 3, 4, 5, 6, 7. Then the holes-to-bits relationship is simply inverse for the high-order four and direct for the low-order four.

96-column card hole		EBCDIC bit	
D	=	0	This notation means that if there are D, C, B, A holes, the 0, 1, 2, 3 bits respectively are 0.
C	=	1	
B	=	2	
A	=	3	
8	=	4	If there are 8, 4, 2, 1 holes, the 4, 5, 6, 7 bits respectively are 1.
4	=	5	
2	=	6	
1	=	7	

Now suppose for some reason, it is required to swap the hole patterns for code positions ([1]) and ([2]).

	EBCDIC bit pattern	Card hole pattern	
		Before swap	After swap
Code position ([1])	0010 1101	D C A 8 4 1	D C B A 1 2
Code position ([2])	0000 0010	D C B A 2	D C A 8 4 1

Consider the situation in going from card hole patterns to EBCDIC bit patterns. As exceptions to the general translation equations above, additional circuitry must be added to

1. detect hole-pattern DCBA2 and generate bit pattern 0010 1101,
2. detect hole-pattern DCA841 and generate EBCDIC bit-pattern 0000 0010.

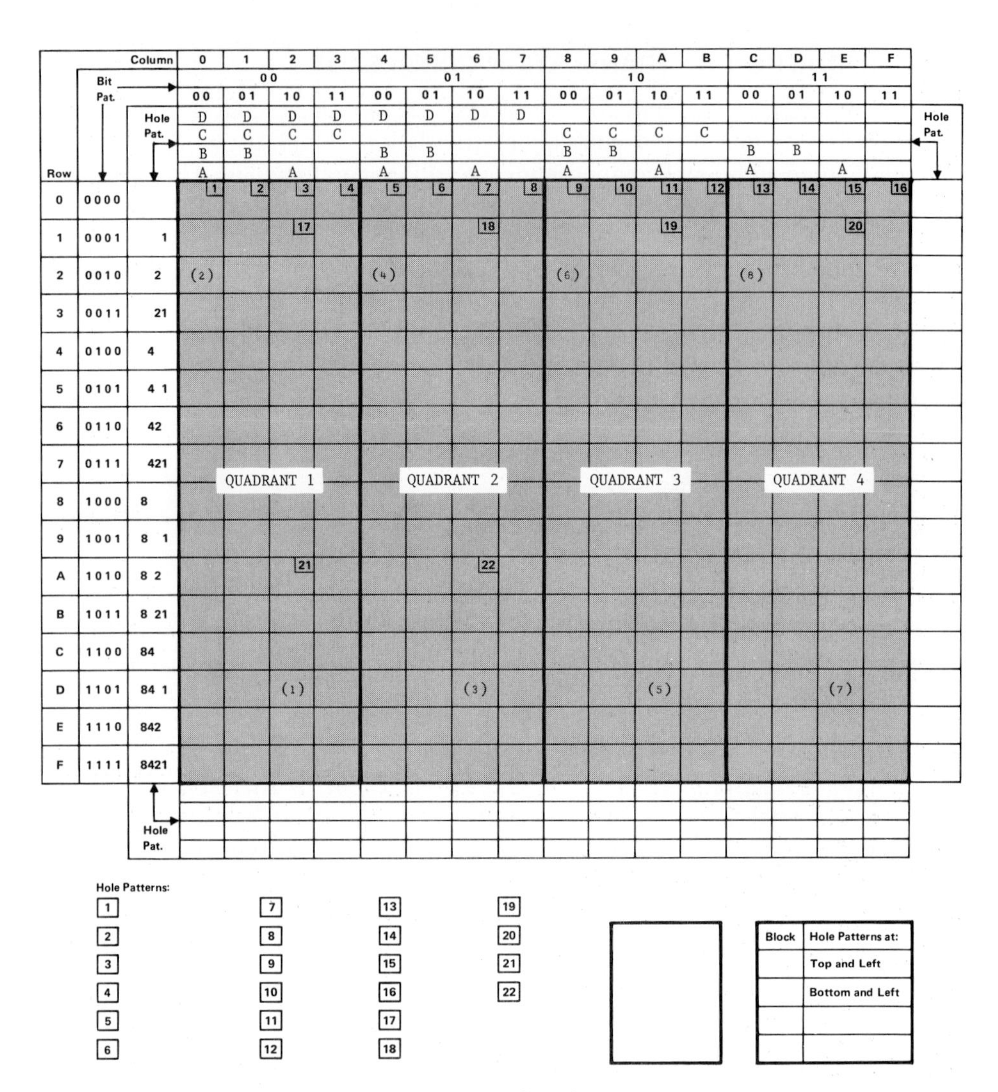

Fig. 27.13 Optimal 96-column card code

Figure 27.13 is shown divided into four quadrants. The quadrants are distinguished one from another by two high-order zone bits 0 and 1, or, equally, by the two high-order zone holes, D and C. The swap that was proposed above took place wholly in quadrant 1.

What would be the result if analagous swaps were made in quadrants 2, 3, and 4? That is, swap (^{3}) and (^{4}), (^{5}) and (^{6}), (^{7}) and (^{8}).

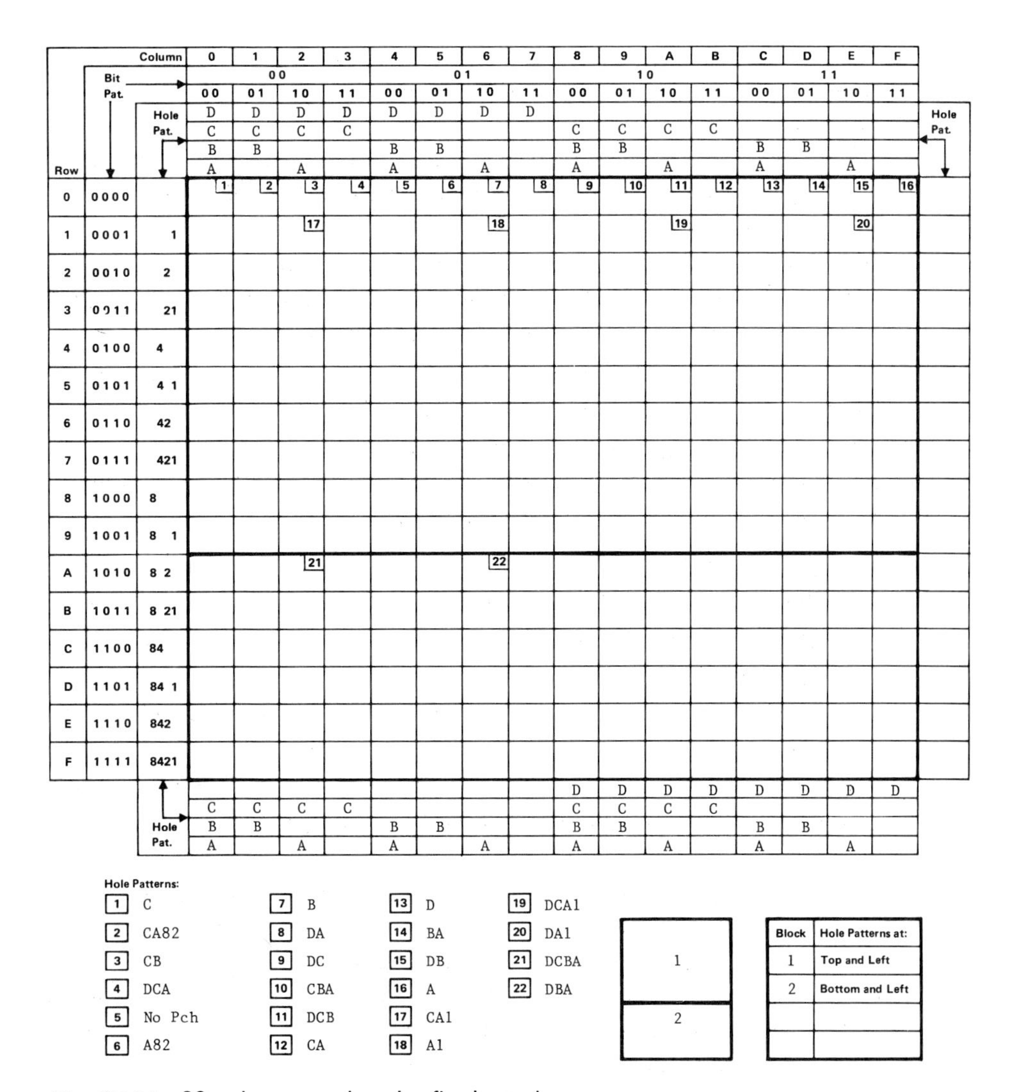

Fig. 27.14 96-column card code, final version

The interesting result is that the detection circuitry can now ignore zone-holes D and C. That is to say, instead of having to detect and analyze all eight card rows, D, C, B, A, 8, 4, 2, 1, only six, B, A, 8, 4, 2, 1 have to be detected and analyzed. Similarly, in going from bit patterns to hole patterns, the two high-order zone bits need not be detected and analyzed. In short, by forcing the exception translation of quadrant 1 into quadrants 2, 3, and 4, detection and analysis circuitry have been reduced; the translation complexity has been reduced.

It is also a fact that if the exception translation of quadrant 1 had been forced into only one other quadrant, the translation complexity would have been reduced, but not reduced as much as it would have been if the exception had been forced into all three other quadrants.

With respect to the 96-column card code, given that certain translation exceptions were required for reasons already given, the translation relationships were consequently complicated. By exercising their art to force further exception translations, the engineers did simplify the translation complexity considerably. In Fig. 27.12, six translation exceptions are shown. Before the 96-column card code was finished, 22 exceptions were forced, as shown in Fig. 27.14.

27.17 SIMPLIFICATION OF TRANSLATION COMPLEXITY

A glance at the translation equations shown in Fig. 27.15 for the 96-column card, Version 3 (Fig. 27.12) and in Fig. 27.16 for the 96-column card, Final Version (Fig. 27.14) reveals that considerable simplification took place in the equations for E0, E2, and E3, as shown below in the total counts of connectors.

	Version 3	Final Version
Common expressions	27	26
Equations	51	34
Total	78	60

27.18 SUMMARY

Criteria 1, 2, 3, 4, 5, 6 were met. Criterion 7 was not met by the hole pattern for numeric 0, but numeric 0 had only one hole in its hole pattern. The objective of Criterion 7 was to minimize the number of holes in the hole patterns for numerics; this objective was achieved.

Criteria 3, 5, 6, and 7 led to exception translations. Given these exception translations, Criterion 8 was met as well as possible.

Common expressions

$$X = \bar{8} \wedge \bar{4} \wedge \bar{2} \wedge \bar{1}$$
$$Y = \bar{8} \wedge \bar{4} \wedge \bar{2} \wedge 1$$
$$Z = 8 \wedge \bar{4} \wedge 2 \wedge \bar{1}$$
$$K = \bar{D} \wedge \bar{C} \wedge \bar{B} \wedge A \wedge \bar{4} \wedge \bar{1}$$
$$F = [\bar{4} \wedge (8 \veebar 2)] \veebar [\bar{8} \wedge 4]$$
$$G = 8 \wedge \overline{(\bar{4} \wedge \bar{2})}$$
$$H = 8 \wedge \overline{[\bar{4} \wedge \overline{(2 \wedge 1)}]}$$
$$j = X \wedge D \wedge \bar{C} \wedge B \wedge A$$

Equations

$$\mathrm{E0} = \{X \wedge [\bar{D} \wedge \overline{(\bar{C} \wedge \bar{A})}] \veebar [(D \wedge \bar{C}) \wedge (B \veebar A)]\} \veebar \{Y \wedge [D \equiv (\bar{C} \wedge \bar{B} \wedge A)]\} \veebar \{(F \wedge \bar{D}) \veebar (G \veebar D)\}$$
$$\mathrm{E1} = \bar{C}$$
$$\mathrm{E2} = \{\bar{B} \wedge \{\{X \wedge \{C \veebar [\bar{C} \wedge (D \veebar A)]\}\} \veebar [Z \wedge \overline{(\bar{D} \wedge \bar{C} \wedge A)}] \veebar [L \veebar F \veebar H]\}\} \veebar \{(B \wedge \bar{C}) \wedge \overline{(\bar{D} \wedge A)} \wedge X\}$$
$$\mathrm{E3} = \{\bar{A} \wedge \{\{X \veebar [C \veebar (D \wedge \bar{C} \wedge \bar{B})]\} \veebar \bar{X}\}\} \veebar \{[A \wedge \bar{D} \wedge \bar{C}] \wedge [X \veebar (\bar{B} \wedge Z)]\}$$
$$\mathrm{E4} = [8 \wedge \overline{(K \wedge Z)}] \veebar J$$
$$\mathrm{E5} = 4$$
$$\mathrm{E6} = [2 \wedge \overline{(K \wedge 8)}] \veebar J$$
$$\mathrm{E7} = 1$$

Fig. 27.15 96-column card, Version 3

Common expressions

$$X = \bar{8} \wedge \bar{4} \wedge \bar{2} \wedge \bar{1}$$
$$Y = \bar{8} \wedge \bar{4} \wedge \bar{2} \wedge 1$$
$$Z = 8 \wedge \bar{4} \wedge 2 \wedge \bar{1}$$
$$F = [\bar{4} \wedge (8 \veebar 2)] \veebar [\bar{8} \wedge 4]$$
$$G = 8 \wedge \overline{(\bar{4} \wedge \bar{2})}$$
$$H = 8 \wedge \overline{[\bar{4} \wedge \overline{(2 \wedge 1)}]}$$
$$I = \overline{(\bar{D} \wedge A)}$$
$$J = D \wedge B \wedge A$$
$$K = X \wedge J$$
$$M = \bar{D} \wedge \bar{B} \wedge A \wedge \bar{4} \wedge \bar{1}$$

Equations

$$\mathrm{E0} = \{X \wedge (D \veebar A)\} \veebar \{\{Y \wedge [D \equiv (\bar{B} \wedge A)]\} \veebar \{(F \wedge \bar{D}) \veebar (G \wedge D)\}\}$$
$$\mathrm{E1} = \bar{C}$$
$$\mathrm{E2} = \{\bar{B} \wedge \{[X \wedge A] \veebar [Z \wedge I] \veebar Y \veebar F \veebar H\}\} \veebar B \wedge X \wedge I\}$$
$$\mathrm{E3} = \{A \wedge \{[X \wedge \overline{(D \wedge B)}] \veebar [Z \wedge \bar{D} \wedge \bar{B}]\}\} \veebar \{\bar{A} \wedge \bar{X}\}$$
$$\mathrm{E4} = \{8 \wedge \overline{(M \wedge 2)}\} \veebar K$$
$$\mathrm{E5} = 4$$
$$\mathrm{E6} = \{2 \wedge \overline{(M \wedge 8)}\} \veebar K$$
$$\mathrm{E7} = 1$$

Fig. 27.16 96-column card, Final Version

Glossary

The following reference terminology is used in these definitions:

Contrast with

Refers to a term that has an opposed or substantively different meaning.

Synonym for

Indicates that the term has the same meaning as another term, which is defined.

Synonymous with

Identifies terms that are synonyms for the term being defined.

Acronym for

An abbreviation generally consisting of the first letters of the words of a term.

See

Refers to multiple-word terms that have the same last word and are defined.

See also

Refers to related terms that have a similar, but not synonymous, meaning.

Deprecated term for

Indicates that the term should not be used. It refers to a preferred term, which is defined.

A-Bit A bit in the A position of a Byte whose bit positions are named B, A, 8, 4, 2, 1 from high order to low order.

Absolute Numeric A numeric with neither a negative nor a positive associated sign. *Contrast with* Positive Numeric, Negative Numeric, Signed Numeric.

Accented Alphabetic An alphabetic, as in the French or Italian alphabets, with an associated accent, such as a grave accent, an acute accent, a circumflex accent. *See also* Diacritic Alphabetic.

Alphabet A set of all the Alphabetics used in a language, including Diacritic Alphabetics and Accent Alphabetics. *See also* Cyrillic Alphabet, Duocase Alphabet, Latin Alphabet, Monocase Alphabet, Non-Latin Alphabet. *Contrast with* Katakana Symbols.

Alphabetic A letter in the Alphabet of a country. Generally taken to mean a letter of the Latin Alphabet, but sometimes must be particularized, as Latin alphabetic, Cyrillic alphabetic, Greek alphabetic, Hebraic alphabetic, and so on. *See also* Small Alphabetic, Capital Alphabetic.

Alphabetic Character An Alphabetic together with its associated Bit Patterns or Hole Pattern.

Alphabetic Extender Positions Positions reserved in EBCDIC for Graphics particular to a country. *See also* National Use Positions.

American National Standard Code For Information Interchange A Coded Character Set consisting of 128, 7-bit Characters. There are 32 Control Characters, 94 Graphic Characters, the Space Character, and the Delete Character.

ANSI *Acronym for* the American National Standards Institute.

AND A logic operator with the property that if *A* and *B* are Binary Variables, *A* AND *B* is 1 if both *A* and *B* are 1, and is 0 if *A* is 0 and *B* is 1 or if *A* is 1 and *B* is 0 or if *A* is 0 and *B* is 0.

ASA *Acronym for* the American Standards Association, now the American National Standards Institute.

ASCII *Acronym for* the American National Standard Code For Information Interchange.

Baudot Code *Synonym for* CCITT #2.

B-Bit A bit in the B position of a Byte whose bit positions are named B, A, 8, 4, 2, 1.

BCD *Acronym for* Binary Coded Decimal.

BCD Code A Code that has the characteristic that the low-order 4 bits of the Bit Patterns of the numerics are Binary Coded Decimal.

BCDIC *Acronym for* the BCD Interchange Code.

Binary Pertaining to a selection, a choice, or a condition that has two possible values or states.

Binary Coded Decimal A coding representation in which the low-order 4 bits of the Bit Patterns of the numerics are the binary equivalents of the decimal digits of the numerics.

Binary Digit In the binary system, one of the digits 0 or 1. *Contrast with* Decimal Digits. *Synonym for* Bit.

Binary Variable A variable that can take two possible values, or represent two possible states.

Bit *Synonymous with* Binary Digit. *See also* A-Bit, B-Bit, C-Bit, D-Bit, Parity Bit, Zone Bit, 0-Bit, 1-Bit, 2-Bit, 4-Bit, 8-Bit.

Bit Code A set of Bit Patterns and associated Graphic and Control Meanings.

Bit Combination *Synonym for* Bit Pattern.

Bit Name The name of the position of a Bit within a Byte. *Synonymous with* Bit Number.

Bit Number The number of the position of a Bit within a Byte. *Synonymous with* Bit Name. *See also* 0-Bit, 1-Bit, 2-Bit, 4-Bit, 8-Bit.

Bit Pattern An ordered set of Bits, usually of fixed length. *Synonymous with* Bit Combination, Bit Representation.

Bit Representation *Synonym for* Bit Pattern.

Bit Sequence The binary sequence of the Bit Patterns of a code, from 000 · · · 0 to 111 · · · 1.

Bit Stream A string of Bit Patterns without regard to grouping by Bit Pattern.

Bit String A string consisting solely of Bits.

Block A string of characters for technical or logical reasons to be treated as an entity.

Byte A Bit Pattern of fixed length.

Byte Size The number or count of Bits in a Byte.

Capital Alphabetic The alphabetics A, B, C, . . . , Z. Also includes the capital Diacritic Alphabetics. *Contrast with* Small Alphabetic.

Card Code A set of Hole Patterns and associated Graphic and Control Meanings.

Card Column On a punched card, a vertical Column.

Card Row On a punched card, a horizontal Row.

Cartridge *See* Chain Cartridge.

C-Bit A Bit in the C position of a Byte whose bit positions are named D, C, B, A, 8, 4, 2, 1.

C.C.I.I.T. *Acronym for* Comité Consultative International Telegraphique et Telephonique.

CCITT #2 A 58-character, 6-bit Shifted Code, used nationally and internationally on telegraph lines.

Chain Cartridge A cartridge holding a chain for a Chain Printer, allowing easy and simple replacement.

Chain Printer An impact printer in which the type slugs are carried by the links of a revolving chain. *See also* Train Printer.

Character A Bit Pattern and its associated Meaning. *See also* Alphabetic Character, Control Character, Delete Character, Escape Character, Graphic Character, Null Character, Numeric Character, Space Character, Special Character.

Character Set *Synonym for* Coded Character Set.

Clocking Track A track on which a pattern of signals is recorded to provide a timing reference.

COBOL (Common business-oriented language.) A programming language designed for business data processing.

Code *Synonym for* Coded Character Set. *See also* Baudot Code, BCD Code, Fieldata Code.

Code Form A general term, including, for example, Coded Character Sets, Packed Numerics, Binary data, Bit String.

Code Meaning The meaning assigned to a Bit Pattern of a Coded Character Set.

Code Name The name assigned to a particular coded character set, such as ASCII, BCDIC, EBCDIC, PTTC.

Code Position *Synonym for* Code Table Position.

Code Table A compact matrix form of Rows and Columns for exhibiting the Bit Patterns or Hole Patterns and assigned Meanings of a Coded Character Set.

Code Table Position The position or location of a Character in the Code Table for a Coded Character Set. There are two common conventions. For ASCII, the position is given as x/y, where x is the Code Table Column Number, and y is the Code Table Row Number. For EBCDIC, the position is given as mn, where m is the Hexadecimal Code Table Column Number, and n is the Hexadecimal Code Table Row Number. *Synonymous with* Code Table Location.

Code Table Location *Synonym for* Code Table Position.

Code Table Column A vertical Column in a Code Table. *Synonymous with* Table Column.

Code Table Row A horizontal Row in a Code Table. *Synonymous with* Table Row.

Coded Character Set A specific set of Bit Patterns or Hole Patterns to which specific Graphic Meanings and Control Meanings have been assigned. *Synonymous with* Code.

Collating Number A number assigned to the Characters of a Coded Character Set, running from 0 to 63 (for BCDIC) and from 0 to 255 (for EBCDIC). The collating numbers give the Collating Sequence of the coded character set, from low to high.

Collating Sequence An ordering assigned to the Characters of a Coded Character Set.

Column A vertical column of a Coded Character Set either in a Code Table or on a punched card. *See also* Code Table Columns, Card Column. *Contrast with* Row.

Column Number The number assigned to a Column of a Code Table. *Contrast with* Row Number.

Comparator Hardware circuitry that compares the relative magnitudes of two bit patterns and indicates the results of that comparison.

Compiler A program that transforms source-language statements of a programming language into computer-oriented language.

Contiguous Alphabet A characteristic of a Code (such as ASCII) such that the Bit Patterns assigned to the Alphabetics have no gaps in the binary sequence of the Bit Patterns. *Contrast with* Noncontiguous Alphabet.

Control An action that initiates, modifies, or suppresses an operation.

Control Character A specific Bit Pattern with an assigned Control Meaning. *Contrast with* Graphic Character.

Control Meaning A particular operation that controls either a hardware or software function.

Cyrillic Alphabet The Alphabet of Slavic languages.

D-Bit A bit in the A position of a Byte whose bit positions are named D, C, B, A, 8, 4, 2, 1.

Data Stream A variable-length string of Bit Patterns representing the data of a data processing application.

Decimal Digit In the decimal system, one of the digits 0 through 9. *Contrast with* Binary Digit.

Delete Character A control character used primarily to obliterate an erroneous or unwanted character, particularly in perforated tape.

Device Control Character A character to control a device (as "On" or "Off") or to control functions within a device.

Diacritic A symbol (such as diaeresis, ¨) used with a letter to indicate pronunciation.

Diacritic Alphabetic An Alphabetic with a Diacritic. *See also* Accented Alphabetic.

Digit A Graphic that represents an integer. *See also* Binary Digit, Decimal Digit.

Digit Punch In punched cards, the 1-Punch, 2-Punch, 3-Punch, . . . , 9-Punch.

Digit Row In punched cards, the Card Rows for Digit Punches.

Dual The mapping of more than one meaning to a single Bit Pattern or Hole Pattern.

Duocase Pertaining to a keyboard machine (such as a typewriter) which can shift from one case to another.

Duocase Alphabet An Alphabet with both Small Alphabetics and Capital Alphabetics.

EBCDIC *Acronym for* Extended BCD Interchange Code.

Eight-Punch *Synonym for* 8-Punch.

Eight-Row *Synonym for* 8-Row.

Eleven-Punch *Synonym for* 11-Punch.

Eleven-Row *Synonym for* 11-Row.

Escape Character A code-extension Character used with a sequence of one or more succeeding Characters to indicate that the Characters which follow the sequence are to be interpreted according to a different Coded Character Set.

Exclusive OR A logic operator with the property that if *A* and *B* are Binary Variables, then *A* Exclusive OR *B* is 1 if either but not both variables are 1, and is 0 if both are 1 or both are 0.

Extended BCD Interchange Code A 256 character, 8-bit Coded Character Set.

Extender, Alphabetic *See* Alphabetic Extender Positions.

FIELDATA Code A 7-bit Coded Character Set developed by the United States Army for military communications system.

Format Effector Character A Control Character to control the formatting of data on a printed or displayed page.

FORTRAN (Formula translation.) A programming language primarily used to express computer programs by arithmetic formulas.

Five-Punch *Synonym for* 5-Punch.

Five-Row *Synonym for* 5-Row.

Four-Punch *Synonym for* 4-Punch.

Four-Row *Synonym for* 4-Row.

Graphic A printed, typed, or displayed symbol to represent an Alphabetic, a Numeric, or a Special.

Graphic Character A specific Bit Pattern or Hole Pattern together with an assigned Graphic Meaning.

Graphic Meaning The Graphic associated with a Graphic Character.

Graphic, Special *See* Special.

Hex *Synonym for* Hexadecimal.

Hexadecimal Pertaining to a selection, choice, or condition that has sixteen possible different values or states. *Synonymous with* Hex.

Hole Pattern The pattern of holes within a single vertical Column of a punched card.

Hollerith Card Code A 256-character, 12-row card code.

IDENTITY A logic operator with the property that if A and B are Binary Variables, A IDENTITY B is 1 if both A and B are 1 or if both A and B are 0, and is 0 if A is 1 and B is 0 or if A is 0 and B is 1.

INCLUSIVE OR A logic operator with the property that if A and B are Binary Variables, A INCLUSIVE OR B is 1 if A is 1 and B is 0 or if A is 0 and B is 1 or if both A and B are 1, and is 0 if both A and B are 0.

ISO *Acronym for* International Organization for Standardization.

Katakana Symbols A set of phonetic symbols used in Japan to represent the Japanese language

Latin Alphabet The Alphabetics of the languages of English, Spanish, Portuguese, French, Italian, German, Swedish, Norwegian, Danish, and Finnish speaking countries. *Contrast with* Non-Latin Alphabet.

Lower Case Alphabetic *Deprecated term for* Small Alphabetic.

Lower Case Letter *Deprecated term for* small letter.

Meaning The sense, significance, or understanding intended to be conveyed by a Graphic character or a Control Character. *See also* Code Meaning, Control Meaning, Graphic Meaning.

Mode Change Character A Control Character that sets or changes some particular mode of operation.

Monocase Alphabet An Alphabet with Capital Alphabetics only or with Small Alphabetics only.

National Use Positions Positions in the ISO 7-Bit Code reserved for graphics particular to a country. *See also* Alphabetic Extender Positions.

Negative Numeric A Numeric with an associated negative sign. *Contrast with* Absolute Numeric, Positive Numeric, Signed Numeric.

Nine-Punch *Synonym for* 9-Punch.

Nine-Row *Synonym for* 9-Row.

Noncontiguous Alphabet A characteristic of a Code (such as EBCDIC) that the Bit Patterns assigned to the Alphabetics have gaps in the binary sequence of Bit Patterns. *Contrast with* Contiguous Alphabet.

Non-Latin Alphabet The Alphabetics of languages such as Russian, Greek, Hebraic, which are not Latin Alphabetics. *Contrast with* Latin Alphabet.

Null Character The Character whose Bits are all zero bits.

Numeric One of the digits zero through 9. *See also* Absolute Numeric, Negative Numeric, Positive Numeric, Signed Numeric.

Numeric Character A Numeric together with its assigned Bit Pattern or Hole Pattern.

One-Bit *Synonym for* 1-Bit.

One-Punch *Synonym for* 1-Punch.

One-Row *Synonym for* 1-Row.

OR *See* EXCLUSIVE OR, INCLUSIVE OR.

Packed Decimal Representation of a decimal value by two contiguous 4-bit BCD Bit Patterns within an 8-bit Byte.

Packed Numeric *Deprecated term for* Packed Decimal.

Paper Tape And Transmission Code A 111-character, 6-bit Shifted Code that is used on paper tape for data transmission.

Paper Tape And Transmission Code For BCD Environments A 111-character, 6-bit Shifted Code for use with computers with BCDIC as the internal code.

Paper Tape And Transmission Code For EBCD Environments A 111-character, 6-bit Shifted Code for use with computers with EBCDIC as the internal code.

Parity Bit A check Bit appended to a string of Bits to make the sum of all the Bits, including the Parity Bit, always odd or always even.

Pattern *See* Bit Pattern, Hole Pattern.

Position *See* Code Position, Code Table Position, National Use Position.

Positive Numeric A Numeric with an associated positive sign. *Contrast with* Absolute Numeric, Negative Numeric, Signed Numeric.

Printer *See* Chain Printer, Train Printer.

PTTC *Acronym for* Paper Tape And Transmission Code.

PTTC/BCD *Acronym for* Paper Tape And Transmission Code For BCD Environments.

PTTC/EBCD *Acronym for* Paper Tape And Transmission Code For EBCD Environments.

PTTC/6 *Deprecated acronym for* PTTC/BCD.

PTTC/8 *Deprecated acronym for* PPTC/EBCD.

Punch *See* Digit Punch, Zone Punch, 0-Punch, 1-Punch, 2-Punch, 3-Punch, 4-Punch, 5-Punch, 6-Punch, 7-Punch, 8-Punch, 9-Punch, 11-Punch, 12-Punch.

Representation The physical form or manner in which the Characters of a Coded Character Set are recorded or transmitted on some medium, such as magnetic tape, magnetic card, magnetic disks, magnetic core, paper tape, punched card, data transmission line.

Row A horizontal row of a Coded Character Set either in a Code Table or on a punched card. *See also* Code Table Row, Card Row, 0-Row, 1-Row, 2-Row, 3-Row, 4-Row, 5-Row, 6-Row, 7-Row, 8-Row, 9-Row, 11-Row, 12-Row. *Contrast with* Column.

Row Number The number assigned to a Row of a Code Table. *Contrast with* Column Number.

Sequence *See* Bit Sequence, Collating Sequence.

Seven-Punch *Synonym for* 7-Punch.

Seven-Row *Synonym for* 7-Row.

Shifted Code A Code in which the meaning of a Bit Pattern depends not only on the Bit Pattern itself but also on a particular preceding Bit Pattern in the string of Bit Patterns, the preceding Bit Pattern being called a "precedence Character" or a "shift Character."

Signed Numeric A Numeric with either a positive or negative associated sign. *Contrast with* Absolute Numeric, Negative Numeric, Positive Numeric.

Small Alphabetic The alphabetics a, b, c, . . . , z. Also includes small Diacritic Alphabetics. *Contrast with* Capital Alphabetic.

Space Character A Graphic Character that causes the print or display positions to move one position forward (that is, to the right) without producing the printing or display of any visible graphic.

Special A Graphic other than an Alphabetic or a Numeric. *Synonymous with* Special Graphic.

Special Character A Special together with its associated Bit pattern or Hole Pattern.

Special Graphic *Synonym for* Special.

Stream *See* Bit Stream, Data Stream.

String *See* Bit String.

Subset A Coded Character Set, each Character of which is a Character of a larger Coded Character Set.

Table Column *Synonym for* Code Table Column.

Table Row *Synonym for* Code Table Row.

Three-Punch *Synonym for* 3-Punch.

Three-Row *Synonym for* 3-Row.

Track The portion of a moving data medium such as drum, disk, or tape, that is accessible to a given reading or recording head position.

Train Printer An impact printer in which the type slugs are carried by a revolving train. *Contrast with* Chain Printer.

Transmission Control Character A Control Character to control intercommunications on data transmission lines.

Twelve-Punch *Synonym for* 12-Punch.

Twelve-Row *Synonym for* 12-Row.

Twelve-Row Card A punched card with twelve punchable Card Rows.

Two-Punch *Synonym for* 2-Punch.

Two-Row *Synonym for* 2-Row.

Upper-Case Alphabet *Deprecated term for* Capital Alphabetic.

Upper-Case Letter *Deprecated term for* capital letter.

USASCII *Deprecated term for* ASCII.

USASI *Acronym for* the United States of American Standards Institute, now called the American National Standards Institute.

Zero-Bit *Synonym for* 0-Bit.

Zero-Punch *Synonym for* 0-Punch.

Zero-Row *Synonym for* 0-Row.

Zone Bit For BCDIC, one of the two high-order Bits; for EBCDIC, one of the four high-order Bits.

Zone Punch On a punched card, a 12-Punch, an 11-Punch, or a 0-Punch for some Hole Patterns.

0-Bit A Bit whose value is zero.

1-Bit (a) A Bit whose value is one. (b) In the BA8421 or the DCBA8421 nomenclature for bit positions, a Bit in the 1 position.

2-Bit In the BA8421 or DCBA8421 nomenclature for bit positions, a Bit in the 2 position.

4-Bit In the BA8421 or DCBA8421 nomenclature for bit positions, a Bit in the 4 position.

8-Bit In the BA8421 or DCBA8421 nomenclature for bit positions, a Bit in the 8 position.

0-Punch A punch in the 0-Row of a punched card. *Synonymous with* Zero-Punch.

1-Punch A punch in the 1-Row of a punched card. *Synonymous with* One-Punch.

2-Punch A Punch in the 2-Row of a punched card. *Synonymous with* Two-Punch.

3-Punch A punch in the 3-Row of a punched card. *Synonymous with* Three-Punch.

4-Punch A punch in the 4-Row of a punched card. *Synonymous with* Four-Punch.

5-Punch A punch in the 5-Row of a punched card. *Synonymous with* Five-Punch.

6-Punch A punch in the 6-Row of a punched card. *Synonymous with* Six-Punch.

7-Punch A punch in the 7-Row of a punched card. *Synonymous with* Seven-Punch.

8-Punch A punch in the 8-Row of a punched card. *Synonymous with* Eight-Punch.

9-Punch A punch in the 9-Row of a punched card. *Synonymous with* Nine-Punch.

11-Punch A punch in the 11-Row of a punched card. *Synonymous with* Eleven-Punch.

12-Punch A punch in the 12-Row of a punched card. *Synonymous with* Twelve-Punch.

0-Row The horizontal Row in a punched card that receives 0-Punches.

1-Row The horizontal Row in a punched card that receives 1-Punches.

2-Row The horizontal Row in a punched card that receives 2-Punches.

3-Row The horizontal Row in a punched card that receives 3-Punches.

4-Row The horizontal Row in a punched card that receives 4-Punches.

5-Row The horizontal Row in a punched card that recieves 5-Punches.

6-Row The horizontal Row in a punched card that receives 6-Punches.

7-Row The horizontal Row in a punched card that receives 7-Punches.

8-Row The horizontal Row in a punched card that receives 8-Punches.

9-Row The horizontal Row in a punched card that receives 9-Punches.

11-Row The horizontal Row in a punched card that receives 11-Punches.

12-Row The horizontal Row in a punched card that receives 12-Punches.

ABOUT THE AUTHOR

Mr. Mackenzie graduated from the University of Manitoba, B. Comm. (Honors) in 1942, and B. Sc. (Honors) 1949. He took postgraduate courses at Columbia University in 1950, 1951.

He joined the IBM Corporation in 1952, leaving in 1954. He rejoined the IBM Corporation in 1959. He worked as a Product Planner on the IBM 1650, 7040, 7090, System/360, and System/370. He is currently a Senior Planner.

He has served as

Consultant to ANSI/X3L2	July 1962 to May 1963;
Member of ANSI/X3L2	May 1963 to July 1967;
Chairman of ANSI/X3.2.3	June 1963 to December 1967;
Consultant to ANSI/X3L2	July 1967 to current date.

He was a member of the USA Delegations to ISO/TC97/SC2, ISO/TC97/SC2/WG2, ISO/T979/SC4/WG1 from October 1963 to October 1971. Currently, Mr. Mackenzie is a member of ISO/TC97/SC2/WG4.

Index